Laura

NOTES AND QUERIES
ON
ANTHROPOLOGY

NOTES AND QUERIES
ON
ANTHROPOLOGY

Sixth Edition
revised and rewritten by

A COMMITTEE
OF THE ROYAL ANTHROPOLOGICAL INSTITUTE
OF GREAT BRITAIN AND IRELAND

ROUTLEDGE AND KEGAN PAUL LTD
Broadway House, 68 Carter Lane
London

*This sixth edition published in 1951
by Routledge and Kegan Paul Ltd
Broadway House, 68–74 Carter Lane
London E.C.4
Printed in Great Britain
by Western Printing Services Ltd., Bristol
First Edition published 1874
Second Edition published 1892
Third Edition published 1899
Fourth Edition published 1912
Fifth Edition published 1929*

PREFACE TO THE SIXTH EDITION

In 1936 the Committee set up by Section H of the British Association for the Advancement of Science to prepare the sixth edition of *Notes and Queries on Anthropology* began its work. Sub-committees to supervise the work in the various sections were appointed and members nominated. An editorial committee composed of the chairmen of various sections with the addition of Dr. A. C. Haddon and Dr. C. G. Seligman was appointed. Unfortunately, owing to the death of several of the members and to the subsequent outbreak of war this plan was not carried out. In 1947 a General Committee with Professor H. J. Fleure as Chairman reviewed the situation. It was found that the only sub-committee that had functioned was that on Social Anthropology and its work was almost complete. Mr. T. K. Penniman and Miss B. M. Blackwood had revised the section on Material Culture; work on the other sections had not begun. The General Committee undertook the completion, revision and editing of the complete volume and appointed B. Z. Seligman editor. In the fifth edition the contributions were unsigned, and it was decided to carry on this practice.

Professor W. E. Le Gros Clark has kindly supervised Part I, Physical Anthropology, and I have to thank Drs. N. Barnicot, A. E. Mourant and J. S. Weiner for contributions. Besides the work done by Mr. T. K. Penniman and Miss B. M. Blackwood in Part III, contributions by the late James Hornell, Mr. A. Digby, Miss Helen H. Roberts and Mr. R. U. Sayce are gratefully acknowledged.

For Part IV, Field Antiquities, I am indebted to Dr. S. A. Huzayyin for a contribution, and much valuable help and advice from Professors D. A. E. Garrod and F. E. Zeuner.

Part II, Social Anthropology, deals mainly with the sociology of non-literate peoples, though the methods described are also suitable in general principle to studies in an advanced society. The advance in Social Anthropology has been so marked since the fifth edition of *Notes and Queries* in 1929 that it was found necessary to rearrange this part, expanding it and presenting in it some topics previously dealt with in the sections on Arts and Sciences and Nature Lore. The general scheme and list of contents on Sociology were decided in committee. Definitions of technical terms were also discussed in committee, a list was prepared and sent to all members for approval. In order to save space and avoid repetition it was found necessary to divide and rearrange some of the material specially contributed. This has meant, in effect, that many sections are the work of several hands.

I wish to acknowledge the co-operation of all members of the Sociological Sub-committee who have contributed freely both in writing new articles and in making valuable suggestions and criticisms. I must mention specially Professor A. R. Radcliffe-Brown, Professor Daryll Forde and Professor Meyer Fortes, who, in addition to their initial contributions, have been through all the final drafts, have assisted me in consultation on difficult questions and have provided notes and suggestions. Besides the contributions from the members of the Sub-committee I gratefully acknowledge contributions from Miss M. M. Green, Dr. E. R. Leach, Professor S. F. Nadel and Dr. F. B. Steiner. Professor R. O'R. Piddington, Professor I. Schapera, Dr. A. N. Tucker and the late Professor Ida Ward also gave valuable assistance.

Previous editions of *Notes and Queries on Anthropology* were edited for the British Association by a committee of Section H. In 1949 the work of completion and publication of the sixth edition was transferred to the Royal Anthropological Institute, which will be responsible for all future editions. The kindly co-operation of all three committees has made the publication of this edition possible. I am also grateful to Dr. D. B. Harden, Recorder of Section H, for his support.

BRENDA Z. SELIGMAN

LIST OF COMMITTEES

GENERAL COMMITTEE
(*Section H of British Association*)

Professor H. J. Fleure, M.A., D.Sc., F.R.S. (*Chairman*)
Professor W. E. Le Gros Clark, M.A., D.Sc., F.R.C.S., F.R.S.
A. Digby, M.A.
Professor E. E. Evans-Pritchard, M.A., Ph.D.
Professor D. Forde, Ph.D.
Professor J. H. Hutton, C.I.E., M.A., D.Sc.
Mrs. Brenda Z. Seligman (*Secretary*)

SUB-COMMITTEE ON SOCIOLOGY
(*Section H of British Association*)

Professor A. R. Radcliffe-Brown, M.A. (*Chairman*)
Professor E. E. Evans-Pritchard, M.A., Ph.D.
Professor R. Firth, M.A., Ph.D.
Professor D. Forde, Ph.D.
Professor M. Fortes, M.A., Ph.D.
Professor J. H. Hutton, C.I.E., M.A., D.Sc.
Professor Sir John L. Myres, O.B.E., M.A., F.B.A., F.S.A.
Mrs. Brenda Z. Seligman (*Convener and Editor*)

COMMITTEE OF THE ROYAL ANTHROPOLOGICAL INSTITUTE
FOR THE PUBLICATION OF *Notes and Queries on Anthropology*
(Sixth Edition)

The President, Honorary Editor, Honorary Treasurer and Honorary Secretary of the Institute (*ex officio*)
Professor W. E. Le Gros Clark, M.A., D.Sc., F.R.C.S., F.R.S.
A. Digby, M.A.
Professor D. Forde, Ph.D.
Professor J. H. Hutton, C.I.E., M.A., D.Sc.
Mrs. B. Z. Seligman

CONTENTS

PREFACE	page v
LIST OF COMMITTEES	vii

PART I. PHYSICAL ANTHROPOLOGY

METHODS OF PHYSICAL ANTHROPOLOGY	3
BLOOD GROUPS	16

PART II. SOCIAL ANTHROPOLOGY

I.	INTRODUCTION	27
II.	METHODS	36
	Scope and Aims	36
	Techniques of Investigation	40
III.	SOCIAL STRUCTURE	63
	Introduction	63
	Territorial Arrangement	63
	Sex and Age	66
	The Family	70
	Kinship	75
	Lineage and Class	88
	Social Stratification	93
IV.	SOCIAL LIFE OF THE INDIVIDUAL	98
	Daily Routine	98
	Training and Education	101
	Life Cycle from Conception to Marriage	104
	Sexual Development	107
	Marriage	110
	Old Age, Death and Disposal of the Dead	124
V.	POLITICAL ORGANIZATION	132
	Political Systems	132
	Law and Justice	144
	Property	148

CONTENTS

VI. ECONOMICS ... 158
    Introduction ... 158
    Production ... 160
    Distribution ... 168
    Exchange ... 169
    Consumption ... 171

VII. RITUAL AND BELIEF ... 174
    Introduction ... 174
    Religious Beliefs and Practices: ... 175
        Beliefs concerning Man ... 176
        Beliefs concerning Supernatural Beings and Agencies ... 179
    Forms of Ritual ... 187
    Magical Beliefs and Practices ... 188
    Witchcraft and Sorcery ... 189
    Ritual in Medicine and Therapy ... 189
    Ritual and Beliefs concerned with Physical Phenomena ... 190
    Ritual and Beliefs concerned with Economic Activities ... 191
    Ritual and Beliefs concerned with Social Structure ... 191

VIII. KNOWLEDGE AND TRADITION ... 195
    Recording and Communication ... 195
    Reckoning and Measurement ... 195
    Cosmology, Seasons, Weather and Calendar ... 198
    Geography and Topography ... 199
    Vegetation ... 199
    Man and the Animal Kingdom ... 200
    Medicine and Surgery ... 201
    History and Myths ... 204
    Stories, Sayings and Songs ... 206

IX. LANGUAGE ... 208
    Gesture, Sign-language and Signals ... 208
    Spoken Language: ... 210
        Phonology ... 211
        Grammar ... 215
        Semantics ... 217

PART III. MATERIAL CULTURE

Introduction	221
Status of the Craftsman	222
Personal Care and Decoration	223
Clothing	234
Habitations	236
Fire	240
Food	241
Stimulants and Narcotics	246
The Food Quest	247
Tools and Mechanisms	257
Weapons	259
Receptacles for Food, Drink, etc.	270
Basketry	272
Pottery	276
Glass	279
Stone, Wood and Metal Work	279
Mining and Quarrying	283
Salt	284
Skins and Fabrics	284
Spinning and Weaving	287
Dyeing and Painting	296
Travel and Transport	297
Art	308
Music	315
Dancing	331
Drama	333
Games and Amusements	334
String Figures and Tricks	335

PART IV. FIELD ANTIQUITIES 343

APPENDICES

Photography	353
Cinematography	359
Collecting and Packing	361
Preservation of Bones	364
Paper Squeezes	365
BOOK LIST	369
INDEX	387

LIST OF ILLUSTRATIONS

Blood Groups	23
Genealogical Method: Diagrams	53
Plaited Basket Work	272
Coiled Basket Work	275
Diagram illustrating the Principles of Weaving	289
Decorative Art	311–314
"Ti Meta," String Figure from Torres Strait	338

PART I

PHYSICAL ANTHROPOLOGY

The Methods of Physical Anthropology

It is a familiar fact that human populations living in different geographical regions differ to a greater or lesser extent in various anatomical features such as skin colour, hair form and the proportions of the body. One of the chief aims of physical anthropologists during the last hundred years has been to record these differences as accurately as possible, and then, by a comparison of the populations thus described, to deduce their origins and inter-relationships. This section is concerned with what may be termed the classical methods of physical anthropology as opposed to the more recently introduced serological methods which are discussed in the next section. The methods dealt with have also been used in work on growth, nutrition, and in the study of so-called constitutional types; although all these fields can be studied from an anthropological point of view, limitation of space prevents a full discussion of them.

The characters chosen for measurement in work on the comparative anatomy of living peoples have been mainly those which are external, because these are the most obvious and the easiest to record on large numbers of living subjects. Physiological variables, such as blood pressure and basal metabolic rate, have also been studied, but for reasons dealt with later these require particularly critical and elaborate investigation before the results can be correctly interpreted.

The characters commonly employed may conveniently be summarized as follows:

(1) *Metrical characters*: these include all measurements of the dimensions of the body. At present these characters allow the most satisfactory quantitative work, but the measurements are subject to various inaccuracies which are discussed below.

(2) *Pigmentary characters*: (skin, hair and eyes). These are more difficult to record quantitatively, at least with existing field-techniques.

(3) *Qualitative characters*: Examples of these are presence or

absence of the epicanthic fold or the sacral spot, the shape of the nose profile, and the hairiness of the body. They are grouped as qualitative because, although they are subject to quantitative variation, satisfactory quantitative methods for recording them do not exist.

No human population is composed of identical individuals; indeed the variability within a population is very often greater than the average differences between populations, so that comparisons necessitate the use of statistical methods. Taking as an example differences of eye colour, and neglecting for a moment certain difficulties of measurement, the distinction between two populations will take the form of a difference in the frequency of a particular colour in the two; in one case there may be 5 per cent of individuals classifiable as blue-eyed, and in the other 10 per cent. If a metrical character such as stature is examined it will be found that individual heights may come out at any value between certain limits; in other words the character shows continuous variation. The populations must then be characterized by the mean value, and the variability of each may be expressed as the standard deviation.[1] Statistical tests must be applied to see whether any observed difference between the means, having regard to the variability and size of the samples, can be attributed to chance. It is impossible to deal here with the statistical methods which are useful in this type of work, but a number of books which the reader may find helpful are given in the footnote.[2]

It is very important that any technique employed shall be as far as possible objective, that is to say that the results will not be significantly influenced by undetermined idiosyncrasies of the observer. If a worker groups hair colours into a number of categories, say, dark, medium and light, he alone knows what he means by these categories, and he may not mean the same thing on different occasions: still less can it be hoped that another observer using the same grouping will mean the same. If, as in

---

[1] The standard deviation (S.D.) is

$$\sqrt{\frac{(\text{sum of deviations from the mean})^2}{\text{Number of individuals in sample}}} \quad \text{or} \quad \sqrt{\frac{S(x-\bar{x})^2}{N-1}}$$

where $\bar{x}$ = the mean, and x = the observed value of the variable. Division by $N-1$, the number of degrees of freedom, yields a more accurate estimate than division by N, the sample size. For a normal distribution it equals the values on either side of the mean within which 68 per cent of the observations lie.

[2] E. G. Chambers, *Statistical Calculation for Beginners* (Cambridge University Press. 1943). K. Mather, *Statistical Analyses in Biology*, 2nd ed. (Methuen, London, 1946), G. W. Snedecor, *Statistical Methods* (Iowa State Coll. Press, 1940).

the case of pigmentary characters, no simple technique for direct quantitative estimation exists, an attempt is made to avoid subjective errors by using standard colour-scales and matching against them; even so, observers differ in their accuracy of matching and errors may be so great as to obscure real differences between populations.[3]

Data based on subjective groupings can only give crude and preliminary information, and even where matching techniques are used caution is evidently needed in drawing conclusions from the records of different workers. Certain characters, for example hair form, are extremely difficult to record objectively. Categories such as woolly, wavy and straight are easily recognized in their extreme condition, but intermediates occur; also, in this instance, workers disagree in their definition of the more commonly recognized types, and no standard matching-scale is available; nor indeed would it be easy to construct an effective one. The amount of hair on the body is even more difficult to assess accurately. The overall appearance probably depends on many factors, such as the length, colour, thickness and shape of the hairs, and the number of hairs per unit area, and all these may vary in different body regions. In cases like this a qualitative impression may not be without value, but its limitations must be appreciated.

Body measurements, although they involve only a comparison with a simple linear scale, are also subject to various types of error. If an observer repeats his measurements on the same subjects, he will find that his readings show a scatter around a mean value. The size of these random errors (which may be measured as the S.D. of the mean value) will vary according to the observer's skill and the intrinsic difficulty of particular measurements. It is much to be desired that every worker should estimate his random error by repeating his measurements on, say, fifty subjects.[4]

If his error is not estimated and deducted from the total variability of the sample it will increase the size of sample needed to show statistically significant differences. Clearly, if the observer finds his error to be large in relation to the differences he anticipates, he must either improve his technique or discard it. In addition to random errors a measurer may make systematic

---

[3] S. W. Grieve and G. M. Morant, "Records of Eye-Colour for a British Population and Description of a New Eye-colour Scale", *Ann. Eug.*, **13**, 1946-7.

[4] M. L. Tildesley, "Choice of the Unit of Measurement in Anthropology", *Man*, **47**, 72, 1947.

errors, so that his mean result is always too high or too low in comparison with some other (presumably more accurate) worker, or with a more objective standard if such exists. Systematic errors may arise from a variety of causes. The common anthropometric measurements are taken between certain points on the body; unless these points are accurately and unambiguously defined and every worker agrees and adheres to a standard procedure, errors will obviously arise which may make the comparison of results useless. Unfortunately no such general agreement has yet been achieved and it is therefore very important for every worker to state which authority he has followed in his work, and to record very carefully any deviation from standard procedure which he may have been obliged to make.

It is worth while pointing out that workers should carefully preserve the records of their individual measurements and observations, even though it may not always be possible to publish these in full, together with statistics derived from them. New knowledge and problems, and advances in statistical technique, may at any time necessitate the calculation of new statistics from the data, and in many cases this can only be done if the individual figures are available.

It will be appreciated that the collection of sufficiently accurate data presents many pitfalls, and a worker intending to make field observations, particularly if he has no anatomical experience, would be well advised to consult an experienced anthropometrician before embarking on the project. A number of standard works on anthropometry are given below.[5]

The Interpretation of the Data

The comparison of populations is simply an extension of the classical method of comparative anatomy to the recent variants of man. The reasoning, in its simplest form, is that similarity of form indicates closeness of relationship. It is worth while considering this method more critically, particularly in the light of genetical knowledge. If the comparison is to yield anything of evolutionary interest, the characters studied must depend mainly on genetic constitution. To take an extreme example, if it were shown that differences in head form between two groups were due

---

[5] R. Martin, *Lehrbuch der Anthropologie* (Fischer, Jena, 1928). A. Hrdlička, ed. T. D. Stewart, *Practical Anthropometry* (Wistar Inst., Philadelphia, 1947). M. F. Ashley Montagu, "The Location of the Nasion in the Living", *Amer. J. phys. Anthrop.*, 20, 87, 1935.

to artificial deformation, the character would be valueless for deducing biological affinities. The situation is usually rather more subtle. Stature, although it is under genetic control, requires adequate nutrition to reach its maximal development; hence differences of stature between populations cannot be accepted as genetic unless the relevant environmental factors are comparable in the two cases. Again, negro skin pigmentation is overwhelmingly genetic; no environmental change, except some rare diseases, can alter it very much: but, in general, southern Europeans possess more skin pigment and also develop more pigment under the influence of sunlight than do many northern populations.

In the cases of physiological variables, the situation may be particularly complex; since many of them relate to regulatory mechanisms, they are particularly susceptible to immediate environmental influences so that it may be difficult to obtain constant readings on a single individual. The blood-pressure, for example, must be taken under strictly standardized conditions since it is readily affected by a variety of factors such as exercise and emotional stress. Even when care is taken to standardize technique it is still not easy to decide whether observed differences between peoples are due to genetic differences or to acclimatization, dietary habits, or chronic disease. Variables such as basal metabolic rate, the level of the blood constituents, or the capacity for work at high temperatures and humidity, require very close scrutiny before assigning genetic causes to population differences. Even when postnatal environmental influences can reasonably be disregarded, the genetic determination of a character may be complex, so that one cannot specify the number of genes involved or determine precisely what will occur in hybrids. Unfortunately almost all the morphological characters used in anthropology are determined by more than one gene, and do not behave in a simple Mendelian manner in crosses. It is one of the merits of the serological characters that the genetic situation is clear and populations can be characterized by the frequency of the genes concerned. A closer study of hybrids between some of the more distinct living peoples may throw further light on the genetics of anthropological characters, but more precise techniques of measurement will be needed before real progress can be expected. Without more extensive observations we are not in a secure position when we infer that features in different peoples which look the same are due to the same genes. Since, as far as is known, the common anthropological characters depend on different genes which segregate independently, there seems little reason to suppose that it will be possible to discern the "parental types" in a hybrid population, particularly as the parent populations would themselves be variable.

Another difficulty of interpretation that may arise is that two populations may have come to possess common characteristics because these characters are adaptive and the populations have been exposed to the same selective influences of the environment, and not because they are particularly closely related. This may, for example, be true for the skin colour and hair form of the African negroes and Asiatic negritos. It must be realized that this issue is distinct from the previously discussed action of the environment on the individual; in the case of adaptive characters we are concerned with traits determined by genes which are subject to natural selection. In general, if the trait increases fitness in the Darwinian sense the genes concerned will increase in frequency in the population. The rate at which this occurs will depend on various factors which cannot be discussed here.[6] In the case of the ABO blood-group system, at least, it is generally thought that the genes concerned are little influenced by selection so that their frequency will not change unless hybridization with a population with a different frequency occurs.[7] If this assumption is correct the blood-group genes provide, in this respect, a very good means of tracing the more remote relationships of peoples. In the case of genes for adaptive characters the frequencies may be expected to change more rapidly so that original similarities will be obscured. The whole question of the adaptive value of the observed differences between ethnic groups is of great theoretical and also practical interest, but at present we have very little reliable evidence on this matter. It is generally thought that the skin pigmentation of some tropical peoples protects them from injury by ultra-violet radiation, but although the skin pigment, melanin, undoubtedly absorbs these radiations strongly, one could not assert on existing evidence that it plays an important role in this respect under natural conditions. Similarly, the relative breadth of the nose has been shown to correlate highly with

[6] T. Dobzhansky, *Genetics and the Origin of Species* (Columbia Univ. Press, New York), 2nd ed., 1941.
[7] W. C. Boyd, "Critique of the Methods of Classifying Mankind", *Amer. J. phys. Anthrop.*, **27**, 333, 1940.

latitude in some regions of the world, but the exact physiological significance of this finding is quite obscure. It may be that some of the commonly observed differences between peoples are themselves non-adaptive, but that they are correlated with other more significant, but at present undetected, actions of the same genes. The observations of the physical anthropologist, working in conjunction with the climatologist and physiologist, may therefore contribute not only to evolutionary knowledge, but to various practical problems in the field of medicine and hygiene.

Anthropometric Techniques

The instruments. Many modifications of the common instruments exist, and there are also various instruments for special purposes which cannot be mentioned here. The list given includes the most common types, which suffice for the measurements dealt with.

At the time of writing all instruments are in short supply, and none are being made in quantity in this country. According to a recent communication[8] supplies have become available from Switzerland.

(1) *The anthropometer.* A metal rod, 200 cm. long, made in four equal detachable sections, and graduated from zero to 200 cm. with a millimetre scale. The rod bears a fixed holder at one end into which can be fitted cross-pieces of various shapes. A cursor bears a similar holder. The 50 cm. section carrying the fixed and moving cross-pieces can be used separately as a sliding caliper, so that the instrument has a wide range of application.

(2) *Large sliding caliper* (Hrdlička type). The scale is 60 cm. long. The main feature of the instrument is the long cross-pieces (26 cm.) which are also much broader (3·5 cm.) than those of the anthropometer.

(3) *Small sliding caliper* (Martin type). The scale is 25 cm. long. Pointed cross-pieces on opposite sides. Suitable for facial measurements.

(4) *Small spreading calipers* (head calipers). With a maximum span of 30 cm. The scale may either be curved and rigidly fixed, or straight and hinged so that it can be folded back (Martin type).

[8] W. M. Krogman, "Anthropometric Instruments", *Amer. J. phys. Anthrop.*, N.S. 6, 507, 1948.

Anthropometric Measurements

The number of different measurements which can be taken on the body is virtually infinite and those selected must obviously be dictated by preliminary inspection of the material and the purpose of the investigation. If one is concerned with differentiating adult samples of various ethnic groups, there is no general rule as to the measurements which should be taken, since those which reveal differences in one case may not do so in another. Nevertheless the measurements given below have been used frequently in this kind of work and may, perhaps, serve as a guide.[9] Preliminary visual inspection of the material, and an acquaintance with previous work on the populations under study, may help the worker to select the most useful battery of measurements.

Size of sample. The size of sample required will depend on the magnitude of the differences between the samples for any particular measurement and the accuracy of the techniques. In general the worker should try to measure between 100 and 500 individuals of each sex. The individuals should be chosen from the young adult group, say from 25 to 35 years of age, to avoid cases of uncompleted growth and senile change. It is desirable, but not always easy, to record the age of each individual. It is possible to obtain accurate measurements from standard photographs provided these are taken at a sufficiently long working distance to avoid serious perspective distortion. A working distance of 25 feet, with a 10- to 12-inch focal length lens is suitable. The method has not, however, been applied to racial studies on a wide scale. The measurements obtainable on photographs will not, in general, correspond to those commonly taken by direct instrumental methods.[10] The following definitions are based on Martin's *Lehrbuch*, 1928 ed.

Head and Face

(1) *Head length* (spreading caliper). From the glabella, which is the most forward projecting point in the mid-line between the eyebrows and above the root of the nose, to the opisthocranion, which is the most posterior point on the occiput; many workers take this latter point in the midline, but others take it to the most posterior point even if this is to one side of the midline.

(2) *Head breadth* (spreading caliper). The distance between the

---

[9] See M. L. Tildesley, *Man*, 1950, 4.
[10] See J. M. Tanner and J. S. Weiner, *Amer. J. phys. Anthrop.*, N.S.7, 145, 1949.

most laterally projecting points (eurya) on the sides of the head, above the level of the ears. According to most workers the bony ridges overlying the attachment of the ears should be avoided. The tips of the caliper should be in the same horizontal and frontal planes.

(3) *Head height*. This measurement is less frequently taken and various methods and terminal points have been employed. Anthropometric textbooks should be consulted for information.

(4) *Minimum frontal breadth* (spreading caliper). The minimal distance between the fronto-temporal bony crests above the outer ends of the eyebrows. Care should be taken to measure from the bony ridges and not the temporal muscles situated behind them.

(5) *Bizygomatic breadth* (spreading caliper). The distance between the most laterally situated points on the bony zygomatic arches. The tips of the caliper should be in the same frontal plane.

(6) *Bigonial breadth* (spreading caliper). The distance between the two gonia, which are the most laterally situated points at the angle of the lower jaw. The angle may be rounded which makes the points difficult to determine; care should be taken to avoid the masseter muscle in front and the depression which may lie between it and the gonion.

(7) *Nasal height* (small sliding caliper). The distance between the nasion, which is the point of junction of the nasal and frontal bones in the midline, and the subnasal point, which is the point at which the lower end of the nasal septum meets the upper lip. Neither point is easy to determine, although the nasion is an important landmark. In many cases the naso-frontal suture cannot be felt below the skin. It is usually a few millimetres above the deepest point of the depression at the nasal root. According to Martin it lies on a line joining the inner ends of the two eyebrows, which may, however, not end sharply. According to Ashley Montagu[11] a line drawn tangentially to the fold of the upper eyelid (superior palpebral fold) crosses the nasion in most cases, but this has been disputed. At least this method provides a repeatable and definitive landmark. The worker should state clearly by which method the nasion has been located. The subnasal point is indeterminate if the attachment of septum and lip is rounded. Hrdlička recommends pressing in the point of the caliper along the line of the septum if this difficulty arises, but this may be resented by the subject.

[11] M. F. Ashley Montagu, *op. cit.*

(8) *Upper face height* (small sliding caliper). The distance between the nasion and the prosthion, which is the lowest point on the gum between the upper central incisor teeth.

(9) *Total face height* (small sliding caliper). The distance between the nasion and the gnathion, which is the lowest and most anterior point on the lower border of the lower jaw; the point is generally taken in the midline but there may be a slight depression here. The gnathion is not a sharp point, and when it is covered by thick soft tissues, it is not easy to locate accurately.

Body Measurements

(1) *Stature.* The subject stands erect, in bare feet, looking straight forward with the arms straight by his sides and heels together. The measurement is then taken from the vertex, which is the highest point on the head, to the ground. In the laboratory some form of stadiometer with a fixed base and sliding scale is generally used. In the field the anthropometer may be used, taking care to keep the instrument vertical; a small detachable footplate may assist in this. Alternatively a board one metre in length and graduated in centimetres and millimetres, as described by Hrdlička, can be attached to a vertical surface at a suitable height, and the stature then determined with a set-square of two broad pieces of wood attached at right angles. In this case the buttocks and often the shoulders of the subject touch the vertical plane. The measurement is obviously liable to be affected by posture and every effort should be made to standardize this. It is also stated that a decrease of stature, amounting to two or three centimetres, occurs in the course of the day.

(2) *Sitting height.* The distance between the vertex and the plane on which the subject is sitting with the thighs horizontal. The measurement is considerably decreased if the lumbar curvature is not maintained and the subject droops forward. This is liable to occur if the thighs are raised above the horizontal, so that it is advisable to use a stool some 30 cm. high, on which boards of known thickness can be placed to get each subject into the correct posture. Either the anthropometer or the wall-scale mentioned above may be used. It is common practice to obtain leg length by subtracting sitting height from stature, since direct measurement of the leg length is somewhat difficult; the value obtained is clearly less, to a varying extent, than the true leg-length, since the head of the femur lies some distance above the

ischial tuberosities in the sitting position. The measurement is widely used in growth studies.

(3) *Arm length*. The distance between the acromial point, which is the most laterally situated point of the acromial process of the scapula at the shoulder, and the tip of the longest finger. The arm is stretched to its greatest length, at the subject's side, during the measurement. The anthropometer is the most convenient instrument. The practice of obtaining all measurements to the ground level, for example acromial point to the ground, and then subtracting the distance, for example, of the finger tip to ground to obtain arm length, is liable to lead to error owing to the arm not being accurately in the vertical plane.

(4) *Forearm length*. The distance between the uppermost (most proximal) point on the head of the radius and the lowest (most distal) point of the styloid process of the ulna. The head of the radius can be felt rotating in the region of the olecranon process (point of the elbow) if the hand is alternately pronated and supinated. A more accurate lower terminal for the forearm is, perhaps, on a line between the styloid processes of the radius and ulna, and the total hand length can also be measured from this line. The upper arm length is taken from the acromion to the radial point.

(5) *Lower limb lengths*. The upper landmark from which the total limb length is measured is the upper edge of the great trochanter, felt below the skin at the widest point of the hips (in the male); the measurement is taken from the trochanter to the ground. This is a difficult point to locate accurately. The thigh length may be measured from this point to the upper edge of the tibia on the outer side of the leg. The leg length is measured from the upper edge of the tibia on the inner side of the knee joint (tibiale) to the lowest point on the medial malleolus at the ankle. A knowledge of skeletal and surface anatomy is clearly required if these and the foregoing arm lengths are to be determined with sufficient accuracy.

(6) *Shoulder breadth* (anthropometer or large sliding caliper). The distance between the two acromial points—*v*. (3)—with the subject standing in normal posture. The measurement is usually decreased if the shoulders are allowed to sag forward or are braced backward excessively.

(7) *Bicristal breadth* (anthropometer or large sliding caliper). The distance between the most lateral points on the iliac crests.

In obese subjects pressure may be needed to palpate the crests.

(8) *Chest breadth* (anthropometer or large sliding caliper). Methods of taking this measurement vary considerably, both as to the level at which it is taken and the subject's posture. Some take it as the maximal horizontal breadth at the nipple level, in the male, others at the level of the fourth costo-sternal junction, or again at the level of the xiphoid cartilage. The arms may be held above the head, raised at an angle of about 45° to the side, or at a right angle forward with the elbows flexed. It is clearly necessary for the worker to detail his procedure. Measurements should be taken at the end of a normal inspiration and also at the end of expiration.

(9) *Chest depth* (anthropometer or large sliding caliper). This is an even more difficult measurement than (8). It should be taken in the same horizontal plane. Curved cross-pieces to the anthropometer can be used with advantage so that the measurement can be taken in the midline both in front and behind; otherwise the straight cross-piece must be laid tangentially across the back. Measurements should be taken in inspiration and expiration as for (8).

Circumferential Measurements

Instruments. Tapes graduated in centimetres and millimetres are employed. Cloth tapes require periodic checking in case stretching has occurred. Some workers also use narrow metal tapes.

Many circumferential measurements, for example those on the limbs, include a high proportion of muscle and adipose tissue and are therefore liable to change with nutritional state, exercise, etc. They have, indeed, been used rather extensively in nutritional work. The unreliability of nutritional assessment by clinical methods, and the laborious nature of chemical methods, have made it difficult to evaluate the anthropometric work in this field. The first requirement would seem to be reliable methods of assessing nutrition by means more direct than anthropometry, so that body measurements could then be examined in relation to this standard. For this reason only a few circumferential measurements are listed.

(1) Chest circumference at the nipple level, in males.
(2) Abdominal circumference at the umbilical level.
(3) Maximum hip circumference.

(4) Circumference of the upper arm, and of the forearm at its maximum just below the elbow.

(5) Maximum thigh circumference.

(6) Maximum calf circumference.

Weight. It is a frequent practice in nutritional work to utilize the relation between weight and height. In the field, records of weight may be obtained with a variety of light platform spring-scales. These should be checked repeatedly for accuracy. Adults should be weighed to the nearest pound, preferably nude. If this is not possible, a deduction for clothing weight should be made. The weighing should be at least two hours after a meal.

Pigmentary characters. The most frequently used matching scales do not appear to be obtainable commercially at present. For this reason, and because of the generally unsatisfactory nature of matching techniques, the common scales are listed without detailed comment.

According to the most accurate investigations using the Hardy spectrophotometer which records the intensity of light reflected over the whole visible spectrum, skin colour depends on the relative quantities of four pigments, melanin, haemoglobin (oxygenated or reduced), melanoid, which is related to melanin, and carotenoids. The last two make only a minor contribution. In very dark-skinned peoples the haemoglobin is effectively masked.

Skin colour. Perhaps the best-known scale is that of F. von Luschan; it consists of 36 tints made of opaque glass. The colours do not fade but reflections make matching difficult. Von Fritsch's series of 48 tints in oil-colour are considered to match skin texture better. A more elaborate instrument is the Bradley-Milton colour-top, in which the four colours are rapidly rotated and a match obtained by regulating the proportions of these colours.

Hair colour. Spectrophotometric results indicate that the whole range of colours is due to the pigment melanin, but there is some doubt in the case of red hair, which may possibly contain other pigments.

The best-known matching-scale is E. Fischer's box of 30 "hair" samples made of cellulose.

Eye colour. This is probably the most complex and difficult of the pigmentary characters from the point of view of objective recording; for one thing the colour of the iris is often not uniform, but the colour is partly distributed as a variable pattern. Marked variation in iris colour is restricted almost entirely to the European region.

The most widely used matching-scale is Martin's box of 16 glass eyes. Workers in various parts of Europe have not always found the eyes provided suitable for the particular population they were examining. A later version of the set by Martin and Schultz included 20 eyes. The Saller matching scale consists of 40 iris types reproduced on black paper. It is doubtful whether the arrangement of the individual eyes close together on the page, each being surrounded by a white ring, and on a black background, provides the best conditions for accurate matching.

Blood Groups

There are three well-established systems of blood grouping, the ABO, Rh and MN systems, which are independent of one another in their inheritance and so give rise to three independent classifications of humanity.

It will simplify matters if we consider first the ABO groups which are always to be understood when the term "blood groups" is used without qualification. There are four principal groups, AB, A, B and O. It is further possible by more refined tests to subdivide groups A and AB into A_1 and A_2, and A_1B and A_2B, respectively. None of the six groups so defined is peculiar to any one section of mankind and most human communities contain some members of all the groups. The frequencies of the various groups, however, differ very widely from one population to another and some populations are known consisting entirely of group O or of groups O and A. Where, as in Great Britain, large numbers of persons can be examined, the frequencies in different regions can be determined very accurately and are a delicate index of mixture of heritage. Differences of blood group frequencies probably always indicate differences of ancestry or the mixing of stocks in different proportions. Similarity of blood groups by no means always implies similarity of ancestry, but in a given small area it almost invariably does so. It is probable that by suitable statistical techniques the study of mixing of stocks can be made quantitative.

In North Wales, moreover, the blood group frequencies of people with Welsh names differ significantly from those of people with English names. As the analysis was based on records made previously, only names and not physical characters could be used as an indication of ancestry. If, however, two populations originally differing both in blood group frequencies and in certain other physical characters come to be closely associated as a mixed community in a given area, there will at first be a marked correlation in that area between blood groups and those physical characters. This correlation will diminish as genetical equilibrium is reached through intermarriage but it may persist for many generations if there are social or other barriers to intermarriage.

The results of an ABO survey can be expressed in terms of the frequencies of three allelomorphic genes, A, B and O. Two independent variables thus define the composition of a population. If the sub-groups of A and AB have been determined, four genes, A_1, A_2, B and O, and three independent variables, are needed.

The Rh, or Rhesus, system is so named because it was originally based on the use of test serum from rabbits immunized against rhesus monkey blood. The most widely available testing serum divides people only into Rh-positives and Rh-negatives, but something like fifty different groups are theoretically distinguishable with the full range of testing sera though some must be extremely rare. The results of testing a given population are expressible in terms of the frequencies of about fifteen gene complexes, of which about seven are sufficiently common and show sufficient variations in frequency to be of value for anthropological purposes. Considering the wide range of these variations in the relatively few non-European populations so far examined, it is clear that the scope for future work is very great. Since Rh-negative persons are much more common in Europe than elsewhere and are mainly confined to populations originating in Europe, Western Asia and Africa, a mere classification into Rh-positive and Rh-negative of populations in other parts of the world will tend to give an index simply of the amount of European, Negro or Indian admixture.

Incompatibility between the Rh groups of a mother and her baby gives rise, occasionally, to haemolytic disease of the newborn, a severe and often fatal anaemia usually accompanied by jaundice. In such cases the mother is usually Rh-negative and the baby Rh-positive. The disease is thus much more frequent among Europeans than among other races, but it is known to arise at times from Rh incompatibilities of kinds which are likely to occur in non-European populations. It is therefore desirable that an Rh survey should be accompanied by enquiries as to the occurrence of deaths of new-born infants from anaemia and jaundice. The blood of mothers, fathers and surviving sibs of such suspected cases would be of special interest. If a diseased child has been born less than two years previously and it is possible to perform a venepuncture on the mother, her serum should be examined for Rh antibody.

Three groups, M, MN and N, are definable on the classical MN system. The results of testing a population are expressible in terms of the frequencies of two allelomorphic genes, and thus the composition of the population is mathematically defined by a single independent variable. Variations in frequency from one population to another are on the whole less than for the ABO and Rh systems. Large numbers of determinations are therefore as a rule necessary in order to establish significant variations. High frequencies of M are found throughout most of Asia and among the American aborigines, including the Eskimos, while the frequency of N is high in most of the Pacific Islands, including Australia. Recent work has shown the existence of sub-groups within the MN system, defined by the new anti-S serum. The very little anthropological work which has been carried out with this serum suggests that its use, when it becomes widely available, will greatly add to the anthropological value of MN testing.

Several other independent systems of blood grouping are known, each giving rise to a simple classification into positives and negatives. Reliable sera for testing for these are very scarce and confined to a few specialized laboratories, but some such sera may become more readily available in the course of a few years and an examination of a variety of populations with them is much to be desired.

Blood Group Tests

Blood group tests are liable to many kinds of error and are best carried out in a specialized laboratory. It is essential that every time tests are set up they should be controlled by the use of bloods of known groups. It is also essential that the person carrying out the tests should have received training at a laboratory where large numbers of tests are carried out, and should have performed numerous tests under supervision using the exact technique which he proposes to use in the field.

If specimens can be sent by air or otherwise so as to arrive at such a laboratory within a few days the results will always be more reliable than if carried out in the field, unless the field worker is himself a specialist with long experience of blood grouping.

With the increase in complexity of blood group tests and with the extension of air transport to many places of great anthropological interest, much highly specialized blood group work is now being carried out on people who a few years ago were accessible only with great difficulty.

(1) *Collection of Blood for Despatch to a Laboratory*

If the field worker thus has access to a laboratory he need merely take and dispatch samples of blood. It need hardly be said that anthropometric observations should be made on the same individuals as are blood-tested.

(*a*) The best method of obtaining blood is to perform venepunctures, taking 2 to 5 c.c. of blood into a dry sterile tube without preservative or anti-coagulant, and sealing hermetically or with a waxed cork. A Baeyer "Venule" can be used conveniently for this purpose. No one, however, who is not medically qualified should perform venepunctures, particularly under field conditions, unless he has had considerable experience of performing them under medical supervision and is fully aware of the dangers of sepsis.

Red cells remain well preserved inside the clot and a saline suspension of them is made at the testing laboratory. The supernatant serum is of use as a cross check on ABO groups and may also be tested in special cases for abnormal agglutinins such as anti-Rh.

Considerable experience of testing of specimens collected and transported in this way shows that the most important single factor determining their condition on arrival is that of sterility. Though "Venules" or other similar tubes are expensive it is even more expensive to pay the cost of air freight on specimens which are unfit for testing. The use of these containers is therefore to be recommended in all cases.

(*b*) If venepunctures are not performed the best method is to collect a few drops of blood from a finger- or ear-prick into a tube of sterile anti-coagulant.

R. T. Simmons and J. J. Graydon, who have grouped large numbers of specimens sent to them from the islands of the Far

East, describe their method of preparing such a solution as follows (*Med. J. Australia*, **2**, 326, 1945).

"*Method of Preparation of Modified Rous and Turner Glucose and Citrate Solution*

"The reagents used were as follows: sodium citrate, neutral [$2(C_3H_4OH(COONa)_3) + 11H_2O$] (Merck); glucose, pure, reagent quality (several brands have been used and all were found satisfactory).

"Two solutions were prepared, 3·8% sodium citrate solution and 5·4% glucose solution in distilled water, and these were autoclaved separately at 110°C. for fifteen minutes. After being cooled, the two solutions were mixed in the proportions used by Rous and Turner, 720 millilitres of the glucose solution to 297 millilitres of the sodium citrate solution, and 1·0 millilitre of a solution of 1% merthiolate was added to give a final concentration of 1/100,000. The mixing was performed in a specially fitted bottle which enabled the mixture to be transferred to sterile two-millilitre vials under rigid conditions of sterility. Immediately after one millilitre of the mixture had been added to each vial, it was stoppered with a sterile rubber stopper. The vials were labelled, and numbered on both label and stopper, and brushed with transparent lacquer to prevent loss of labels should the vials become wet in transit. Two drops of blood were added to each vial when the specimens were collected. This slightly modified Rous and Turner solution has now been in use here for six years, and we believe that, by means of this preserving solution, it is possible to obtain blood samples from any part of the world, provided the samples are stored in ice and air transport is available for the return of the specimens to the laboratory."

A suitable 1 per cent stock solution of merthiolate is the following: merthiolate, 1 part; borax, 1·4 parts; sterile distilled water to 100 parts.

All solutions containing merthiolate should be kept in coloured bottles or in the dark and it is recommended that they should be used within a month.

The ear-lobe or finger-tip of the subject, which should be clean, is rubbed with cotton-wool moistened, if available, with ether, or with spirit (methylated or any strong distilled alcohol—an absolute alcohol content of 70 per cent gives the maximum sterilizing effect). The pricker, either a triangular surgical needle or a "blood-gun" is best kept in a tube of spirit and rubbed with spirit on cotton-wool before and after use and between successive subjects. A prick is made and the part "milked" so as to get two good-sized drops into the tube, which is then sealed, and labelled. The skin is again wiped over with spirit on cotton-wool and if bleeding continues a small piece of cotton-wool is applied for a few minutes.

While this method is widely used, it should be realized that it carries a small but definite risk of transmitting a virus disease. Approximately one English person in 200 is a healthy carrier of the virus of homologous serum jaundice. This is a liver disease with a mortality rate of over 1 per cent. The virus is not killed by ordinary alcohol sterilization, and is sometimes transmitted by surgical needles. The frequency of the disease in non-Europeans is quite unknown. Ideally, therefore, needles should be boiled or flamed each time they are used. Flaming will of course in time spoil their temper.

Whichever method of collecting blood is used, the resulting specimens should be sent to the laboratory by the quickest means possible. It is extremely desirable that they should be refrigerated on the journey at a temperature just above 0°C. *Freezing solid, even for a moment, will render them useless for testing.* Next to bacterial infection this is in practice the commonest cause of loss of specimens. The precise method of refrigeration must be left to local resources and ingenuity, but it is sometimes possible to use a thermos flask and to replenish it with ordinary ice at intervals. Solid carbon dioxide ("dry ice" or "Drikold") *must not be used* as this would freeze the specimens solid, nor must salt be added to ordinary ice.

The length of time for which the specimens will keep depends greatly on the completeness of their sterility but in general it should not be assumed that they will keep for more than three days at atmospheric temperature, or two weeks in a refrigerator.

(2) *Blood Grouping on the Spot*

If tests have to be carried out in the field it will hardly ever be possible to do anything but ABO grouping. Only workers with wide experience of blood grouping should attempt anything more and they will not require instructions such as are given here. It is almost essential to obtain dried grouping serum. This is already available in small quantities and in a few years will probably become generally available. Every effort should be made to obtain dried serum, but if the liquid product must be used tubes should be discarded as unreliable if they become cloudy or develop a smell.

Cases may rarely arise where no prepared serum is available but the expedition includes a doctor and persons known to be of groups A and B. Their serum, provided it has been previously tested for potency in a laboratory, may be used for grouping purposes. It must be remembered that anti-A serum is obtained from a B person, and anti-B from an A.

Whatever the source of the serum it should be tested at frequent intervals against known A and B bloods to guard both against loss of potency and against the development of false positive reactions which can be produced by serum contaminated by bacteria.

If dried serum is used one or two pipettes should be available, marked with the volume of water needed for reconstituting one ampoule. The ampoule is opened and the given volume is added of water which has been boiled and then cooled. Distilled water is only essential if the local water is extremely hard or brackish. The ampoule is then gently shaken until a uniform solution is obtained. This may take some minutes. If the dried serum does not dissolve but persistently forms a jelly it should not be used. Dried serum reconstituted and not used the same day should be discarded.

Two kinds of serum, anti-A and anti-B, are used for testing. Not the slightest contamination of one of these by the other is to be allowed—separate pipettes, stirring rods, etc., must be used for the two.

In order to obtain blood for testing, the finger or ear is pricked as described above. It is best, if possible, then to mix the blood to be tested with about twenty times its volume of physiological saline (0·9 per cent sodium chloride in distilled water) and supplies of this fluid should be arranged for. One large drop of blood, or enough to give a deep red mixture, is then added to about 1 c.c. of saline in a small tube and well mixed.

A drop of each serum is placed, most conveniently by a Pasteur pipette with a rubber teat, on a porcelain or opal glass tile, or a microscope slide. The drops are labelled with a grease pencil. If labelling is done carefully it is possible to test several bloods on a single tile.

One drop of the blood-saline mixture is then dropped from a clean Pasteur pipette into each drop and stirred into it with a glass rod or platinum loop. If a rod is used there must be a separate rod for each kind of serum. If a platinum loop is used it must be flamed and cooled between successive drops.

Alternatively a very small drop of blood may be transferred with a rod or loop directly into each drop of serum, but results are less reliable, especially in unpractised hands, than if a blood-saline mixture is used.

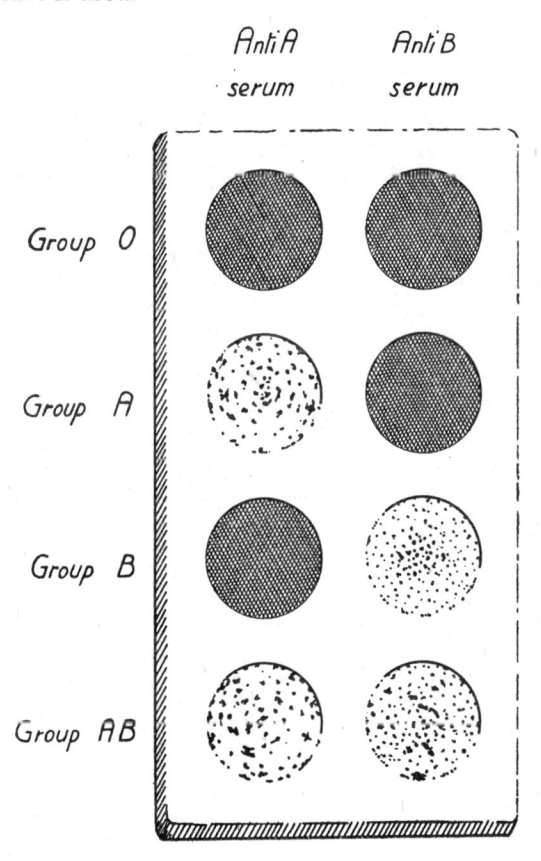

	Anti A serum	Anti B serum
Group O	●	●
Group A	∴	●
Group B	●	∴
Group AB	∴	∴

The tile or slide is then rocked gently for not more than five minutes and the result observed. Either the blood will remain uniformly mixed with the serum or the red cells will agglutinate, i.e. come together to form clumps 1 to 2 mm. across, leaving the rest of the serum clear. If only anti-A causes agglutination the blood group is A. If only anti-B causes agglutination the group is B. If both sera produce agglutination the group is AB, and if neither the group is O.

If the weather is hot and the serum evaporates during the course of the test some granularity may occur which is not true agglutination. This can be dispersed if a drop of physiological saline (0·9 per cent sodium chloride) is added, whereas true agglutination is unaffected.

For the most reliable and complete diagnosis of ABO blood groups it is necessary to test not only for the antigens in the red cells but also for the agglutinins in the serum of the person tested. For this purpose blood must be taken into a tube, allowed to clot and the serum tested against known A and B cells (or better, A_1, A_2 and B).

The serum of an O person contains anti-A and anti-B.

The serum of an A person contains anti-B.

The serum of a B person contains anti-A.

The serum of an AB person contains no blood group agglutinin. It will however very rarely be possible to carry out this test outside a laboratory.

A. S. Wiener, *Blood Groups and Transfusion*, 3rd edition (Thomas, Springfield, Illinois and Baltimore, Maryland, 1943), gives a very full account of the technique involved in blood grouping, and of the state, at the date of publication, of anthropological investigations on the blood groups. An exhaustive account of anthropological blood group investigations up to the date of publication is given by W. C. Boyd, "Blood Groups", *Tabul. Biol.*, 1939, **17**, Pt. 2. ABO blood group testing is fully described in "The Determination of Blood Groups", *Medical Research Council War Memorandum No. 9* (His Majesty's Stationery Office, 1943). A full account of the theory and practice of Rh blood group testing is given in "The Rh Blood Groups and Their Clinical Effects", by P. L. Mollison, A. E. Mourant, and R. R. Race, *Medical Research Council Memorandum No. 19* (His Majesty's Stationery Office, 1948). A full account of all the known blood group systems will be found in "Blood Groups in Man", by R. R. Race and R. Sanger, *Blackwell Scientific Publications, 1950.*

Apart from the ABO groups and in a few cases the MN and Rh groups, no part of the world has been adequately studied and any reliable data are worth obtaining and publishing. Little or no work, even on the ABO groups, has been published on N.E. Africa (except Egypt), Burma, continental Malaya, Persia, and many of the less accessible areas of South America and of Asia.

PART II

SOCIAL ANTHROPOLOGY

CHAPTER I

INTRODUCTION

THE object of this book is twofold. It is intended to be a handy *aide-mémoire* to the trained anthropologist doing field work, and also to stimulate accurate observation and the recording of information thus obtained by anyone in contact with peoples and cultures hitherto imperfectly described. No people or culture has been exhaustively described, so any observer may expect to find some useful work in any area in which he may be.

In every anthropological inquiry it is essential to distinguish clearly between observation and interpretation. In notes and reports, theory and fact should not be merged. The observer who wishes to give a theoretical construction to his material should consider this separately after recording his facts.

The value of theory as a stimulus to relevant observation is fully recognized by anthropologists. Every trained investigator is guided in his field work by his working hypotheses. He may be influenced by a particular school of thought, and this must affect his line of approach and his selection of problems for enquiry, the complete study of the nature and activities of a society being beyond the scope of a single observer. A framework of coherent ideas is a necessary basis for field work; but this should not bias the observation or the record of facts. Amateurs untrained in anthropology interested in the areas in which they find themselves and wishing to devote their leisure to anthropological observation are apt to assume that they are free from bias. This, however, is far from the case: every person is himself the product of a particular cultural tradition and upbringing, and is thus already socially and psychologically conditioned. Unless he is scientifically trained his observation will certainly be hampered by preconceived attitudes of mind. Moreover, it is so usual to regard as "natural" the habits and customs normal in one's own *milieu*, that the observer is apt to consider some behaviour unworthy of record when similar, and abnormal when markedly different from, that which is customary in his own culture. This is especially the case when moral or religious ideas are involved. It is in order to overcome these obstacles that the notes and questions in this volume have been framed, as well as to indicate lines of enquiry worthy of investigation, and the method of obtaining and recording relevant facts.

It is strongly urged that all who intend to undertake anthropological investigation should take a thorough course of anthropological training. It is not suggested that the instructions given in this book can take the place of training.

Before embarking on his enquiries, the field worker would be well advised to make himself thoroughly familiar with the existing literature on the area. If it is adequate and accurate it will save him both time and effort and assist him to formulate any special problem he may wish to investigate, and serve as a check on his observations. If it is inaccurate or scanty his attention will be drawn to facts which others have neglected or misunderstood.

A great deal can be learnt, also, from a careful study of some of the best anthropological monographs on other parts of the world (v. Bibliography) both as to methods of inquiry, the formulation of problems and the presentation of records.

The word *native*[1] is used in this volume because it is presumed that investigations will be made in the vast majority of cases among peoples in their native habitat, The terms primitive, savage, and aboriginal are avoided, because these words have all been used too loosely. It would be difficult to define what any writer means by primitive, but there is one distinction that is obvious and important, that between literate and non-literate peoples. Most of the latter and many of the former are also distinguishable from Western Civilization by their ignorance of experimental natural science and their lack of the technological equipment based on it (v. section on Material Culture and Economics).

This volume is mainly concerned with the non-literate peoples, but the aims and general principles expounded here are applicable to both kinds of societies. For striking as is the diversity of human cultures it can no longer be maintained that this indicates fundamental differences in human mentality. Recent advances in psychology have established the basic identity of human mental structure in all branches of the human race. Sociologists have discovered similar principles at work in all types of society and have drawn attention to the existence of differences between one social

---

[1] It is unfortunate that this word is frequently used in a derogatory sense by Europeans in non-European areas. However no European should feel resentment at being referred to as a native of his home country, and it is in this sense that the word is used in this volume.

stratum and another in the same society which are as great as the differences distinguishing one society from another. Moreover, as the dominance of Western civilization in the world extends, so-called primitive peoples are showing a capacity to acquire the customs, beliefs, technology and languages distinctive of our civilization. Cultures change and develop, as historians have frequently stressed, but this does not imply changes in the physical or mental constitution of the peoples concerned.

Definite rules as to making contact with non-literate peoples cannot be laid down, except that a sympathetic understanding must be reached before reliable information can be obtained. The methods of reaching such a relationship must vary according to the character and capacity of the investigator and the familiarity of the natives with Europeans. The attitude of the native to the European observer must inevitably be influenced by the type of contact he has already had with Europeans, and it must be remembered that such contacts have often been very unfortunate. The investigator must take this into consideration and adapt his behaviour accordingly. The unsophisticated native is often suspicious of all strangers. If a stranger comes with attendants who can be regarded as an armed guard, he may expect a hostile reception, and should he consider it necessary to carry a weapon, he should do so unostentatiously. A sporting rifle and a shotgun are, however, of great assistance in many districts where the natives may welcome extra meat in the shape of game killed by their visitor.

Apart from extreme shyness and agressive hostility, there are many other attitudes requiring great patience and tact. Complete indifference to strangers, blended with a feeling of superiority, is sometimes found in areas where the European is no longer a novelty but has so far done very little to alter native life, and where such needs as the native has for imported objects are supplied by non-European traders. Such an attitude is often a shock to the European, who consciously or unconsciously expects to be respected and at least supposes that the material culture he has to offer must be appreciated. Perhaps more embarrassing are those natives who are officiously attentive in order to exact favours. The idea that natives will say anything to please the investigator, and will invent information, is often found among Europeans who have dealt with the people mainly as their employers. Experience suggests that such views are generally

B *

much exaggerated, and the investigator who establishes a sympathetic understanding with the people and develops a system of checking his material need not be deceived by individuals of this type.

Observers for whom anthropological work is a secondary interest must take into account the type of contact that they personally or their respective professions have had with the natives. They may start from a vantage point, or they may actually be at a disadvantage as compared with a complete stranger. To put the matter in a crude form, the impression made on the native by the civil servant may be primarily that of a tax collector, by the missionary that of one who disapproves of dancing and of all native practices, by the planter that of a harsh procurer and master of labour, by the trader that of a profiteer. On the other hand, the government official may be in favour as the protector against slave raiders and blackbirders, the missionary as protector and benefactor, the trader as a purveyor of valued goods, and labour may be regarded as a means of obtaining exchange goods or even as an adventure. Though the issue can seldom be so clear cut, the resident observer cannot entirely dissociate himself from his profession, but he will do well to face his position clearly when he wants to make a new kind of contact with his native neighbours. In the same way, the anthropologist who is a newcomer must realize that he is a European, usually under government protection, and he must make the most or the least of his privilege as seems best to him.

The sex of the investigator makes some difference to the scheme of work. A woman will seldom be suspected of hostile intention; the mere presence of a woman sometimes inspires confidence, both on this account and where natives are suspicious of the stranger's interference with their own womenfolk. Among very unsophisticated natives, and among those with whom an interpreter is necessary (a woman interpreter being rarely available), a woman may find that she is regarded primarily as a stranger and is given the status of a male, and she has to make her enquiries among the male population. Or she may have to choose whether she will make her investigations mainly among the women—thus obtaining information that would be withheld from a man—or whether, if she requires more general information, she will associate chiefly with men. In non-European cultures unmarried women have not as a rule any honourable status (except possibly as members of a religious cult); natives may fail completely to understand the status of an unmarried European woman, and it may be necessary to invent a husband and children waiting for her in her own country.

As already stated, no single individual can investigate the whole field of human activities; even in a small community he will not have time or opportunity to observe the habits of daily life of both sexes of all ages; all economic, ritual, and recreational activities; all laws, customs and institutions.

Each investigator is likely to be interested in some special aspect. He will be guided by his predilection, training, capabilities and the opportunities for study that the particular community affords. Residents in a district who do not aspire to a complete sociological study may be interested in definite aspects of the culture around them. Their observations will be valuable if systematically recorded.

It is necessary to add a warning that long residence in a community, whether as missionary, government official, medical officer, trader or settler, does not by itself qualify anyone to speak with authority about its social activities. It is true that a missionary or a doctor, through his close and intimate contact with individuals, may be in a position to make valuable observations on such topics as the family life or religious and magical ideas of the natives; an administrative officer, similarly, may have unusual opportunities for getting information about law and political organization, or an agricultural officer for the study of the local economic system. These opportunities must be used in order to make deliberate and systematic inquiries over a period of time. Information derived by casual methods and in a haphazard way has ceased to be regarded as having anthropological value.

An anthropologist is often "adopted" into the tribe that he is visiting. This courteous gesture on the part of the natives is of great advantage and should be accepted; it will give the investigator a status. It should, however, be realized that it is a gesture; the investigator, though he may be accepted as a participator, cannot really be incorporated into native society without altering the social situation considerably both as regards his reactions *vis-à-vis* the natives and their attitude towards him. There may be some hardy individuals who can undertake to live as the natives do, but for most investigators, especially in tropical areas, this is not practicable. Further, it must be pointed out to those who aspire to take this course that it involves psychological and sociological adjustments far more difficult than that of eating unaccustomed food. Firstly, it precludes note-taking; then, the investigator must become a participator instead of an observer; this cannot but influence both his emotional and his intellectual outlook, and completely change his methodological approach. He who throws over his own social status will succeed only if he is given an honourable status among his hosts. It is possible to achieve this among a literate people, or one which—though the majority may be illiterate—has a literate class. Among such people the best course to adopt may be to get oneself accepted as a member of a native household, and those who are able to do this should obtain excellent records.

Care must, however, be exercised in the choice of a household. It must be realized that wherever there is rank, and division of labour, there is likely to be social snobbery of one kind or another. Hence the investigator who hopes to gain a wide view of the culture of a given area must avoid mixing too exclusively with one group. It seems superfluous to add that, while he should not impose his own ideas of social status upon the natives, he must pay deference to theirs. He must first accept the social position for himself that the natives are willing to give him; as a rule this will be an honourable one—especially where there are chiefs—but he must not be annoyed when he is treated as an equal, or even less, though this may come as a shock to him. On the other hand, he must not prejudice his reception by associating at first with a group that the people he wishes to investigate consider pariahs. He should pay attention first to that group which is considered "the best people", and find his informants among them; it will always be easy to work lower in the social scale afterwards, while the reverse may prove impossible.

The investigator should neither regard non-literate peoples as entirely different from, nor as just the same as, the "civilized"; he should consider their performances and their behaviour in relation to both their cultural background and their mental capacity. He should be aware that his own categories are not universal, and he should avoid forcing these on his informants.

It is important that not even the slightest expression of amusement or disapproval should ever be displayed at the description of ridiculous, impossible or disgusting features in custom, cult or legend.

An interest in language, technical processes, history, music, art, games and string figures or photography, is unlikely to be regarded with suspicion and may often prove a better beginning for sociological work than direct investigation, and always an observer who can take an intelligent interest in crafts or games will have an advantage. He should join the natives in their activities, and be ready to contribute himself, but he must not expect his European contributions to be more admired than the native products; he must assume the attitude of a learner, not of a teacher.

The investigator will need to account to the natives for his presence, and this is best done by telling the truth, by explaining to them that the customs of different peoples vary very much and that a study of the peoples of the world is being made. The investigator must be ready to tell them of the habits of Europeans in their own *milieu*, without suggesting that these present a model to be copied. Patriotic flattery may be useful: if told that the customs of neighbouring tribes have been recorded but those of their own people are unknown, most people will react favourably. It is sometimes useful to pretend incredulity to induce further information, or to relate customs of neighbouring peoples and to elicit comment and comparison. A recent investigator found it useful to tell his informants that he was writing a book in which their descendants would read about their customs, so that if they deceived him they would be deceiving their own posterity.

It is generally worth while to have a good supply of simple drugs, antiseptics, and dressings. Simple medical help can be a useful means of establishing cordial contacts. A supply of small presents, including tobacco, cigarettes and sweets (it is advisable for the giver to partake of unknown foods when offering them to unsophisticated natives) is always useful, and among some people money gifts can be used. Toys, ostensibly shown to amuse children, will often interest adults and prove very useful. Salt is much appreciated in some parts of Africa. As a rule beads, cotton cloth and coloured handkerchiefs are valued inasmuch as they are already local articles of trade; preferences can be discovered from the traders in the nearest market town. Novelty *per se* is seldom attractive, and women can be just as offended by the offer of (to them) unsuitable beads as are European girls if given presents suitable for elderly women.

Every culture has its own conventions, rules of conduct and

etiquette. These, and the common forms of greeting, should be learnt as soon as possible. When visiting a sacred place or when present at a ceremonial, the utmost care must be taken not to do or say anything to offend the susceptibilities of the people. It is best to make these intentions clear at the outset, and it is probable that any unwitting mistake may be overlooked or regarded as ignorance on the part of a stranger. It is better to be over-scrupulous about such matters than to run the risk of offending by some act that may be regarded as sacrilegious.

The Use of Literary Evidence

The fieldworker in non-literate societies will not have literary sources other than material collected by himself from informants. In societies which possess the art of writing or of making other kinds of records he may have literary sources such as religious scriptures, historical documents, poetry and fiction as well as inscriptions on stone, metal and wooden objects and memorials. Such sources may provide valuable data especially on earlier periods. If the fieldworker has to rely on translations he must endeavour to obtain versions by different hands, as the difficulty of finding exact equivalents in a European language for key concepts in non-European languages leads to considerable variations in the interpretation of literary sources. Consultation with experts in regard to the meaning of concepts which have undergone changes from period to period is advisable. Literary sources may be of special value in throwing light on the relations between a literate class and the illiterate mass of the people in areas where literacy is restricted to a political, religious or cultural élite. In interpreting literary data, the fieldworker must remember that literature often deals with ideal behaviour, not with actual custom. But ideals thus set out may be a powerful factor in actual behaviour, especially in religious matters, and comparison of the ideals with the realities of behaviour may be of great value. Both religious scriptures and documents that purport to be historical often contain a great deal of commentary, legend and myth and these must be distinguished, if possible, from the records of fact. Literary records also tend to stress the unusual rather than everyday occurrences and cannot be used to give a picture of everyday custom without critical evaluation.

Natural Environment

No sociological study of a community can be undertaken without an understanding of the natural environment within which it exists and from which it draws its subsistence. The investigator should make himself familiar, therefore, with the geographical, geological and meteorological features of the district, and obtain some knowledge of the flora and fauna. The important bearing of such factors on questions of distribution of population (*v.* Demography), type of dwelling (*v.* Technology), economic life (*v.* Economics), and material culture is sufficiently obvious. As the anthropologist cannot be expected to be competent on all these subjects he must seek information from those who are.

The material features of the area, including altitude, and the character of the soil, must be noted as well as the vegetation, especially with regard to pasturage and the supply of wild food, wood, and other useful products, and to the support both of dangerous carnivora or destructive animals, and those hunted for food. The presence or absence of fish as a source of food in sea, lakes, or rivers, the rainfall, and the prevalence of insects as pests, as well as food, should be noted.

The presence of minerals should be noted, and whether these are worked by the natives for themselves, or for Europeans.

Means of communication largely dependent on geographical features and utilized by the native population should be noted. Those dependent on the presence of people of a higher culture, such as railroads, motor roads, steamers on watercourses, and air-lines, may prove important, not so much because the native population may make use of them, but owing to the economic effect that they may have on the population.

The customary attitudes and religious observances of a people to the climatic, physical, and geographical features of the environment are significant both with reference to the economic use they make of their environment and as an index of their empirical knowledge of natural laws (*v.* Ritual and Belief, Economics, and Knowledge and Tradition).

CHAPTER II

METHODS

Scope and Aims

Social Anthropology deals with the behaviour of man in social situations. Sociological generalizations can only be formulated from the careful study of social activities and institutions among specific peoples or in definite areas or cultures.

For any given culture or area material must be collected by (1) direct, and (2) indirect observation; the two methods must be continually integrated. The questionnaire method at one time advocated for anthropology has only a limited utility. In this volume a number of questions will be given under separate headings, but it is not suggested that they should be put directly to the informants. These questions indicate the subjects for investigation, and have been framed as the result of collaboration by a number of fieldworkers from many areas, so as to bring out the range of variations found in the customs and usages of different societies. Examples of well-defined customs are occasionally given, to guide the investigator in his researches.

Direct observation supplemented by immediate interrogation is the ideal course; it is most satisfactory to begin an investigation into any particular subject by way of direct observation of some event, and follow it up by questions as to details, variations, similar events, etc. This may not always be possible. For instance, no death may have occurred in the neighbourhood during the investigator's visit. He must then resort to the indirect method of obtaining information; he should ask how the dead are disposed of, and whether he may visit the site of burial (or other method of disposal of the dead). He should find out whether it is usual to make any offering when visiting such a site, and if so he should provide it. He should watch the behaviour of his informant at the site, and from this he should develop his general investigation.

The purist, who will only record what he has himself seen, will give just as incomplete a picture of social life as he who relies entirely on recording information given in answer to direct questions. Both methods must be followed. When writing up results it should be noted whether the event described has been observed or not; and if not, whether corroboration from reliable witnesses was obtained.

A single observation is rarely sufficient to yield reliable information. The investigator should endeavour to witness the kind of social behaviour he is studying on several occasions. He will not only obtain fuller detail in this way, but will be able to establish also which elements of a native institution or custom are fixed or stereotyped, and which parts of it are liable to vary from occasion to occasion. Many societies have definitely formulated patterns of behaviour associated with the critical events in the life of the individual (e.g. birth, weaning, initiation, marriage, deaths, homicide, etc.), or of the community (e.g. ceremonies regulated by the calendar, or institutions connected with recurrent economic activities). In other societies customs may be less formalized. This may render them more difficult to study; but they are not less significant than the formalized customs to which anthropologists in the past have as a rule devoted most attention.

Similarly, information obtained by the indirect method must always be checked with several informants to ensure accuracy. The accounts given by different informants may diverge considerably owing, firstly, to the differences in personality between informants—their relative reticence or expansiveness, their carelessness or conscientiousness—and, secondly, to social factors that limit or specialize an informant's knowledge: sex, age, rank, or other forms of social differentiation. Such social factors should be noted.

(1) The first rule in all investigations is to advance from the concrete and tangible to the abstract. Social events must be recorded as they happen. Accounts of how natives "think" or "feel" are of little value without information as to how they actually behave in concrete situations.

(2) It is important to remember the difference between theory and practice among ourselves, and to realize that such differences are likely to exist in other cultures. A certain procedure may be related as normal, and this may actually be the ideal or correct procedure, yet in practice it may never be carried out exactly as described.

(3) Whatever the type of social or individual behaviour under investigation, several examples should be recorded, so that the normal and the deviation from the normal can be observed and

the difference between the normal and the accepted ideal can also be discovered.

Remembering these three rules, the investigator will observe differences between stereotyped and spontaneous behaviour; local, family, and individual variations of custom should be noted, as well as the differential observance of customs according to age, sex, rank, temperament, etc. The observer must beware of misusing broad classifications. If he has vague ideas about totemism, animism, "primitive communism", etc., he may "discover" such institutions where they do not exist, and even where they do exist, fail to record their specific features or functions. A number of definitions will be given in this section for technical terms and words used in a technical sense,[1] and their use is recommended. These words are printed in italics.

These definitions represent a classification of social phenomena, not a one-sided interpretation of cultural facts, and thus form an important part of the approach to the study of society. The investigator is warned, however, that he must constantly analyse the phenomena under observation with reference to the local situation; it is not sufficient to label them. In any case he should say exactly what is meant in the context in which any such classification is used.

Modern social anthropology lays particular stress on the interdependence of the different aspects of social life in a given society. No sociological study of a particular problem can be complete without investigating its connection with the social structure, the economic system, religion, language, and technology. Again, to understand the social life of a people, a knowledge of their environment, their history, and the extent to which contact with other peoples has occurred, is essential. The interdependence of the various aspects of culture is theoretically important, and must fundamentally affect the method of investigation. Direct investigation into any subject will never disclose the whole field. For instance, the investigator must be ready to follow up sociological clues that he may discover when studying the technology of fishing or house-building: final appreciation of kinship duties may only be fully brought to light in the study of religion, of chieftainship or of ritual.

The fieldworker may undertake his work with any of the following objectives:

[1] The page on which a word is defined can be found in the index.

(1) The study of native culture as it is at the present time.

(2) The study of culture contact and change as specific processes. In a culture where differentiation has occurred, intra-group influences may be found which resemble influences of culture contact. e.g., the existence of political and/or religious centres, or trade centres, may create variations in the culture expressed in changing values and modes of life.

(3) The reconstruction of history and the tracing of migrations in the past.

(4) All three approaches may be used, with the aim of discovering universally valid social laws.

While it is obvious that no study of the contemporary culture of a people can be made without direct observation of it, much work in the other categories has been marred because it has been undertaken without adequate observation of the present as a foundation. It must be stressed therefore that the study of present conditions is the first necessity whatever the ultimate objective may be, and it is with this aim in view that this volume is planned.

Even in a study of present conditions some investigators tend to concentrate entirely on the so-called "truly native elements", and so misconstrue the actual state of affairs. Although interest in the "uncontaminated native culture" is legitimate, it must be stressed that an accurate record cannot be made by arbitrary selection of the material, nor can a sound understanding of any culture be reached by observation of one locality only. The selection of the most suitable centre for intensive work is a matter of some importance. If there are not too great difficulties to overcome in the way of language or of the hostility of the natives, it is best to choose for the first centre of study a locality well removed from the main focus of foreign influence. But the investigator would distort the picture he draws if he did not observe the reactions of this so-called "untouched" society to European influence. Even if direct contact has not yet occurred, there will have been intercourse between some members of this community and other members of the same tribe who have been in closer contact with European civilization. Moreover, the establishment of a government, with the consequent prohibition of head-hunting, human sacrifice, cannibalism, feuds, infanticide, etc., or the existence of a centre where money passes in exchange for goods and labour, is bound to affect native institutions, even if control has not yet been directly undertaken, and money is still not used as an internal medium of exchange. If European contact is in an early stage, inquiry from the older members of the community may give a picture of the purely native society as it formerly existed. Other centres in the same cultural areas should be chosen where contact is closer, and the effect of direct European influence on native institutions and customs can be observed. Further, the part played by higher cultures of non-European origin must be observed, as well as such factors as the acquisition of foreign goods (*v.* Material Culture) by trade, ceremonial exchange, or pillage.

The methods described in this book are applicable both for the study of more or less intact native societies and of those which are changing rapidly owing to culture contact or for other reasons. Naturally anyone more especially interested in problems of the latter type will have to select his area of fieldwork accordingly.

As we have emphasized, the historical study of a native society must begin from a careful analysis of its present constitution. Where authoritative historical evidence is available, the investigator must evaluate it in accordance with the canons of the historian. Great care must be exercised in interpreting oral traditions historically. Frequently evidence about the past of a society can be derived from an analysis of present-day custom, from the material culture, archaeological and linguistic data, and the physical anthropology of the people.

The establishment of general sociological laws is the task of comparative theory, which does not fall within the scope of this book.

Techniques of Investigation

The Personal Situation

The fieldworker, after taking stock of general conditions, his own training and temperament, and the aim of his investigation, must be guided by common sense in setting up his living quarters. He must remember that his methods of working will be affected by his decision. He may live (*a*) outside the native community, or (*b*) within it.

(*a*) He may occupy a rest-house or a house or camp specially put up for him at a short distance from the community that he wishes to study. His visits to the community will be known, and more or less timed; his informants' visits to him may come to be regular interviews. These disadvantages may be overcome by creating friendly relations with a number of informants, who will give him news of any activities of interest in the locality and guide him to them, while at his own quarters by keeping a supply of cigarettes or some kind of refreshment that is appreciated locally an informal welcome can be imparted. By being outside village territory he may be able to take an objective view of the community as a whole.

(*b*) If conditions of climate and sanitation are favourable, he may be able to live within the community. This is often possible nowadays, in areas where convenient communications and the spread of European civilization have led to the introduction of many of the elementary amenities of Western life. The advantages of a more informal approach may be considerable, but the investigator may become a larger influencing factor than is convenient; events that the community does not wish him to see may be postponed, or his dependence on one part of the community within whose territory he is domiciled may affect his relations with other parts of it. Whichever plan he adopts he should be aware of his own emotional reactions to the community which forms his field of study.

The investigator will need an establishment sufficient for his well-being and for the exigencies of whatever equipment he requires. He should, however, keep his household as small as he conveniently can. In remote parts every extra person may be a strain on the food supply of the neighbourhood. Where it is possible to employ natives of the district this should be done; contact with the servants' families will be useful. Town "boys" when taken into unfamiliar and unsophisticated surroundings may not only be a trouble and a responsibility but actually a source of danger; they are apt to consider themselves superior to their less sophisticated neighbours, and may tend to take unpermitted liberties. If it is impossible to have local natives as attendants, it is better to have "boys" who regard the natives as dangerous, or even as cannibals, rather than those who despise them as slaves or inferiors. If the servants are not natives of the district, it may be advisable to camp well away from the village and to allow them to go into the village only if they are on a definitely friendly footing with the natives.

Language as a Medium of Study

Though the best work cannot be done without a knowledge of the native language, it must be emphasized that such knowledge does not guarantee accurate information and is not a substitute for training or for a methodical approach. Good work, especially survey work, can be done by means of interpreters, or—in areas where a certain number of natives are bilingual—by the use of a common second language, such as pidgin English in many parts of Melanesia, Arabic or Hausa in Africa, Spanish in America, etc., etc. It is far better to work with an interpreter than to attempt an independent enquiry with an inadequate knowledge of the language. Unless the enquirer has exceptional linguistic ability, it is improbable that he will in a period of, say, six months, acquire sufficient mastery of the language to enable him to use it as his medium of enquiry.

When an interpreter is engaged, it is essential to test his accuracy in every possible way and to train him to be exact and consistent. In testing and training an interpreter, the genealogical (*v.* p. 50) and other concrete methods are of the greatest service. The genealogical method gives no scope for anything but straightforward answers, and by its use the interpreter gains the habit of accuracy if he does not already possess it. He learns that what is wanted is not an interesting or plausible story, but exact facts. It is surprising what a transformation can be effected in the value of an interpreter by even a few days' use of the genealogical method. Further, the knowledge which results may interest him greatly and one who was at first only an interpreter may become a most valuable fellow-investigator. It is possible, sometimes, to obtain the services of a literate member of the community, both as interpreter and as clerk. With judicious surveillance and due recognition of the fact that such assistants have sometimes been considerably estranged from the culture by their education, very great help can often be rendered by them. They must not be relied upon as informants, but can be useful in making routine records (e.g. of the crops), and in taking down texts from an informant's dictation.

Use of Native Terms

Whenever work is done with an interpreter, or directly by means of a common language, the investigator will soon find it impossible to keep wholly to his own language. It will become clear that there are many native terms that have no English equivalent, and it will be found essential to use the native words. As the work progresses, this use of native terms will probably become more and more necessary, and it often comes about that the language of communication between the observer and the interpreter has become almost unintelligible to a bystander, nearly every noun being native, in a general setting of English. There is no more potent source of misunderstanding than the failure to appreciate the difference of meaning between some native term and the English word which has been taken to be its equivalent. This happens especially when a *lingua franca* such as pidgin English forms the means of communication. An apparently hopeless misunderstanding may at once disappear with the substitution of the appropriate native term for that in use in the jargon.

Choice of Informants

The selection of informants will to some extent be determined by circumstances beyond the investigator's control. Certain hints on the choice of informants may, however, be useful.

It may be well to explain what is meant by an informant. It is not suggested that these are necessarily individuals who come to the anthropologist and give information daily in answer to questions; every member of the society is a potential informant, his behaviour may be observed and his remarks noted. It is, however, extremely useful to train two or three intelligent people so that they understand the method of investigation and are able to act consciously as intermediaries and to give information with precision. Such informants can be specially useful to cultivate close acquaintance with a limited number of men and women of position, through whose agency it will be possible to start inquiries into such subjects as family life, crafts, religion and magic, etc. They may take the inquirer first to their homesteads, show their own crafts, and introduce him to their relatives and friends. They can act as guides and informants on special occasions and during ceremonies.

Although it is advisable to work with the same informants over a considerable period, one individual should not be kept in attendance for too long a stretch,; he may have duties of his own to attend to, and is likely to tire from unaccustomed work before the investigator does. In most areas, besides the chosen informants, other natives tend to drift in when investigation or conversation is going on, and join the circle for a while. These should be encouraged: they may volunteer corroborative evidence, and though they may be unwilling to contradict the main informants their attitude may help in deciding whether the latter are reliable or not. Often it is from among such an informal group that the best informants are found. Sometimes the principal informants take a protective or adoptive attitude towards the investigator, and this may have very advantageous results; on the other hand, if these attitudes tend to prevent the investigator from obtaining reliable information from others, he will be obliged to withdraw tactfully from his position. When secrecy is encountered its cause should be investigated; it may be due to suspicion created by external policy, or it may be inherent in certain aspects of the culture.

Whenever possible, information should be obtained directly from specialists (priests, rain-makers, doctors, iron-workers, makers of special objects, etc.), and these should be visited and suitable presents made to them. They may be flattered by the attention, and willing to give regular information afterwards, but when very esoteric matters are concerned, they may have to be approached with great patience and diplomacy. It is unwise for an investigator to ask a specialist (especially one with religious or magical power) to visit him until a friendly relationship has been established. It is necessary also to obtain the opinion of ordinary people about specialists—their rules, qualifications, character, etc., in order to discover what such specialists mean to the lay community.

Private Information.—At an early stage of the inquiry it will probably be found that there are many matters about which the people will talk readily and freely, not only when with the investigator but also when others are present; while there are other matters about which at first they will not talk at all, and certainly not in the presence of others. The demeanour of witnesses will reveal when a topic is not considered suitable for public discussion, and it is then better to postpone it and tackle it quietly with some special informant. Reticence is not necessarily a sign of secrecy; it may be a matter of etiquette, and its investigations may lead to interesting information on social behaviour.

Remuneration.—A man who comes day after day to give information will need some recompense, but it is inadvisable to pay directly for information. The best plan is to ascertain the amount that a man normally earns in following his usual occupation— or, if he has no paid occupation, the amount which would recompense him for the loss of a day's work—and make this the basis of payment. If a man is hired by the day or half-day, he comes to look on the occupation as regular work for which he is paid as for any other kind of labour, and any idea of direct payment for information is avoided. In some areas, where working for wages is not a recognized practice, informants may attach themselves to the visitor out of curiosity, to gain prestige, or perhaps even to keep a watchful eye on the stranger. Such natives can be very useful to the investigator, and should be shown hospitality in the form of tobacco and food, and given presents in kind or money.

Where chieftainship is well developed, it may be necessary for the investigator to give a present of some value to the chief, and practically to put himself under the latter's protection or to become his guest. He may then be obliged to find his first informants among dependants especially deputed to look after him, and only later choose his own informants.

Essential Types of Documentation
There are four essential types of documentation:
(1) Descriptive notes and records of investigation.
(2) Maps, plans, diagrams, drawings, and photographs.
(3) Texts, etc. (*v.* p. 49).
(4) Genealogical and census data (*v.* p. 50).

(1) *Descriptive Notes, etc.*
It is unwise to trust to memory; notes should be written as soon as possible. The investigator must sense the native attitude to note-taking in public. Many peoples do not object to it, simply regarding it as one of the European's unaccountable habits. Others are suspicious; some may become tolerant if they can be assured that the purpose of the notes is not for taxation assessment, etc., while a few people who are otherwise friendly will never tolerate the practice.

Note-taking falls roughly into three classes:
(*a*) Records of events observed and information given.
(*b*) Records of prolonged activities or ceremonies.
(*c*) Journal.

(*a*) All notes should be headed with date and locality. It should be stated clearly whether the notes record direct observation, or information in answer to questions or given voluntarily. The names of informants should be stated.

It is helpful to use inks or pencils of two colours, one for ordinary use and one for correction, corroboration, or extra information obtained after the note was first made.

Leading questions should be used only with caution, and require corroboration. Questions that suggest their own answer should never be put, i.e. "Do you buy bread from the baker?" suggests both purchase and the existence of a specialist. "How do you get your bread?" does neither.

(*b*) When recording any prolonged activity it is unwise to interrupt the participators by asking them questions; it is, however, difficult to make a correct report merely by recording what is seen. It is frequently possible to visit beforehand with some well-informed person the place where the activity is about to occur. This person should be one of the future participators or some older member of the community who knows the procedure well. All preparations for the work or ceremony should then be noted, the tools or utensils examined and described, plans or diagrams made when necessary, and the names of all the principal participators expected should be noted and in some cases their genealogies recorded. With this information it will be easier to watch and understand the activity, whether this be of an economic, festal, social, or ceremonial nature. During the event it is useful to jot down notes of actions in correct time sequence. Later, these notes may be used to question observers or participants (preferably a group of both parties) and to obtain fuller details and explanations. As a rule, it is necessary to keep closely to the time sequence in the procedure; if some striking feature is discussed out of order it is often impossible to get any information about the events that preceded it.

(*c*) Journal. A journal written daily is the of greatest assistance to the investigator when he comes to work up his notes. Going through his journal should help him to see the native life in perspective, to appreciate the daily routine and the outstanding events, seasonal fluctuations in food, labour, recreation, ritual, etc. It will also help him to a due appreciation of *lacunae*. Suggested daily entry headings are:

> Weather.
> Self.
> Local activities and occupations.
> Food.
> Special events.

(2) *Maps, Plans, etc.*

Maps and plans of the area under investigation are essential to a clear understanding of the economic and social life of a people, although accurate large-scale surveys will almost certainly be beyond the capacity of the average anthropologist. What the anthropologist can and must do can be divided into two main groups, small-scale sketch maps of the whole area, and plans (large-scale maps) of small areas. All maps and plans should be provided with an indication of scale, orientation, legend, and if possible the latitude and longitude of some given point on the map.

The small-scale sketch maps should be used to indicate, within a rough topographical framework, the principal areas of settlement, communications, tribal and cultural distributions.

Traverses with a prismatic compass and cyclometer can be made on journeys through the territory and will be accurate anough for most purposes. Today many areas have been surveyed and mapped, and for all coastal areas in the world British Admiralty or U.S. Hydrographic Office charts are available. While lacking in interior topographic detail the two latter sources are of great value in determining fixed points upon which to base sketch maps, and the former may form the basis on which the anthropologist's distributions are based. A classification of the territory into types of country on the bases of rocks, soil characters, altitude, slopes, drainage and water supply, and types of vegetation will serve as a basis for the study of cultural distributions, but great care should be taken to avoid showing more than two or three distributions on one map. A map which is obscured by too much detail is worse than useless. The symbols employed should be clearly keyed. It is frequently an advantage to prepare an outline base map showing a few major features to provide a framework on which each distribution is plotted separately.

Location of hunting and agricultural land, fishing rights, etc., will probably fall into the category of sketch maps. As far as possible these distributions should be correlated with physiographic features.

Plans of village sites, houses, etc., can be used to note the inmates of every house, their social status and occupation; the location of burial grounds, shrines or sacred places, club-houses or open spaces reserved for communal activities. They may also be of great value in noting complex large-scale activities such as big ceremonies, in which the disposition of important people and objects should be indicated. If such a survey is carried out in conjunction with the genealogical method, a sound basis for all further investigation will be formed.

Photography is an indispensable adjunct to field work. A good collection of photographs serves not only to document a descriptive account but will be found to be invaluable memoranda. With modern cameras every phase of a technological process, a ceremony, an economic activity, etc., can be rapidly and unobtrusively photographed. Photographs of commonplace, everyday activities and events, e.g. cooking, herding cattle, collecting roots or firewood, and so forth, should be systematically collected, and will be of immense assistance when it comes to writing up the daily life of the people. Cinematographic records, especially of customs and activities that are falling into disuse, are well worth making if the investigator is in a position to do so.

For most problems of social anthropology sound records are unnecessary. They may be needed by the student of linguistic or of native music; both these branches of research require special training, which would include training in the use of mechanical apparatus such as the phonograph (*v.* p. 315).

The following data may prove useful in collating information on environmental conditions:

(1) A classification of the territory into types of country on the basis of rock and soil characters, altitudes, slopes, drainage, and water supply and vegetation. A sketch map of the territory showing the extent and distribution of these areas, which will serve as a basis for the study of cultural distributions. Traverses with a prismatic compass and cyclometer sufficiently accurate for sketch mapping of these distributions can be undertaken in the course of journeys through the territory.

(2) The extent to which the land types so determined are recognized by the people and have particular significance attached to them should be investigated. In the course of ethnographic inquiry, investigate the relation of land types to distribution of settlement, to agricultural pursuits and to the movements of population. From these data an analysis of the occupation of group territories and of the various aspects of land utilization can be attempted, and the results embodied as far as possible in a series of sketch maps of the areas of representative portions. Particular attention should be paid to the following:

(*a*) Pattern, size and stability of settlements which may be distinguished as nucleated or dispersed; factors affecting the location and stability of settlements including both physical conditions, such as water supply, soil exhaustion, and need for bush fallowing or grass burning and pressure on land, and cultural values such as mythological and religious ideas tending to impede communities from migrating.

(*b*) Pattern of land utilization for cultivation and/or grazing, hunting and fishing, including seasonal and annual sequences of occupation of different tracts within the territory. An attempt should be made to estimate the area exploited in one season or year, the aggregate areas of productive land and any reserves of undeveloped territory controlled by or available to the community. These conditions must be considered in relation to both physical conditions and economic organization (*v.* Economics and Land Tenure).

(*c*) Attention should be paid to communications and the mobility of population in relation to physical conditions in which there may be marked seasonal changes, to available means of transport and to the location of productive land in relation to settlements and to the development of internal and external markets. Consideration of the maintenance of paths, roads and bridges in difficult country will involve inquiry into the organizations (e.g. age sets, corvees, etc.), developed or adapted for these purposes.

(3) Information should be sought on native criteria of distinction between soils and any differentiation between them in agricultural or other uses. Particular adaptations to soil conditions, e.g. draining, mound construction, transport, and mixing of soil should be noted and native practice and theory concerning soil fertility and exhaustion should be recorded.

(4) In connection with work on technology the character, location and associations and abundance of any rocks and minerals, e.g. pottery clay, grindstones, flint, semi-precious stones used should be investigated. It should be noted whether these are utilized by the native population or exploited by an alien culture.

(3) *Texts*

The writing of texts, so valuable for obtaining linguistic material, gives important data, and cultural facts as well. Complete texts may be taken down from dictation by an informant who has been asked to relate some incident in his own daily life, some process in which he is interested, a story, myth, or event in family or tribal history. Such texts should be amplified by direct questioning; they then become valuable anthropological data. Further, texts should be made of everyday speech, of children's talk, of talk between kinsfolk, fellow workers, etc. Unless the investigator has a very good knowledge of the language he should try to have every text translated at once.

(4) *The Genealogical Method*

The genealogical method has proved of such value in anthropological research that it is now considered an essential technique in sociological investigation. It is commonly found that genealogical knowledge plays an important role among non-literate peoples; ancestry is often traced back several generations, and a large number of collaterals is known by name. It is clear that this knowledge has a functional value; genealogical data are used in the regulation of marriage, inheritance of property, succession to chieftainship, etc. It seems almost a truism to state that the investigator must understand a principle which is in constant use in his own field of study. Yet such understanding is not so simple as might appear to the uninitiated; not many people trouble to analyse the principle underlying genealogical kinship in their own culture so that it is not surprising that they should find themselves at sea in alien cultures. Before going into the technique of taking genealogies more must be said of the uses of the method.

In a small community it is often possible to take the genealogies of all the inhabitants, and this census can then form the basis not only of sociological work but also for investigation on population and migrations. The data in the genealogies will not only give the investigator the names and relationship to one another of all those whom he will meet in daily work, but will further give him information about individuals not present in the community. Such knowledge is a great asset. There are few people who are not flattered by the personal attention that is shown to them when greeted by their correct names; the skilled fieldworker will use the data he has gained from a few informants to make many more personal contacts. The study of kinship (*v.* pp. 75–88), so necessary to social anthropology, can only be adequately undertaken by means of the genealogical method. When recording the daily and the ceremonial life of a group in which the genealogies have been recorded, the observer will be able to follow the grouping that habitually takes place, whether persons who associate in various activities are genealogically related, and if so, how. He will be able to discover exactly which members of the family are allowed free entry to the house, which are treated ceremonially. When any important event occurs the investigator will know who the individuals are who render assistance, etc. On the occasions of ceremonies connected with birth, marriage, and death genealogical data concerning the principal participants are invaluable. In collecting genealogies the investigator will find corroboration, or new information which he may not have expected, with regard to the remarriage of widows, special marriage customs observed among chiefly families, etc., etc. Thus, both from the point of view of gaining exact information and as an actual introduction to the group with whom work is to be done, the collection of genealogical data affords a sound basis and should be begun as soon as possible.

In most places it will be found that there are certain members of the community having special genealogical knowledge who may be used as informants. The evidence of young men in genealogical matters must be accepted with caution, except for their own generation, for knowledge of this kind is acquired slowly, in most cases through the teaching of the older members of the community. While collecting the genealogies there will, of course, be much overlapping; a family referred to in the ancestry of one man's father will appear again in that of the mother of another, and of the wife of a third, and thus ample means of corroboration and of ascertaining the trustworthiness of different witnesses is provided.

Certain precautions must, however, be taken. The natives must not be allowed to suppose that the work is actually a government census, which might be used for purposes of collecting taxes; any idea of the kind that may get about must be contradicted, and confidence must be gained on that score. There may be specific cultural reasons that make the imparting of genealogical data a roundabout process; the investigator must never try to override such obstacles, but should recognize them as social traits and investigate them as such; then in a sympathetic manner he is sure to find a way of circumventing the difficulties arising from such traits.

One difficulty is the existence in some culture of taboos on names, especially on those of the dead and of certain relatives; for this reason it may even be necessary to collect each genealogy from persons who do not themselves appear in it. Other difficulties arise from the practice of adoption and of exchanging names, while either the paucity or the plurality of personal names may also be a source of confusion. When, however, such sources of difficulty are once recognized, they become merely new instruments for understanding the social conditions of the people; thus, when it is found that adoption exists (*v.* Adoption, p. 73), detailed inquiry should be made and concrete information obtained which will enable a complete study to be made of the very practice that caused the difficulty.

Technique of the Genealogical Method.—In recording genealogical tables it is convenient to write the names of males in capital letters, and those of the females in ordinary writing. The names of the social divisions, villages, etc., may be written in red ink, to be replaced by italics in printing. In recording a marriage, the name of the husband may be put to the left of that of the wife. Whenever a large mass of genealogical material has been collected, it is most convenient to write on one sheet descendants in one line only, and to give cross-references to descendants in the other line, citing the genealogies in which they are to be found. The line chosen will depend on whether the people stress patrilineal or matrilineal descent. Thus with patrilineal descent and a family of sons and daughters, the children of the sons will be given on one sheet, while the children of the daughters will appear in the genealogies of their husbands. When a person has died unmarried, it may be well to indicate whether death has taken place in infancy or in adult life, by the abbreviations *d.y.* (died young), and *d. unm.* (died unmarried). The names of the living should be underlined or distinguished in some other way, to enable a genealogical census of the population to be taken.

The following symbols are recommended (*Man*, May 1932): Triangles have been substituted for squares, as triangles have been widely adopted for males in anthropological work.

△   for males.
○   for females.
◇   for sex unknown.
⟁ ⌀ ⌀   may be used for deaths in infancy.

METHODS 53

These symbols may be hatched, cross-hatched or dotted to indicate clan or any other identification that it is convenient to show in a diagram.

The relationship between siblings should be indicated in the usual way by using a vertical descent line attaching the symbol to the horizontal sibling bar. This bar should always be above the symbols.

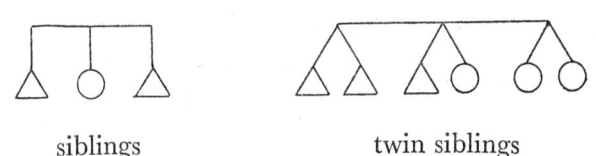

siblings   twin siblings

It has been customary to indicate marriage by the symbol =. Another method is suggested which has considerable advantages. The marriage relationship may be indicated by a horizontal coupling bar joined by a vertical line to the base of the sex symbol. This bar is always below the symbols.

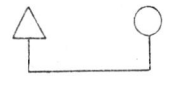

A dotted line indicates an illegitimate union. The offspring of marriages and illegitimate unions are indicated in the usual way by a vertical line from the horizontal coupling band. If a marriage coupling bar and a vertical descent line need to cross in a diagram a loop should be made in the marriage coupling bar where the section takes place. By this method polygynous and polyandrous marriages can be conveniently shown, and should one party be married to two or more spouses belonging to different generations their offspring can be shown on one level by lengthening the vertical descent lines. The order of the marriages can be shown by a number on the marriage coupling bar.

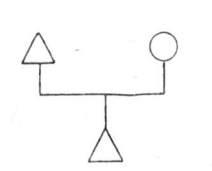

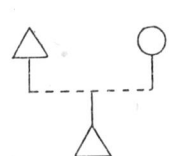

In diagram 1, A has first married b, who is unrelated to him; as a second wife he has married c, his mother's brother's widow, and as third he has taken d, his wife's brother's daughter. The offspring of these three marriages who are half-siblings, are all shown on one level. Their sibling relationships can be traced by

c

following the vertical descent lines to the marriage coupling bars of their respective parents.

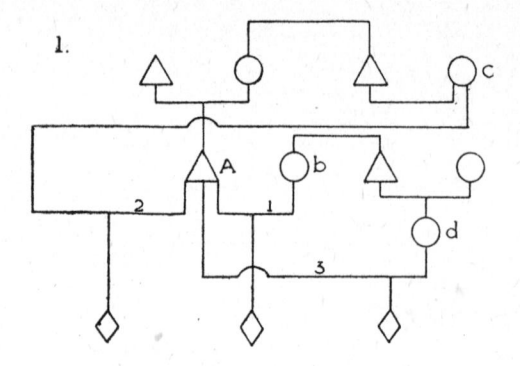

1.

Diagram 2 shows cross-cousin marriage with matrilineal descent. It can be seen at a glance that with this marriage a man belongs to the group both of his father's father and of his mother's mother.

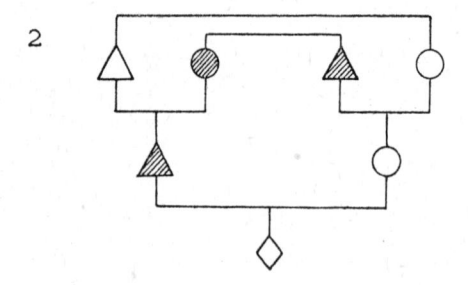

2.

In collecting genealogies, accurate information can be obtained with a minimum knowledge of the language. The following method is necessary both for those employing an interpreter and for those who are familiar with the native language, because it is only by means of careful investigation into the kinship system that the exact significance of such apparently simple words as "mother", "father", "brother", and "sister" can be ascertained.

First the informant should be asked the name of his mother, the woman from whose womb he was born, then he should be asked the name of the man she married, the one who begot him.

Then how he addresses each of these, or what the native word for this relationship is. Then enquiry should be made as to how each parent addresses the informant. Having thus obtained the terms for father, mother and child, and having recorded the names of the informant's father and mother, the question should be put: "Had so-and-so (mother) and so-and-so (father)—both by name—other children besides yourself? What are their names?" Having obtained the names and sex of the other children of the informant's father and mother (his own full brothers and sisters), the next inquiry should be how the informant addresses them, and reciprocally how they address the informant. Thus a genealogical record of an *elementary family* (v. p. 70) is obtained, with the terms for father, mother, child and *sibling* (v. p. 71). Using the native terms that he has ascertained, the investigator is in a position to continue the genealogy and to record other persons related by kinship and affinity (v. p. 76) and their terms of address.

The informant should be asked whether he is married, and, still by means of examples, the native words for husband and wife will be obtained, then the names of the children borne him by his wife should be added.

Having obtained the terms for husband and wife, the investigator can now use these five terms—father, mother, husband, wife, child—for gathering all further information. "She who bore", or "he who begot" should be added to the terms for mother and father, because the informant may be using these terms in a wider sense than the investigator. With these five terms the most complicated genealogical tables can be drawn up; the terms for brother and sister, cousin, uncle and aunt, must be avoided, as the sense in which they are used varies greatly (v. Kinship Structure, p. 76).

The next step is to ask "Did so-and-so (the father, by name) have wives other than so-and-so (the mother, by name)? Did these two (by name) have children?" "Did your mother (by name) have another husband? Did these two have children?" Thus the half-sibling relationship is made clear.

The names of the parents of the informant's parents should be recorded, their children other than the informant's parents, their wives and children, in the same way, and so on, for as many generations as the informant can remember. This will give one genealogy as far as it can be traced. The genealogy of the informant's wife may be recorded next.

Sampling

Most forms of social inquiry aim at generalizations, that is, the assertions made are claimed to be "typical" or "normal" for a particular group of people. In many forms of inquiry such assertions are only useful if they can be reduced to quantitative terms so that both the mean and the range of variation are expressed as numbers. This is particularly the case where the subject of inquiry concerns such matters as human fertility, wealth and income distribution, crop yields, trade, food consumption, etc. It is important that every field worker should understand what procedure is necessary to ensure that a set of numerical data bears a known relationship to the observed facts. For many workers in the past, "a typical random sample" seems to have meant nothing more than the miscellaneous selection of facts that happen to have been jotted in a notebook. This is not good enough; numerical data are only reliable if they are derived from a *systematic* set of observations.

The Necessary Essentials of a Useful Sample

Where facts need to be assessed in quantitative terms it is frequently impossible to observe in detail 100 per cent of the field of available data. Inferences drawn from a sample set of data will be statistically valid or invalid according to whether the sample is representative of the whole field or not. It is impossible to ascertain whether or not the sample is representative unless the individual items that comprise the sample have the following characteristics:

(1) They must be evenly distributed throughout the whole field.

(2) They must be selected by some mechanical process which eliminates all personal bias on the part of the observer.

(3) They must include all relevant variants in approximately the same proportions as they occur in the total field.

It follows that in every case a preliminary census is necessary to delimit the total field and assess its range of variability. *Without such a census it is impossible to assess the reliability of a detailed sample.*

Sampling in a Homogeneous Field.—The anthropologist will frequently be dealing with material sufficiently homogeneous to permit him to ignore condition (3) above. In such cases the most practical sampling technique is to allot a number in consecutive series to each item in the total field. Then give the same series of numbers to a stack of cards or counters, thoroughly shuffle the stack and then pick out individual numbers at random to form your sample. Alternatively, if, say, you require a 10 per cent sample, take the shuffled stack as before and pick out every tenth card. Examples of the type of statistical problem for which a sampling technique of this kind is adequate are: the yield from a large number of apparently similar plots of land; the number of occupants in each of a large number of apparently similar houses.

Sampling in a Non-homogeneous Field.—In statistical work in civilized communities, variations of wealth, social class, and occupation are so great that for most purposes it is necessary to subdivide the total field into a number of categories before the process of sampling is applied. It is then possible to ensure that the proportions of each category occurring in the sample correspond to the proportions in the total field; if this device is not adopted then the sample can only be made representative by being made unduly large. Non-homogeneous fields of this kind also occur in anthropological data, as for instance in cases where the age, wealth or occupation of individuals are significant variables. The following example illustrates the principle involved. Consider a community of 1,000 households of which 950 are cultivators, 30 are fisherfolk, and 20 are traders. If the sampling procedure recommended in the previous paragraph were applied without preliminary separation of the field into occupational categories then a 10 per cent sample (i.e. 100 households) might quite possibly exclude all the fisherfolk and all the traders. Clearly for certain types of study the inferences from such a sample would be highly unreliable. If however a preliminary census has been taken the proportions of the basic occupations in the total population are known and one can therefore ensure that the sample is composed of 95 cultivators, 3 fishermen, and 2 traders. The selection of individual cultivators, fishermen and traders must be by the technique described in the previous paragraph.

Methods for the *analysis* of quantitative data cannot be covered in this book and the reader is referred to works on elementary statistical calculation. E. G. Chambers's *Statistical Calculation for Beginners* (C.U.P., 1940) is suggested. In the field, however, the technique of calculation is less important than the technique of sampling. It cannot be too strongly stressed that if the original data are not collected in accordance with a correct sampling procedure, then no amount of mathematical manipulation later can produce results having scientific value.

Size of Samples.—The reliability of a sample increases only in proportion to the square root of its numerical size; other things being equal, a sample of 200 items will be only twice as reliable as a sample of 50. The purpose of sampling is to save effort and the collection of large samples is usually a waste of time. If the total field is really efficiently categorized, so that the sample, as it were, becomes an exact scale model of the whole, it is occasionally possible to obtain accurate results from samples which are as small as 0·01 per cent of the total field. However, the field worker who is not himself a statistical expert is advised not to claim any general validity for inferences based on samples which are less than 5 per cent of the total field, even when the normal tests for reliability appear to hold good. There is no optimum size for a sample. Where possible avoid samples containing less than 30 observations. Where small samples are unavoidable, do not neglect them as useless, but remember that they will require special treatment in the final statistical analysis.

Demography

A classified enumeration of population is not only intrinsically valuable in providing data for statistical estimations of density of population, sex ratios, fertility rates and other indices; it is also essential to an understanding of social structure since the organization and functions of kin groups, associations, and other social groups will be found to be related to their numerical strength and to be modified with changes in that strength. The term clan (*v.* p. 89) is for example applied to unilineal kin groups of widely varying character ranging in size from quite small territorially compact groups to very large aggregates of individuals dispersed through a number of local communities. Data for the study of the relation between demographic conditions and social institutions are urgently needed. A census should therefore be planned to afford statistics of value for the elucidation of the particular sociological conditions as well as for the calculation of ratios and rates of change of more general interest to the vital statistics.

Ethnographers and others attempting a demographic survey among primitive peoples will often have to rely for obtaining statistics of population on their own unaided efforts confined to a relatively small and selected area. An intensive and exhaustive statistical record of even the smallest selected area or of a single village is usually of far greater value than a wide and extensive enumeration of the population of a whole country undertaken with less accuracy, method, and detail, and omitting much of the information which may afterwards be required. When collecting demographic records it is essential that the data under each heading should be complete for the group and/or time period in question. As omissions will invalidate conclusions the greatest care should be taken that no items are overlooked. The group and the period of time to which data refer should always be clearly defined.

Two basic difficulties are frequently encountered in collecting and analysing vital statistics among primitive peoples. First, data cannot as a rule be obtained for the same group at a number of regular intervals of time and changes in the composition of the population have to be inferred as far as possible from the results of a single enumeration. It is therefore necessary to pay particular attention to the collection of data of value in estimating trends. Second, exact information as to the actual age of individuals is not usually obtainable. This renders important the fullest use of native criteria of age and seniority. It is usually possible to make use of socially recognized stages in the individual life cycle to obtain sufficiently standardized criteria for distinguishing between adults and minors and for establishing the age of death of deceased members of groups. Where *age sets* (*q.v.*) are formed at fairly regular intervals it is possible to attempt a classification of adults by age. When the population is too large to allow a detailed investigation of the community as a whole, it is possible to estimate ratios and population trends by sampling. For this purpose the field worker should attempt to obtain as full a record as possible of closed groups, i.e. groups for which particulars of all offspring of the adult members of one sex (preferably the women) can be obtained. When the population is reasonably homogeneous with respect to living conditions, a co-residential group will usually be the most convenient for this purpose. To be of any value a sample should include at least 200 living adults of one sex. Data from primitive peoples are rarely adequate for the calculation of reproductive ratios by the methods employed in the analysis of census data from European areas, but the ratios of minors of either sex to the total population of that sex in a closed group will afford some indication of the net reproductive capacity. Indices of mortality by sex and at different ages can be estimated from data on the number and order of births and the subsequent history of the children born. For this purpose a series of females and their children is desirable. In a patrilocal polygynous society it may be difficult to obtain data on all the births of an adequate series of females who have completed their reproductive period. In such circumstances an estimate may be based on the births to wives of the men of a group. Distinction must be made between live and still births, death while being suckled, death before puberty, death during later minority, and death when adult. Sociological and psychological attitudes to birth and death should be investigated in order to eliminate errors due to discrepancy between native and European standards. This is particularly important with reference to the deaths of newly-born infants. It is important that no babies dying soon after birth should be omitted and also that abortions and still births should be distinguished from live births and that cases of infanticide should not be reckoned among the former.

A record of births and deaths in an enumerated population over a delimited period is very valuable. Data should be obtained for one or more complete years and the observer in constituting himself a registrar of births and deaths should ensure, either by personal investigation or by the reports of assistants, such as reliable heads of kin groups, that the record is complete. Where such a record is taken the enumeration of the population should refer to the middle of the period of the record.

Where it is impossible to enumerate the entire population an estimate may be obtained by taking a complete census of adult males and multiplying the figures for adult females and minors of each sex, etc., in the sample selected for detailed investigation by the ratio between the total number of adult males and the number of adult males in the sample. To estimate density of population it is necessary to find the area of the territory and its subdivisions by a survey of the group territory, mapping boundaries and settlement sites. Where more refined methods are impossible a useful estimate may be obtained from a cyclometer and prismatic compass survey. Attention should be paid to evidence of any recent changes in the areas of the group territory and of expansion or contraction of settlement sites in the preceding period.

Detailed vital statistics can usually be obtained in connection with genealogical inquiries and the work can best be begun with a household census of the group in which kinship is being studied by the Genealogical Method (*q.v.*).

An attempt should be made to obtain separate ratios for the different elements in populations which include markedly distinct racial stocks. It is, however, exceedingly difficult to identify racial characters in a mixed population, and care must be taken to discriminate between actual racial (somatological) and cultural distinctions. The former can only be utilized as criteria in a demographic, as distinct from an anthropometric, survey when the contact between the racial types is recent and the differences between them are great, e.g. Chinese and Polynesians in the Pacific. It may sometimes be possible in such cases to demonstrate the decline of one racial element within the total population. By a combined genealogical and census inquiry the extent and progress of miscegenation should also be investigated. Where social stratification such as a caste system is observable within the population inquiry should be directed to the relative strength and fertility rates of the different strata.

It will be clear that no definitive list of points for inquiry can be laid down but the applicability of the following should be considered when planning a census:

Enumeration of all living members of the community or of a selected group with particulars for each individual: (1) name with household and territorial location; (2) origin, i.e. whether born a member of the community or an immigrant, and if the latter, age at time of immigration; (3) sex; (4) age (actual or by categories); (5) names of parents and whether living or deceased with estimate of age and cause of death; (6) kin and/or status (caste) group affiliation; (7) association (societies) membership; (8) marital condition including names and kin groups and origins of spouses living, divorced or deceased; (9) children living (distinguished as adult or minor) and deceased (indicating age at death) by each spouse.

Where the above cannot be obtained for the whole community and a random sample alone is investigated in detail a simple enumeration of the entire group by sex, distinguishing adults and minors, or, failing that, of adult males should be attempted.

Such data will make it possible to work such classifications and ratios as the following:

    1.—Total population and population of social groups classified

c*

by sex and age. Where ages in years cannot be obtained the age distribution of population can be usually given to show proportions within the following periods (or stages) which should be determined: pre-puberty, minority (below about 20 years, this threshold is particularly valuable), adulthood, old age, i.e. approximately 50 years and over; the percentages of minors in each sex.

2.—Sex ratios: masculinity is usually expressed by the ratio of females per 100 males. Masculinity should also if possible be determined within different age categories.

3.—Birth and death rates: the mean number of births and of deaths of live-born offspring classified by age and sex for a sample of women who have completed the reproductive period.

4.—Morbidity: ratios of various causes of death by age categories among the deceased members of a sample series within a given period of time.

5.—Marital indices: ratios of bachelors, monogynous and polygynous males. Ratios of marriages with members of the various kin and/or territorial groups for males and females. Ratio of divorces to adult married males and females.

6.—Effective mating ratio, i.e. ratio of number of males to females at the respective average nuptial ages.

7.—Sterility ratio: the proportion of sterile women among those who have reached the menopause or some definable age limit which approximates to it.

8.—Migration ratio: the proportion of immigrants among, and emigrants from, the sample co-residential group by sexes.

CHAPTER III

SOCIAL STRUCTURE

Introduction

An essential prerequisite for the study of any community is a sufficient knowledge of its social structure. By the social structure is meant the *whole network of social relations in which are involved the members of a given community at a particular time*. It defines on the one hand the forms in which people are grouped for social purposes in that society, and on the other the socially recognized ties reflected in the behaviour of individuals to one another and to their social groups.

In regard to every social group the following characteristics should be noted:

(*a*) Local or territorial limitations, if any.

(*b*) Composition (*v*. Demography).

(*c*) Mode of acquiring and losing membership, e.g. by birth, adoption and marriage (voluntarily or not) on the one hand; by death, by expulsion on misdemeanour, by marriage, by voluntary departure, etc., on the other.

(*d*) Form, i.e. the constitution and forms of behaviour distinguishing it from other groups, including the rights and duties and the modes of conduct in general prevailing between members of different groups; and the relations of groups of the same kind to one another, and of different kinds of groups to one another.

(*e*) Function: the relation of its activities to those of other groups and of the total community.

With regard to relations between individuals it should always be ascertained:

(*a*) whether they are temporary or permanent;

(*b*) how they are established, e.g. by birth, marriage, exchange of goods, etc.; or broken up, e.g. by voluntary renunciation, by repudiation, through lapse of time, etc.;

(*c*) what rights and duties and modes of conduct in general hey imply.

Territorial Arrangement

In studying the social structure of a community it will be found convenient to begin with the forms of local grouping, starting with the smallest, the homestead, and going on to the largest territorial group. *A homestead may be defined as a single habitation, the occupants of which constitute a household.* Genealogical inquiries (*v.* Genealogical Method) often yield data on the constitution and determining factors of local groups. Similarly in inquiring about the homestead, information concerning the type of family will be received (*v.* p. 70). Further, the type of domicile (temporary shelter, tent, single hut, or composite dwelling) will be seen in relation to the economic life of the people (*v.* Economics).

A homestead may be a single dwelling or a group of dwellings with or without a common yard, either open or enclosed; the use of each hut or room should be discovered both as regards occupants and other purposes—cooking, sleeping, storehouse, guest-room, etc. It should be noted whether part of the homestead is reserved for the use of women or children. The presence of shrines (*v.* p. 130), graves (*v.* p. 127), and sanitation (*v.* p. 99), position and arrangements, should be noted. The kinship and other social and economic ties between all members of the homestead should be ascertained.

Homesteads may be isolated or grouped together to form villages. It should be ascertained whether the occupants of a village are related by kinship or other ties. *A village may be defined as a territorially separate collection of homesteads, which is regarded as a distinct unit, and of such a size that its inhabitants can all be personally acquainted.*

Homesteads may be grouped in larger units than villages. If such a unit is compact, yet of such a size that any of its inhabitants can have personal contact with only a section of his fellow-members of the group, it may be called a *town*. A town may be culturally or economically heterogeneous, as often happens if it is a centre of manufacture or of commerce; but the term implies an administrative unit under a single political authority. Villages or towns may be very compact and stockaded, or may be extensive settlements stretching the length of the cultivated area.

From the administrative point of view (*v.* Political Structure), a *local group—that is any aggregate of people, who inhabit a single clearly demarcated locality and who regard themselves and act as a unit in relation to other local groups—may be subdivided, and these subdivisions are best referred to as wards.*

A village, or a town, is a permanent or relatively permanent settlement. In some localities economic requirements may necessitate seasonal migrations in search of water or pasturage. The population may be concentrated in permanent settlements during one season and may be dispersed in scattered households or different groupings in another part of the country at another season. Nomadic peoples, or those dependent principally on hunting or collecting for their food supplies (*v.* Economics) have no permanent abodes. But it will generally be found that the groups which habitually move about together tend to be precisely defined in terms of tracts of territory with which they are specially associated. A *horde is a group of nomads claiming exclusive hunting or grazing rights over one or more particular defined areas, within which its wanderings are as a rule confined.*

In analysing the structure of any local group, it should be ascertained how far it behaves as a unit and how far its members show a sense of unity. Has it a political head? (*v.* Political Structure). Is there coordination within the unit for work, feasting, or hunting, or defensive action? (*v.* Warfare). All occasions on which the group acts as a unit should be recorded. Is grazing or cultivated land, or are hunting or fishing rights, held in common by occupants of a village or members of the local group? Is there a meeting place, club-house, guest-house (*v.* p. 100) or dancing ground, within the local group? All such should be described. Is marriage between members of a local group permitted?

In some parts of the world there are long-houses, and a number of families live each within its own section in each house; these houses may form both local and kin groups.

If there are *seasonal migrations*, it should be ascertained whether these are determined by conditions of pasturage, etc., whether each household moves separately, or whether the local group moves together and settles together in the temporary quarters, and the local limits within which such migrations are fixed.

Local groups may be based on occupation, e.g. blacksmiths may live in a group apart from the rest of the community, or towns may be divided into occupational areas (*v.* Economic Life). All such should be described.

A plan indicating the homesteads and other sites of individual or communal significance, i.e. meeting-places or dancing-grounds, shrines, graves, cattle byres, source of water supply, etc., should be made, as well as more detailed plans of individual homesteads.

Such plans, together with genealogies of the local group, will give a firm basis for any sociological study.

It should be noted whether homesteads are scattered according to individual convenience, or arranged in some organized order; if the latter, who is responsible for the arrangement; if there are look-out houses or other forms of defensive structure, and if so the organization and upkeep should be investigated.

Territorial boundaries are habitually known; it should be ascertained what these are and whether artificial boundary marks are made. Are any agreements made with neighbouring groups for crossing territory, following game put up on own territory into that of another group, etc.? What is the penalty for a stranger crossing territory without permission?

A tribe may be defined as a politically or socially coherent and autonomous group occupying or claiming a particular territory.

Means of communication within a tribe may be kept open; investigation should be made into the presence and upkeep of roads and waterways and any tribal boundaries. (For political organization of the tribe, v. Political Structure, p. 136). Within the tribe there may be organized measures for the upkeep of roads and other means of communication.

Sex and Age

Among all peoples social distinctions are made between men and women; sometimes these distinctions extend through the whole sphere of social life, sometimes they are confined to certain aspects. The investigator should note what discriminations and associated taboos exist between men and women in regard to such things as: dress and decoration, speech, etiquette, food, recreation and freedom of movement; division of labour, ownership of property, participation in political life (e.g. tenure of office, voice in public affairs, etc.), legal responsibility (e.g. the right to sue or be sued directly); participation in tribal activities (e.g. attendance at ceremonies, conducting ceremonies, practising magic, etc.); specific forms of grouping (e.g. clubs or secret societies restricted to one sex only).

Both co-operation and antagonism between the sexes may be given social significance. An extreme form of sexual segregation is found in Muhammadan and Brahman societies, where women are secluded in their own quarters in the home and are allowed outside only subject to severe restrictions.

Where such segregation takes place it should be discovered if any male besides the husband has access to the women's quarters. Often the close relatives of the women or male slaves are allowed such privileges. This may have the effect that children in a patrilineal society grow up in closer intimacy with their maternal relatives than with their paternal relatives.

The presence and duties of eunuchs should be ascertained.

The relative status of the sexes may be seen in their legal rights and capacity to hold office, and in the treatment meted out to both sexes from childhood upwards. Is women's work lighter or harder than men's? Is women's work despised by men?

Women may be always under male tutelage—the father, the husband, and at his death the son or other heir of the husband (*v.* Widows).

Any culturally accepted psychological characteristics of either sex should be investigated. It should not be assumed that there are universal male and female types.

Distinctions between the sexes are frequently correlated with native ideas of the physiology of sex. Thus, many of the restrictions imposed upon women apply to them only during the period when they are capable of bearing children, not to young girls or to women past the menopause. Again, some restrictions apply only during the menstrual period, pregnancy or childbirth (*v.* Knowledge of Physiology of Sex, p. 202, also Magic, p. 188). All such restrictions should be investigated.

Age

Gradations of age and maturity are usually recognized in all societies. Thus, we generally find classifications of people according to age, e.g. infants, children, adolescents, adults, old people. Each stage may have characteristic forms of behaviour, dress, occupation, recreation, etc. The investigator should also ascertain if the transition from one age level to another is made the occasion of ceremonial, either for a whole group or for an individual; or if no formal recognition is given to it.

In many societies persons of the same sex and of about the same age are formally grouped into distinctive sets, which are usually formed at successive intervals. The *age-set is a formally organized group of age-fellows, that is, youths or girls, men or women. Each age-set may pass through a series of stages each of which has distinctive status, ceremonial, military or other activities. Such stages are generally known as age-grades.*

The formal constitution of an age-set frequently involves a collective ceremonial and ritual initiation, accompanied by special teaching of tribal law and custom, instruction in sexual matters, and sometimes by physical initiation (*v.* Circumcision, etc., p. 228), which is the mark of the attainment of maturity (*v.* Life Cycle, p. 108). In warlike societies the age-grade system often serves the purpose of providing for military education of the youths, as well as that of organizing military enterprises (*v.* Military Organization, p. 104).

The bonds of age fellowship may unite the sets of men of the tribe and cut across those of kinship and clan brotherhood. Among many warlike peoples society seems to be organized predominantly on an age basis. The lower sets act as schools in which endurance and restraint are taught, the central sets form the military force, and the upper sets the administrative power. Corresponding to the men's sets may be sets of girls who cohabit with the warriors, and women who marry the seniors. In some cases the men have right of access to the wives of their age-fellows; age-sets may also regulate marriage by means of prohibition. Without definite age-sets there may be social and economic ties between age-mates—they may help one another in work, borrow one another's property freely, and play a part in the choice of a spouse for their age-mate and in the ceremonial of betrothal and marriage.

As age is a factor in rank, conferring rights and privileges in matters of precedence, food, dress, decoration and ceremonial, and as the initiation ceremonies in the sets are usually secret, there is much in common between age-sets and secret societies. The essential character of the age-set system, as opposed to the *secret society*, is that the former is equalitarian; every man is a member of his age-set, and as the set rises in rank he passes through every grade of seniority in his own tribe simultaneously with all its other members. The secret society, on the other hand, is selective, and exercises considerable power over persons who may never be able to enter its ranks. Payments of some kind are often necessary on initiation into each age-set; these may be small contributions to a ceremony or feast usually within the scope of all, and, though they may be made separately, all the members of the set rise in rank together. This feature of the age-set organization breaks down where the initiation fees are heavy.

Age-sets and grades are often named, and these names must not be confused with the terms indicating physiological age; in many systems a person belongs to the same named set all his life, while in others he passes from one named grade to another. The names are frequently topical, referring to some event in connection with the initiation ceremony. They are often arranged in a definite series, when either the same names recur or the sets may be considered as subdivisions of groups that recur. The number of sets to a generation is usually fixed, but may vary even within the tribe; but where the cyclic factor in age-sets is present an alternation of generations is seen—thus a man and his eldest son cannot belong to the same named set. As well as the ceremonies of initiation into a set, there may be larger and more important ceremonies occurring at longer intervals when the government of the country is handed over from one group of grades to another.

One of the most common ceremonies of initiation into age-grade organization is the puberty rite of circumcision or incision, with or without the corresponding female operation of clitorodectomy. Other mutilations, such as knocking out of teeth or scarification, are also widespread. But mutilation is not a necessary part of age-set organization. Other ceremonies occur, such as extinguishing all fires, followed by a ritual rekindling, dramatizations of death and birth, sacrifices, and rites of purification. It may be necessary to forgive an initiate his faults ceremonially, and a criminal may be refused initiation.

Seclusion is very common, and usually involves a definite course of instruction and ordeals, with the imposition of food and other taboos. Occasionally the lower-grade initiation may be common to both sexes.

All ceremonies should be described in detail as far as possible; there may be some difficulty, as the higher grades may be secret. Investigation should be made into the organization of the initiation; whether ceremonies are held for individuals or for groups of initiates at regular or irregular intervals; who is responsible for the ceremonies; the behaviour of the initiates to the opposite sex before, during, and after initiation; the relationship between initiates and initiated. All objects used in initiation should be described, and all legends relating to myth and ritual collected. Where payments are made it should be discovered by whom and to whom they are made, as well as who provides all animals for sacrifice. Special cults and cult objects connected with the ceremonies should be described. Any connection between age-groups and the use of clubhouses should be ascertained.

The age-set may be a tribal institution and its function mainly educational (including sex instruction), military, administrative, religious, or social (as where age-fellows eat together and act as hosts to travelling fellows). Commonly all elements exist, but the emphasis varies.

In many of the simpler societies age confers authority in matters of law and government (*v.* Political Organization), the oldest group having the decisive voice in all matters that concern the regulation of public life. In almost all primitive societies the older men, and sometimes women, play the leading part in ceremonial and ritual activities.

In many societies, too, seniority by age and seniority by generation are distinguished, i.e. a classificatory "father" may take precedence of his classificatory "son", irrespective of the actual age of both individuals

The Family

The need to define the varieties of family becomes evident when the use of the word in English is analysed. In common English parlance the word "family" may be used to mean (*a*) the group composed of parents and children; (*b*) a patrilineal lineage (*v.* definition, p. 71); (*c*) a roughly defined cognatic group (*v.* definition, p. 75) frequently including affines (*v.* definition, p. 76); (*d*) a group of relatives and their dependents constituting one household. Where the sociological background is known, the meaning can be understood from the context; this cannot be assumed in social anthropology, and the first duty of the investigator is to discover by exact methods the type or types of social groups existent in the society that he is investigating. For this purpose definitions are necessary.

The elementary or simple family is a group consisting of a father and a mother and their children, whether they are living together or not. It is normally the basic unit of social structure within which the two primary links of kinship are formed, i.e. those of parenthood and siblingship. *Compound families are (a) polygynous, a group consisting of a man and two or more wives and their children; (b) polyandrous, a group consisting of a woman with two or more husbands* and her children; (c) a group formed by the remarriage of a widow or widower having children by a former marriage.

The family in this sense is based on *marriage, which is defined as a union between a man and a woman such that children borne by the woman are recognized as the legitimate offspring of both partners*. *Monogamy is the institution or custom by which a person (man or woman) is permitted to have only one legal spouse at a time*. *Polygamy has two varieties: polygyny is the institution or custom by which a man is permitted to have more than one wife at the same time; polyandry is that by which a woman is permitted to have more than one husband at the same time*.

Siblings: full siblings are persons of either sex who have the same father and mother. (Brother = male sibling, sister = female sibling.) *Two persons who are children of the same mother and different fathers, or of the same father by different mothers, are half-siblings.*

When dealing with the relationship beyond the elementary family, the question of descent must be considered. It must be noted that social usage always determines the lines by which descent is reckoned. *Socially, descent is the recognized connection between a person and his ancestors. An ancestor is one from whom a person is descended, either through the mother or the father.*

Unilineal descent is a distinct reckoning of descent. When reckoned exclusively through males it is called patrilineal or agnatic descent; when reckoned exclusively through females it is called matrilineal or uterine descent. If *bilateral* descent (i.e. "biological" descent) is reckoned, the number of generations that can be included in a social unit is perforce limited. Descent will be considered further in relation to lineage and clan (v. p. 91).

While, or perhaps because, genealogical relationship is of the greatest importance in primitive society, fictitious relationship is frequently set up by means of recognized social conventions. These fictions will be discussed later (v. Adoption, p. 73; Kinship System, p. 67; Marriage of Widows, p. 117; Marriage of the Dead, p. 118).

Inquiries into the type of family will entail others into the homestead or domicile (v. p. 64). The constitution of the *domestic family*, i.e. the *family customarily occupying one homestead*, should be ascertained. It should be noted whether children of either sex remain in their parents' domestic group after marriage, whether there are occasionally other persons attached to it and who these are—widows, widowers, orphans, infirm persons, etc.—and to which parent they are related (*v.* Homestead).

The matrilineal family is sure to present certain problems that will be plain to the observer, but it should not be assumed that with patriliny the family will necessarily resemble the Western European type; investigation of the family structure is always necessary.

In the polygynous family it should be ascertained whether the wives all live together, or whether each wife has a separate house or apartment within the same homestead, or a complete establishment of her own (*v.* Marriage, p. 112).

The polyandrous family may present rather different problems (*v.* Marriage).

A group may be described as a *joint family* when two or more lineally related kinsfolk of the same sex, their spouses and offspring, occupy a single homestead and are jointly subject to the same authority or single head. The term *extended family* should be used for the dispersed form corresponding to a joint family.

In all types of family the extent of co-operation and cohesion should be observed, whether there is an authoritative head of the family, and who this is. It should be noted whether the family (or the household) forms an economic unit; the economic activities of all members of the family should be investigated and whether religious or other rituals are performed within the family (*v.* Religious Ideas and Practices, p. 191).

Status of Members of the Family. The status of each member of the family can only be understood when social, economic, legal and religious factors are considered. The rights, duties and privileges of each member of the family should be noted throughout the life history (*v.* Life Cycle). It should be ascertained in which member of the family authority is vested and to what extent authority can be used. Has a man right of life and death over any member of his family, or can he sell or pawn any member of his family into slavery? Does authority wax or wane in old age? The position of husband and wife respectively in household management should be ascertained, and the relation of each to totemic, ancestor, or other cults of the spouse. The economic relationship of husband and wife should be inquired into; is either specially responsible for the maintenance of the other, or of the whole family? Can each member of the family hold property separately, or do they hold it in common? Is there any age at which children (male and female) become independent of their parents and guardians? Can children own property to dispose of at their free will? The treatment and status of orphans and posthumous children should be ascertained.

Physiological Parenthood and socially recognized Parenthood. In the foregoing it has been assumed that the married couple are the physiological parents of their children. There are, however, a number of conditions when this is not so (*v.* Marriage, p. 118). When the physiological parent is not the socially recognized parent the roles of both should be investigated.

Prohibited degrees will be treated under marriage regulation (*v.* Incest, p. 113), but it should be noted here that in investigating the stability of kinship ties within the family, correlation, if any, with the prevalent incest rules should be made. Further, complete understanding of the family is not possible without investigation of the sexual unions permitted and prohibited within it, and note should be made of the precautions, if any, that are taken to prevent sexual intercourse within the prohibited degrees.

Adoption. The extent of the custom of *adoption* varies very much. In some cultures it is very highly developed. Adoption may be (1) sporadic, or (2) so regular as to be part of the social organization of a people.

(1) The motives for sporadic adoption will not be hard to find; by this means a home will be found for an orphan, or a child for a childless couple. The need for the latter may be intensified by the belief in an after-life, when a descendant is required to carry on the family cult or where there is property to be inherited.

(2) Among a prolific people who do not practise infanticide such motives would be insufficient in themselves to give rise to a general custom of adoption.

Whenever a case of adoption is discovered, full inquiries should be made, and where it is customary a number of cases should be examined. Often there is so much secrecy in connection with the practice that this is difficult. A full account should be obtained of the whole procedure of adoption from the time it is first proposed to the time when the tie between a child and its real parents is broken and replaced by a fictitious tie between it and the adopted parent. The motives for adoptions should be investigated; the genealogical relationship, if any, between the real and adopting parents; the class or other social groups of both parties;

the social status of parents and adopting parents; whether the real parents voluntarily give up their child, or are compelled by custom to do so; the amount and nature of payment, if any; the degree of secrecy and the methods of enforcing it. If the parents and the adopting parents belong to different social groups does the child change its group by adoption, and what attitude is inculcated towards its old group with regard to (*a*) relationship and marriage regulations; (*b*) cults and ceremonies; (*c*) hostility and blood-revenge? Are such customs binding, or optional?

Can an adopted child inherit from and succeed to his adopted parents? Would a true child always take precedence of an adopted child? The ceremonial of adoption should be recorded; this is sometimes symbolized by a dramatization of rebirth, or of suckling, or blood may be sucked from the adopting parent.

Adoption of Adults. What are the motives for this? Are strangers or captives adopted? If so, is the ceremonial the same as for the adoption of infants? Does the adopted adult abide by the marriage taboos that would be incumbent on him if he were the true son of the family into which he is adopted? Is there any recognized method of adoption or affiliation of illegitimate children? In some societies a domestic slave becomes equivalent to an adopted child?

Fostering. If a mother should die or be unable to suckle her child, what steps are taken to preserve its life? Is an attempt made to rear it on artificial foods, or is it given to a foster-mother or wet-nurse?

If the wife's sister or the co-wife is capable of suckling the infant, would one of these women be chosen in preference to any other?

Are there any conditions, social or occupational, under which individual women, or certain classes of women, are not permitted to suckle their young? In pre-Christian Iceland it was the custom for a landowner to send his children to a married couple of inferior rank who would rear them with their own children. Where fostering, either sporadic or customary, takes place the following inquiries should be made: Is the status of the foster-parents equal to or inferior to that of the real parents? Are the foster-parents connected by ties of kin, clan, or common membership of any other social group, with the real parents? At what age does a child leave its foster-parents, and when it does so is it still bound to them or to their children by any social ties, e.g. the duty of blood-revenge, or artificial kinship? What payment, if any, is made to the foster-parents? Are all the children of one family sent to the same foster-parents, or are they scattered? At what age does a child usually leave its real parents, and when, if ever, does it return to them? Can children inherit from or succeed to their foster-parents? How far are the foster-parents responsible socially and economically for the child? What reasons are given for the custom of fostering?

Family Life. The habitual as well as the ceremonial behaviour of all members of the family towards each other should be observed, both as between parents and children and between siblings and other members of the domestic family. Especial regard should be paid to habits concerning food and sleep, work and play, etc. Demonstrations of affection, authority, approval and disapproval, should be noted between members of the family, as well as the extent, if any, of reticence and modesty displayed. No details of family life are too trivial to record. Such observations should be made in the same way in the polygynous, the polyandrous, and the joint family. It should be noted which relatives besides the elementary family are free to enter the homestead, and whether they are offered food in a familiar or a ceremonial way (*v.* Behaviour between Relatives, p. 84).

The part, if any, played by tradition and family pride should be noted. Is pride fostered within the family circle by repetition of the achievements of gifted or prominent members, either living or dead? Are familial peculiarities and habits encouraged or discouraged? Do certain families take up specialized occupations, or are certain families renowned for special forms of skill? (*v.* Economics, p. 169).

Kinship

The bonds of marriage, parenthood and siblingship (*q.v.*) which connect the members of the elementary family with one another result in a network of relationships by kinship and affinity. *Kinship is relationship actually or putatively traced through parent-child or sibling relations, and recognized for social purposes.* All societies recognize cognatic kinship within certain limits. *Cognates are persons descended from the same ancestor whether through males or females, while patrilineal or agnatic kinship is kinship traced through males only* (Caius: *Sunt autem agnati per virilis sexus personae cognatione juncti*).

Matrilineal or *uterine kinship* is kinship traced through females only.

Affinal relationships result from a marriage. They link a person with his or her spouse's kin, e.g. the relationship of a man to his wife's sister or to his mother's brother's wife.

It is convenient to use the term *kinship system* (short for system of kinship and affinity) to denote the pattern of social usages observed in the reciprocal behaviour of persons who are, or are regarded as being, related by kinship or affinity.

The study of kinship beyond the confines of the family is essential to the understanding of social organization. It is important to examine the behaviour of kin and affines towards one another—in domestic life, economic affairs, legal disputes, initiation and marriage ceremonies, funerals, and other occasions of social life.

In the investigation of a kinship system the first step is to record the *kinship terminology*, i.e. *the set of terms used in addressing or speaking of relatives*. The terms in use in a community should be recorded (*a*) by the genealogical method (*v.* p. 52), (*b*) by direct observation, how relatives address one another and speak of one another in daily life. It will generally be found that the terminology is unlike that of English, and therefore precautions are necessary in using English terms to translate native terms. Such a word as "uncle" should be generally avoided, unless qualified as "maternal uncle" or "paternal uncle". It is generally best to describe the relation exactly, as "mother's brother" or "father's brother". With regard to cousins, it is necessary to distinguish two classes. *Cousins (first cousins in our terminology) are the children of siblings.* Anthropologists distinguish between *parallel cousins, children of two siblings of the same sex,* i.e. *children of brothers or children of sisters,* and *cross-cousins, children of two siblings of opposite sex,* i.e. *father's sister's child and mother's brother's child.*

Paternal and *maternal* should be used in the ordinary English sense as appertaining to or concerning the father and the mother respectively. In either a patrilineal or a matrilineal (*v.* p. 71) group a man may speak of his mother's kin as his maternal kin, i.e. his maternal grandfather is his mother's father; his maternal cross-cousin (*see* above) is his mother's brother's child; his father's sister's child is his paternal cross-cousin. The cross-cousins are not related to him lineally; whether descent is patrilineal or matrilineal they must belong to other lines, but it is convenient to describe them as paternal or maternal relatives. *Patrilateral* and *matrilateral may be used as synonyms for paternal and maternal.*

Our own relationship system has been called by Morgan a *Descriptive System* (*v.* p. 79). This name is misleading, because our terms do not describe the relationships. The characteristic features of our system are (1) the existence of terms which are applied only to lineal relatives, as father, mother, grandfather, granddaughter, son, daughter, grandchild, and (2) the lack of discrimination between collateral relationships through males and through females, as illustrated by the use of the terms uncle, nephew, cousin, etc. Wherever this system, or a similar system, is found in other cultures, not only should the terms of relationship be obtained, but a study should also be made of the mutual behaviour of persons using the terms. It should not be assumed, in those cases in which this system is found, that because the terms are similar to our own therefore family reactions, duties, responsibilities, privileges, and mutual bonds, necessarily follow a pattern similar to ours.

The systems more usually associated with the simpler cultures were termed *classificatory* by Morgan. A *classificatory terminology is one in which lineal relatives (father, son, etc.) are addressed or spoken of by terms which also apply to certain collateral relatives.* For example, the same term may be applied to the father and to the father's brother, or the terms for siblings may be applied to parallel cousins. In a fully developed system of classificatory terminology brothers or sisters and sometimes siblings of opposite sex are regarded for certain social purposes as being equivalent to one another. Where siblings of the same sex and of opposite sexes are equated, a person addresses the brothers of both his father and his mother as "father", and all the sisters of both parents as "mother". Where this is the case there is a small number of kinship terms and they are used in the widest possible manner. More usually, siblings are differentiated by sex, so that the mother's brother and the father's sister are given distinct terms. In both these types of kinship system the terms are used in a wide classificatory manner, because it is not only the parents' siblings (with or without sex differentiation) who are equated with the parents, but the siblings of any antecedent or descendant are equated with him or her in terminology.

Classificatory systems are often, but not necessarily, associated with the existence of clans, lineages (*v.* pp. 88–91), or similar social groups defined by unlineal, i.e. either patrilineal or matrilineal, descent (*v.* p. 71). Within a clan the equivalence of siblings is assumed for an indefinite number of generations, so that all clansmen of the same generation are "brothers", while those of the next ascending generation are "fathers" and the next descending generation are all "sons", although genealogical relationship need not and frequently cannot be traced. On the side of the other parent, however, the equivalence may hold only so long as a genealogical relationship can be more or less accurately traced. Thus a man will have classificatory brothers, fathers, etc., who are his own clansmen and clansmen of each other; he will also have classificatory kinsmen, who are not his own clansmen but are related to him by definite genealogical links through a parent or antecedent who is not of his clan.

In some communities (as in Australia) kinship terms of this sort go beyond clan or even tribal relations; hence strangers, never met or seen, are regarded as potentially belonging to some category of kinsmen. This usage is no empty courtesy; the classificatory terms are applied according to the same rules that are current in the local group. Affinal links to distant classificatory relations are sought, and are used to establish relationship to members of distant hordes or tribes. Once these links are recognized, a stranger receives a status in the local community wherever he may be; should he not be able to do so he would not be accepted in the community, and he might be regarded as an enemy.

Owing to the wide use of terms under the classificatory system, everyone in a community may be found to be related in some way to everyone else; in addition, owing to customary forms of marriage and other social forces, persons may be classed together who stand in separate genealogical relationships, e.g. among some people certain cross-cousins fall into the parent-child relationship.

Most societies have prescribed regulations for marriage and the remarriage of widows. It is sometimes found that these primary and secondary marriages are reflected in relationship terms in use.

The Australian systems have been usually known as "class systems" because the persons addressed by a single term fall into definite (usually named) groups; but to avoid ambiguity it is better to call these groups divisions or sections. All members of

SOCIAL STRUCTURE 79

these sections of the opposite sex are either potential mates or persons with whom marriage is never permitted. The terminology is classificatory, but owing to an indirect system of grouping (*v.* Clans, p. 91) children do not belong to the section of either parent but to that of the father's father. It is immaterial whether the moieties (*v. below*) recognize patrilineal or matrilineal descent, the subdivisions work out in the same way (*v.* p. 92).

Systems of this kind are associated with a division of society into two exogamous halves or moieties, usually known as a Dual Organization (*v.* p. 92).

In some kinship systems there is a tendency to *describe* relationships accurately while still using the terms in a classificatory (non-isolating) manner, i.e. siblings may be called respectively "child of the father" and "child of the mother", and under these terms "child of the father's brother" and "child of the mother's sister" will be included respectively; cross-cousins may then be described as "child of the mother's brother", "child of the father's sister", etc. Descriptive terms may also be used in an isolating sense, as "mother-in-law" in our kinship system. In our own system, brother-in-law, though the word is descriptive, is not isolating; it includes two classes of persons, the husbands of all sisters (m. and w.s.), and the brothers of the spouse (m. and w.s.). In other systems siblings are differentiated according to their order of birth. In all cases inquiries should be made to discover whether these differentiations are associated with customs concerning marriage, inheritance, special duties, etc., and how they are related to the formal kinship terminology.

Application of the Genealogical Method to the Analysis of Kinship Systems. It is necessary to use the genealogical method (for instructions *v.* p. 52) in order to elucidate the relationship system. When a number of genealogies has been recorded the terms of relationship should be obtained by asking the informant what he calls the various other members recorded in his genealogy. The list of terms (p. 81) gives the relationships of most importance, arranged on the basis of reciprocal terms.

When working with a man (whom we will call X and who may represent Ego in the system), he should be asked what he calls his father, mother, brothers, and sisters, the brothers and sisters of his father and mother, and other more distantly related persons in his genealogy, care being taken to speak of these persons by their personal names, not by the names of their relationship to him (*v.* Genealogical Method, p. 55). In every case, after asking what X calls A or B, it should be ascertained what A or B calls X, thus obtaining the terms in reciprocal pairs. If, for instance, a man is asked what he calls a woman who is shown by the genealogy to be the wife of his mother's brother, the term that he gives in reply will be entered opposite this relationship, and then the term applied by this woman to the man will be entered in the corresponding line of the second column, giving the equivalent of the husband's sister's son, and so on, until examples of all the relationships in the list have been sought out in the genealogy. When the lists of relationship terms have been ascertained, it is helpful to make another list, putting the native terms first in one column and then giving the English equivalents; by this means one can see at a glance which relatives are classed together under the same terms.

To obtain the correct terms used by women, at the same time checking the reciprocals given by men when addressing female relatives, it is well to go through the same list of relatives with the informant's sister and wife. Terms should be obtained from at least three different genealogies.

In a population where there is any considerable amount of intermarriage it frequently happens that two persons are related to one another in more than one way. Where preferential marriages are customary, this must be a constant feature, and it must also occur in areas that are definitely restricted for geographical or other reasons. In such cases it should be ascertained by which path the generally stressed relationship is traced, and why, also whether the other relationships are ignored or are occasionally recognized for any particular reasons.

If the genealogies obtained are not full enough to supply named individuals for all the relationships required, the informant may be asked: "If So-and-so (named) had a child, brother, sister, or wife, what would you call that person, and reciprocally what would he or she call you?" Or objects may be placed in a certain order, and the investigator may say to his informant: "Suppose this object is So-and-so, here is his brother, or child," and then proceed to ask for kinship terms.

The following is a list of persons for whom kinship terms should be ascertained genealogically. The reciprocal to the relationship in the first column appears in the second column.

SOCIAL STRUCTURE 81

Father	{Son {Daughter
Mother	{Son {Daughter
Brother[1] (m.s.) ..	.. Brother (m.s.)
Brother (w.s.) ..	.. Sister (m.s.)
Brother (child of own father and another mother)	
Brother (child of own mother and another father)	
Sister (w.s.)	.. Sister (w.s.)
Sister (child of own father and another mother)	
Sister (child of own mother and another father	
Elder brother (m.s.) ..	.. Younger brother (m.s.)
Elder brother (w.s.) ..	.. Younger sister (m.s.)
Elder sister (m.s.) ..	.. Younger brother (w.s.)
Elder sister (w.s.) ..	.. Younger sister (w.s.)
Father's brother ..	.. Brother's child (m.s.)
Father's brother's wife	.. Husband's brother's child
Father's brother's child Father's sister ..	..} Brother's child (w.s.)
Father's sister's husband	.. Wife's brother's child
Father's sister's child ..	.. Mother's brother's child
Mother's brother ..	.. Sister's child (m.s.)
Mother's brother's wife	.. Husband's sister's child
Mother's brother's child	.. Father's sister's child
Mother's sister ..	.. Sister's child (w.s.)
Mother's sister's husband	.. Wife's sister's child
Mother's sister's child	.. {Son's son (m.s.)
Father's father[2] ..	.. {Son's daughter (m.s.)
Father's mother ..	{Son's son (w.s.) .. {Son's daughter (w.s.)

[1] It is unlikely that all these terms for the brother/sister relationship will be found in one system. In many systems there is a reciprocal term for the brother/sister relationship, and another that is used between sisters and between brothers. In some systems difference of seniority between brothers and sisters is very important, while in others a sharp distinction is drawn between brothers and sisters as to whether they are the children of the mother or of the father. In such systems the father's brother's children are usually classed with father's children, and the mother's sister's children with the mother's children.

Where seniority of brothers and sisters is distinguished terminologically, investigation should be made to discover whether this also occurs for the spouses and children of brothers and sisters and for the brothers and sisters of the father and the mother.
   m.s. = man speaking.   w.s. = woman speaking.

[2] In some systems it is important to find out the terms for the spouses of all the grandchildren, and, conversely, for the grandparents of the husband and wife—this is especially important where there is an alteration of generation, when the father's father may be classed with the brother, and the son's son's wife with the wife.

Mother's father	..	.. {Daughter's son (m.s.) / Daughter's daughter (m.s.)
Mother's mother	..	.. {Daughter's son (w.s.) / Daughter's daughter (w.s.)
Husband	.. ..	.. Wife
Wife's father	..	.. Daughter's husband (m.s.)
Wife's mother	..	.. Daughter's husband (w.s.)
Husband's father	..	.. Son's wife (m.s.)
Husband's mother	..	.. Son's wife (w.s.)
Wife's brother ..	..	.. Sister's husband (m.s.)
Wife's sister ..	..	.. Sister's husband (w.s.)
Wife's brother's wife	..	.. Husband's sister's husband
Husband's brother	..	.. Brother's wife (m.s.)
Husband's sister	..	.. Brother's wife (w.s.)
Wife's sister's husband		
Husband's brother's wife		
Son's wife's parents		

The above list includes nearer relatives only. In order to find how far the classificatory principle extends, the terms used for more distant relatives must also be obtained; thus, the terms for the brothers and sisters of the grandparents on both sides, for their wives and children, and for the latter's children, should be obtained, as well as the terms for the more distant relatives of the wife. The terms for the wife and children of the sister's son, and for the husband and children of the sister's daughter, are frequently of sociological interest. Whenever the term "child" occurs in the list it is necessary to ascertain in the same way, whether "son" and "daughter" are called by different terms, and the reciprocals must be obtained. Among primitive peoples, relationship terms, unlike our own which depend only on the sex of the person spoken to, are often equally dependent on the sex of the speaker.

Several terms may be used for the same relative. Thus, often one term is used when addressing a relative, and another term when speaking of this relative to another. Sometimes the former is merely the vocative form of the latter, but often it may have quite a different form. The use of different terms of this kind is especially frequent in the case of the father and mother, but should be inquired into throughout. Sometimes two different sets of terms may thus be obtained, of which one may be much simpler than the other, and a false idea would be given if one set only had been obtained

Very often, indeed usually, terms denoting relationship are given with the possessive, "my father", or "his mother", and it may be that the word is never actually used without a possessive. For this reason it is often convenient to give the terms in the possessive form, and in such a case the possessive used should be noted. The full list of possessives should be obtained, and will sometimes be found to differ from those used for other purposes. Notice whether any linguistic differences are made in the form of the term, when the relative referred to is that of the informant ("my father", etc.), or that of the person spoken to ("your father", etc.), or that of the person spoken to ("his father", etc.). Differences in person are generally indicated by differences in the possessive pronoun, but distinct words may be used instead. Special terms may occur for members of a family according to their order of birth.

It is important to study carefully the way in which the terms are used. It is as a key to the study of social relationships that the value of the genealogical method makes itself felt. It may be found that in ordinary daily life personal names are used rather than relationship terms, but that in certain circumstances a person must be addressed not by his name but by the appropriate term of relationship. The investigation of these circumstances will throw valuable light upon the part played by kinship in regulating the life of the people. There may be occasions, for instance, when the use of a relationship term in preference to a personal name is deliberately designed to remind the person spoken to of the relationship in which he stands to the speaker; or such use may induce him to grant a favour or fulfil an obligation. It is quite common for children to call their relatives by the relationship term, while they themselves are addressed by name. The terms of relationship are not merely forms of speech used in addressing or referring to people; they stand for actual social relationships, i.e. there are certain obligations, privileges, rights, etc., which regulate the mutual conduct of a person and anybody to whom he applies a particular term of relationship. It should be noted how far the use of relationship terms as terms of address is customary, permitted but not enjoined, or compulsory, and the context in which such terms are used should be recorded.

Where terms such as "father", "mother" are applied not only

to one's own parents but also to the father's brother's, mother's sisters, and more distant classificatory "parents" (as is customary in classificatory systems) it should be noted whether any modifying words such as "own" are found, i.e. whether the native distinguishes in terminology between his immediate family circle and more remote relatives, by the use of either qualifying adjectives or by differences in intonation, or other means. It must be emphasized that the terms of kinship do not imply that the native's attitude of mind and his social relations to all the people designated by the same term are exactly identical. Both in the ideas and feelings of the people, and in their customary behaviour, the "own" brothers or parents may occupy quite a different position from the cousins, uncles, and other distant relatives although the former may be designated by the same words as the latter.

The term "father" is often applied to any man older than the speaker, even if the two are not genealogically related; "brother" to any male person of the speaker's generation, and so on. If this is so, it should be discovered whether these "fathers", "brothers", etc., belong to some community or association to which the speaker belongs, or whether the terms are used merely out of courtesy.

Behaviour between Relatives. Having obtained a list of relationship terms, the correct traditional behaviour towards all relatives should be ascertained, including relatives in the classificatory sense, both by genealogical reckoning and by affinity. A certain amount of information can be obtained by direct questioning while dealing with kinship terms, but the bulk will be discovered gradually while investigating the social life of the people in its practical as well as in its ceremonial aspects. Besides the traditional or correct behaviour (such as informants will usually describe), the actual behaviour should be observed; this may be found to vary considerably from the ideal standard. The reasons for the lack of conformity in individual cases should be investigated. Every opportunity should be taken to discover the relationships of individuals co-operating in any activity. It may be discovered that individuals who work together or pass their leisure together are actually related, i.e. that economic and friendship ties are formed within certain kinship groupings.

No behaviour pattern should be taken for granted as "natural", even that between parents and children. As a matter of fact this particular pattern varies considerably in different cultures and in different social strata of the same culture. Though we may speak of behaviour patterns as social norms, it must not be supposed that an accepted pattern rules out the expression of individual sentiment, only that there is a customary mode of behaviour which is more or less generally accepted.

The behaviour between parents and children is frequently formally regulated when the latter reach the age of puberty, and even during infancy the kind of attention given by the respective parents is largely ruled by custom and varies from culture to culture. The pattern of behaviour observed between brother and sister varies greatly and is usually regulated by custom; they may be easy companions, or may treat one another with ceremonial respect or even complete avoidance, or they may be close friends and economic partners yet present a rigid formality in each other's presence on all subjects that are sexually toned. Behaviour between siblings of the same sex may be as between equals or may be regulated by seniority.

The respect in which seniors are held may be shown in a traditional manner, but it is necessary to ascertain which relatives are regarded as seniors, as this does not necessarily tally with age, i.e. a man younger than ego may be classificatory father and hence a senior. The father's brother and the mother's brother may be treated quite differently, and this may have nothing to do with seniority but be dependent on the type of descent recognized. For instance, where the classificatory system is found with patrilineal descent, the father and the father's brother are treated in much the same way; that is to say, although the degree of intimacy may not be the same, the type of deference and the kind of demands made on each other will be similar. The mother's brother, as male representative of the mother's family, may be treated quite differently. With patrilineal descent he frequently exercises no control, but is treated with a degree of intimacy and affection which would not be shown to the father or his brother. With matrilineal descent the position may be reversed, the maternal uncle exercising authority while affection is shown to the father and his brother. Again, where cross-cousin or parallel cousin marriage is customary, that uncle who is the father of the prospective bride or of the girl who theoretically should be regarded as such, may be treated in the manner properly adopted towards the father-in-law. The attitude towards grandparents may be one of respect due to age, though among many peoples persons two generations apart call one another companions and treat one another as such. Grandchildren and grandparents may even have a *joking relationship* (v. below). The behaviour of persons of the opposite sex within the family or larger kinship group is regulated by custom (v. Sex and Marriage, pp. 108, 110).

In many cultures some persons who stand in definite relationships to each other avoid one another formally. These persons are usually of the opposite sex and related by marriage, e.g. the most commonly practised *avoidance* is of the son-in-law and the mother-in-law, but avoidance between members of the same family, father and daughter, and brother and sister, also occurs, as it does between persons of the same sex, e.g. a man and his father-in-law. Avoidance as a mode of behaviour is always said to express respect, and though the rules are mutual they are not equal, e.g. it is the duty of a man to avoid his mother-in-law, and should he fail he has frequently to pay a severe penalty. The degree of avoidance may vary from prohibition to enter the village where the particular person lives, to that of not mentioning the personal name, eating from the same dish, or smoking from the same pipe. The most usual form is the prohibition on son-in-law and mother-in-law being found in the same hut at one time, and the necessity to address one another with head averted and to use extremely polite language.

Avoidance may be practised only between the persons actually related genealogically or by affinity who are addressed by a certain term, or it may be extended to all persons to whom the same term is used, however remotely connected, e.g. the avoidance of the mother-in-law may extend to all those women whom the wife calls "mother" as well as to all those who are called "mother" by the wives of all the men who are regarded as "brother." In such cases the degree to which avoidance is carried is usually regulated by the closeness of the connection. Avoidance may continue throughout life, or in the case of the parent-in-law avoidance it may be mitigated or cease after the birth of the first child, or after certain exchanges of gifts.

The opposite pole to avoidance is extreme familiarity; people who stand in certain relationships may commonly accost one another with buffoonery. This is commonly known as the *joking relationship*. The relationship here again is not equal; one person stands as the butt of the other's wit, and though he can retaliate he must not take offence, and can only respond in a jocular manner. Or a fiction of sentimental interest may be enjoined, or free conversation on marriage or on sexual matters may be permitted between certain relatives.

Wherever ceremonial forms of behaviour are in vogue it should be discovered whether all persons who are addressed by the same relationship terms are treated in the same way, or the ceremonial forms of behaviour apply only to those who are closely related. Such ceremonial modes of behaviour may tend to overshadow the observance of the spontaneous emotional reactions between members of the family or other relationship group. It should not, however, be assumed that the latter are obliterated by them.

Duties and Privileges. It is usually found that definite duties, economic, legal and ceremonial, are associated with the relationship tie that unites two persons. This is, of course, true of every society. Among ourselves the economic duties fall within a small group, and the ceremonial duties, though extending over a much wider sphere, are not very rigid. In the simpler cultures, on the other hand, duties of all kinds appear to be more binding and the sphere to be wider, and these may be found to correspond with the type of relationship system in existence. In investigating all ceremonies, especially those concerning birth, puberty, marriage, and death, the duties of all relatives should be ascertained. Similarly all privileges should be discovered, e.g. among some peoples a man may take the property of his maternal uncle or grandfather without special leave, while among others he may behave in the same way to certain relatives of his wife.

Relatives by marriage or affines fall into two reciprocal groups: (1) those in which the affinal link is the first traced from ego, i.e. kinsmen and kinswomen of the spouse, and (2) those in which the kinship link is the first one traced, i.e. persons who have married kinsmen and kinswomen. Whereas the term "mother-in-law" relates exclusively to (2), "brother-in-law" in our nomenclature belongs to both (1) and (2), and as it is used both by males and females includes four different kinds of relationship; for this reason the English term must not be used, but the relation described, i.e. a man's brother-in-law is (*a*) his wife's brother, to whom a certain type of behaviour is due, (*b*) his sister's husband, to whom another type of behaviour may be due; though this is a reciprocal relationship it is not one of equality. A woman's brother-in-law is (*a*) her husband's brother, whom in many societies she regards as a secondary husband either during her husband's lifetime or at his death, and her relationship to him may be different according to whether he is the elder or younger brother of her husband (*v.* Marriage and Relations between the Sexes, p. 117), (*b*) her sister's husband, whose relationship to her may depend on a man's right to marry the sister of his wife, and vary according to her sister's relative age. It is obvious that these brother-in-law relationships have different social values, and the customary behaviour will be found to be related to the legal and social code of the people investigated. The ties set up by the payment of *bridewealth*, or the services due by a bridegroom and his relatives to the relatives of the bride, as well as the customary marriage laws and rules of inheritance, may all influence the behaviour towards relatives by marriage.

The parents, siblings and other close kinsfolk of the spouse are the relatives-in-law *par excellence* to whom special deference is due (*v.* Avoidance). The economic bonds between these relatives should be fully investigated (*v.* Bride-wealth). It should be discovered whether economic assistance or service is rendered before or after the payment has been completed, and if so on what occasions. Further persons connected by marriage may have special ceremonial duties towards one another, especially with regard to death ceremonies. The custom of *avoidance* shown towards relatives of the spouse has already been mentioned (*v.* p. 86). There is, however, great variety in the treatment of the affines, e.g. the wife's sister and the wife's brother's wife may be treated quite differently. A man usually avoids his wife's parents, and sometimes his wife's mother's brother, but less generally his wife's brother. A woman generally avoids her husband's parents. The avoidance pattern and its duration should be investigated for all relatives by marriage.

The spouses of kinsmen and kinswomen fall into specific position according to the relationship in which these kinsfolk stand to a definite person; thus, the wife of the father's brother may be looked upon as "mother", i.e. other wife of "father". The husband of the father's sister and the wife of the mother's brother frequently hold positions of special importance.

Lineage and Clan

A lineage consists of all the descendants in one line of a particular person through a determinate number of generations. A Patrilineal or *agnatic lineage* consists of all the descendants through males of a single male ancestor; a *matrilineage* consists of all the descendants through females of a single ancestress. Where the living members of a lineage form a recognized social group it may be called a *lineage group*. A large lineage group, i.e. one whose members are separated by several generations from the founding ancestor, may be divided into segments or sub-lineages, each consisting of the descendants of some descendant of the founder of the whole lineage.

In most societies the lineage is exogamous, a notable exception being the Arabs. A lineage may be identified by exclusive common ritual observances of a totemic or other kind, and may have common or joint property rights over land, watering-places or herds. Chieftainship, priestly offices, or specialized crafts may be vested exclusively in a lineage.

The number of generations included in the genealogy of a lineage should be ascertained. In some societies the largest lineage recognized may be that which has a common ancestor or ancestors three generations back, whereas in others, eight, nine, or more generations may be counted. Such genealogies cannot always be taken as a measure of the length of time a lineage has existed in a distinct group. There is a tendency to compress lengthy genealogies in many societies. Moreover, as the lineage ramifies with the process of time, it often tends to split up into smaller lineages. Often there is no wider genealogical and exogamous grouping recognized than the lineage of four or five generations, and when the lineage expands it divides into units of this sort.

The questions that arise in connection with the clan (*v.* below) arise also in connection with the lineage, e.g. whether or not it is a local group, what its economic, political, and ceremonial functions are, its role in the regulation of marriage, etc.

It is often difficult to draw a clear distinction between a lineage and a clan. However, *a clan may be defined as a group of persons of both sexes, membership of which is determined by unilineal descent, actual or putative, with ipso facto obligations of an exclusive kind.* In the fifth edition of this work the clan was defined as an exogamous group. Although exogamy is one of the most usual clan obligations, this characteristic has now been discarded for purposes of definition. Certain unilineal groups, notably the Arab "tribe" and the Polynesian *hapu*, having the other main characteristics of that group but not exogamy, may be conveniently regarded as clans. (For definitions of exogamy and endogamy, *v.* Marriage, p. 115).

A clan may be the largest possible lineage recognized in a society. But frequently the lineage is a subdivision of the clan, in which alone common ancestry can be traced by actual genealogical steps. In such cases the clan may consist of a group of lineages, each genealogically independent of the other but all claiming to be descended from a remote common ancestor, who may be a mythological figure (culture-hero) or a totemic (*q.v.*) ancestor.

Clanship. There is usually a belief that clansmen and clanswomen are descended from a common ancestor, and members of a clan usually address one another or refer to one another by kinship terms. It should be noted that the clansmen (where descent is patrilineal) and clanswomen (where descent is matrilineal) may be grouped together territorially, but this is neither necessary nor universal. Clans may be therefore localized in this sense or dispersed. The rule of exogamy, which in the clan commonly necessitates the presence of spouses belonging to other clans in all local groups where it is operative, also tends to the dispersal of clansfolk over wide areas. The mode of expression of the tie felt between members of a clan should be investigated. Emphasis may be laid on the rule of exogamy (its presence or absence should always be noted), on the common possession of a totem (*v.* Totemism, p. 192), descent from a common ancestor, or even in some cases, common habitation of a village or district. All occasions should be recorded when clansmen combine for work, war, recreation, or ceremonial purposes, as well as those times in the life cycle of the individual when he seeks co-operation with the clan. Notice especially whether any economic obligations obtain among clansmen, e.g. in the sharing of food, garden produce, etc. (*v.* Economic Life, p. 185), and to what extent legal responsibility for actions committed by an individual rests upon his clan as a whole (*v.* Justice, p. 146). What part does the clan-organization play in connection with age-grades, secret societies, men's clubs, or the ceremonial distribution of wealth?

The clan may be a political unit, and a study of the political organization (*v.* p. 136) should be made with regard to chieftainship, councils, warfare, etc. There are frequently, but not necessarily, clan chiefs. It should be noted, however, that where the clans are dispersed clan chieftainship and political chieftainship do not coincide. A clan chief may be a territorial chief over an area where his own clan predominates, but he may claim allegiance from members of other clans in the area. Thus, men following their chief in war may find themselves opposed to some members of their own clan. Inquiries should be made as to correct behaviour in such circumstances.

Clans may differ *inter se* with regard to (*a*) status, (*b*) occupation and technical prerogative, (*c*) ritual and customary behaviour, (*v.* Caste, Class). There may be a chiefly or royal clan, all members of which have peculiar customs, and special privileges. There may be sacred objects, groves, or shrines, peculiar to each clan, and where this is the case there may be legends associating a certain clan with a certain locality. Inquiries should be made concerning the attitude and behaviour of clansmen towards such places, and the attitude of those clans resident in the area that are not associated with such shrines should also be discovered.

The clan may regulate marriage; it is commonly an exogamous group. Where clan exogamy is the rule it should be noted whether breaches of the rule occur and how they are dealt with; what is believed to be the result of such breaches and whether there is any recognized ritual performed to mitigate their effects and regularize the unions.

Bilineal reckoning of descent is incompatible with clan organization as it would not result in mutually exclusive groups and provide the vehicle for the continuation of tradition. But recognition of the line of descent of the parent to whose clan the individual does not belong may be shown in various ways. It may be demonstrated in a sense of kinship to the whole clan without involving membership, showing respect for its traditions, taboos, and marriage regulations. This may be restricted to those members of that clan to whom genealogical kinship is traced.

Beside patrilineal and matrilineal descent there are other methods less obvious but which have considerable significance where they exist.

Indirect descent, by which a child does not belong to the totemic clan of its father or mother but to that of his father's father, is characteristic of Australia (*v.* Genealogical Method, p. 79). Another indirect system has been discovered in Papua, called by the natives the "rope", in which males trace their descent through the mother and females through the father.

Double unilineal descent by which two complementary sets of descent groups are recognized (usually for different functions)—an individual belonging to his father's patrilineal group and his mother's matrilineal group—is also well established.

An *asymmetrical* system may be recognized; although this does not affect the affiliation to named clans, it affects their grouping and the marriage regulations. A descent line is *asymmetric* when it is determined differently by the two sexes. Clan descent is recognized by both sexes in the same line, patrilineal or matrilineal, and this may be called the *dominant* or *overt* line; the opposite line is recognized also by the sex that is not dominant, and is accordingly traced only through members of that sex. This may be called the *submerged* line, e.g. the submerged line may be kept secret. With exogamy and asymmetric descent, a brother and a sister cannot marry with a sister and a brother, and where cross-cousin marriage is allowed this too is asymmetrical, i.e. if marriage with the mother's brother's daughter is allowed marriage with the father's sister's daughter is prohibited. Thus clans must fall into series of at least three. It is possible that other combinations and varieties of descent may yet be discovered, not previously suspected when unilineal descent only was considered.

Foreign influence and alterations in economic conditions may tend to change the system of descent, and may require special investigation. Clans may be grouped together in confederacies or systems of complementary clans. A widespread form of grouping is known as *Dual Organization*, which is found with or without subdivision into classes or sections, the essential feature of the system being that *everyone* is involved in it. If a man of moiety A marries a woman of moiety B, their children will belong to moiety A or moiety B (according to the type of descent recognized for the moieties) and will necessarily have kinsfolk in both groups. The system is frequently associated with enjoined marriages with some definite relative—generally, though by no means necessarily, the cross-cousin. It may frequently be found that one large clan has subdivisions which do not intermarry because of their recognized mutual relationship, and therefore they marry into several other smaller clans. Such a condition may give the appearance of a Dual Organization, but if the members of the smaller clans may themselves intermarry it will be seen that this condition is structurally different from Dual Organization. Dual Organization is frequently correlated with definite systems of ritual, belief, and custom, in which the universe is divided into two opposed parts sometimes associated with land and water, summer and winter, or with totemic beliefs, etc. Attention should therefore be given to any definite customs, traditions, rituals and functions, as well as to the regulation of marriage associated with the two moieties of the Dual Organization.

Whatever the type of descent, forces that tend towards integration and fission should be investigated. A complete list of the clans of the area (or culture) should be made, noting any evidence that others have become extinct. It should be stated whether the clans are exogamous. The mode of descent, totem (if any), locality, chieftainship, etc., as well as legends, myths, and ritual, should be recorded. Splitting, subdivision, and formation of new clans should be recorded, as well as the extinction of clans.

Social Stratification

Rank and Class. Some societies are stratified in *social classes* or, where these are closed, *castes* (*v.* p. 94). The division into slaves and free men is widespread though the permanence and rigidity of slave status vary widely. The division into serfs, commoners and aristocracy (which may include various grades) is much more limited. Social classes entail differences in status and civic rights often conditioned by their descent, in the access to positions of power or influence, in wealth, and also in occupation and habitual modes of living, in apparel and the right to use certain ornaments. Groups so defined must be relatively exclusive and permanent to deserve the name of classes, i.e. the barriers being of a kind not freely overcome by individuals. The degree of mobility between the classes presents an important problem.

Class systems are closely associated with a highly differentiated economy. In some cases they may be associated with ethnic heterogeneity, but this needs investigation. The rights, powers, privileges and obligations of all classes should be investigated. Is any class held to enjoy "supernatural" powers, or are its members subject to taboos or ritual restrictions (*v.* Ritual and Belief, p. 181)? How is membership of a class determined, by heredity, through adoption or initiation?

In some societies certain groups, clans, lineages, or families may be considered superior in rank to others, although there may be no apparent differences in the mode of life of their members. All myth and ritual concerning such distinctions should be investigated, as well as public opinion concerning them.

Apart from recognized classes and hereditary rank, individuals may be held in social esteem; the particular criterion demanded for such prestige should be noted. It may be success in war, skill in some occupation, knowledge of tribal lore or esoteric knowledge, or for some personal quality which is generally admired.

Caste is an institution most highly developed in India where Hindu society is divided into a large number of separate groups, mostly functional or tribal in origin. Social intercourse between castes is very strictly limited and intermarriage is generally prohibited; so to a less rigid extent is the taking of water by one caste from the hands of another, or of food cooked with water. The rules in regard to raw food, or food parched, or cooked with *ghi* (clarified butter) are less rigid. The sharing of tobacco smoke, commonly drawn through water by means of a *hugga* or hubble-bubble, is subject to the same limitations as the taking of water. Many castes are themselves split up internally into sub-castes in which the rules of endogamy are often not so strict as those within castes. Sometimes there is a hypergamous system in which a socially inferior sub-caste can obtain wives from another and higher sub-caste on payment; more often a higher sub-caste obtains wives from a lower sub-caste into which it does not allow its daughters to be married. There is a general theoretical distribution of all castes into four *varna* or "colours" representing firstly the priesthood, secondly nobles or soldiers, thirdly the people generally, cultivators and traders, and fourthly servants and menials. Outside these four groups, the first three of which are known as "twice-born" on account of the symbolic rebirth which forms an essential part of their initiation ceremony, are the outcastes, that is, castes which on account of their occupations or habits are considered so excessively polluting that they are outside the pale of Hindu society. All castes and many sub-castes are endogamous, but within each endogamous unit is a number of exogamous units composed actually or theoretically of descendants from a common ancestor and forming a group within which marriage is prohibited. Theoretically it should be possible to arrange castes in a scale of social precedence, but actually it is impossible to do this as status and custom vary considerably from place to place, and castes over a period of time may succeed in changing their status and even their name, while boundaries between *varna* are vague and indefinable. The system is very conservative and lends great stability to society, and serves admirably to hand on from generation to generation skills and secrets of craftsmanship; but it acts in many ways as a deterrent to the introduction of new and improved methods in industry and results in an economy dependent on the interplay of a large number of segregated and sometimes conflicting interests.

Slavery and other Servile Institutions. In many communities persons exist who have not the status of freemen. All such persons are liable to be described as "slaves". But the condition of such persons is sometimes so unlike "slavery" as it existed in recent times in western countries that the name seems hardly appropriate. When the word "slave" is employed in the following questions it must be understood as applying equally to "serfs" or analogous persons.

Note the native terms for all kinds of servile institutions and persons. Are these words used in other contexts as well? Are any of them regarded as abuse, and by whom? How does the servile person address his master, other members of the master's family, and other freemen; how is he addressed himself? It is important to make out the exact legal and social status of such persons, as shown, for instance, in their relation to the rest of the master's family and their status as regards property, marriage, and inheritance. Are there any tribes, clans, or families who, though they are not slaves, are of traditionally lower status and have recognized duties or tribute to pay to the secular or religious authorities? The following points deserve observation. If slavery exists, is it of long standing or recent introduction? What is the proportion of slaves to the free population? Has it increased or decreased within historic times? Is there evidence for a difference of race between the slaves and the freemen or a myth of such difference? How is the status of slavery created? By birth, capture, conquest, trade, debt, crime, voluntary surrender, or at the wish of the family for debt (a variety of this is "pawning"), or religious dedication, or for any other causes? Are all kinds of slaves treated alike? If not, how are their position and treatment affected by any of these conditions?

All kinds of "slavery" should be noted: how the slaves are used, and if certain occupations are reserved for slaves; if so, what reason is given? Does the kind of employment influence their status and treatment? What are their relations with free workers? Are slaves used in war; in particular, have they the right to bear arms? Can the slave trade for his master, and is the master then liable for all the slave's debts? Is any attention paid to the education of slaves? Does the division of labour according to sex prevailing in the society extend to the slaves; or have male slaves to perform "women's work", and vice versa? Describe the dress worn by slaves, and all badges, brands, mutilations, and peculiar hairdressing; their food, and the mode of their burial or disposal after death. Do slaves live in the master's house, or in his village, or other villages? Are there "slave villages"? Does a difference in this respect affect the status of the slaves? May slaves be eaten? (*v.* Cannibalism, p. 245).

Who may (or do) own slaves? Are any slaves owned by the king or chief, or employed on public works, or in positions of authority or trust? Are slaves owned by temples, oracles, gods, or religious societies, or by the dead? Are slaves presented or distributed by the king as a mark of esteem? May such slaves be sold, married, or killed by the owner? Have commoners to ask the chief's permission when they want to keep slaves? Are any groups excluded from ownership of slaves, and does this coincide with other disabilities?

Ascertain the owner's rights over a slave. Has he the power of life and death, the right to punish, either generally or for attempted escape or suicide? If an owner kills a slave, is he ritually unclean? Has he any special rights over female slaves; any rights to slaves' earnings and savings? Is the treatment of slaves humane or cruel in general? How are old and sick slaves treated? Note limitations of owner's rights. Is there any appeal to any external authority (apart from European intervention), or any penalty for ill-treatment of a slave? Is the limitation of the master's power considered to be caused by a right of the slave or by a right of the party interfering on his behalf? Does the interfering party acquire by its action additional rights over the slave; if not, is there any other compensation? Note all customs relating to the sale of slaves; slave-markets, import and export, professional slave-dealing; formalities of sale; ritual acts; are slaves renamed on being acquired by, or changing masters? Can a slave voluntarily transfer himself from one owner to another? By what means? Are slaves lent, hired, pawned, bequeathed, prostituted? In case of sale, are slave-families dispersed or kept together?

Has a slave responsibility for criminal offences; if so, how is he

SOCIAL STRUCTURE

punished? Has he any capacity to sue or be sued at law? Or to give evidence? Are slave witnesses liable to torture? Can a slave act as a substitute for his master in an ordeal, capital offence, or in cases of substitution for homicide?

Has a slave the right to hold money or other property, including other slaves; any power of testamentary disposition, or of inheritance from the owner, from other freemen, and from other slaves; any capacity for contract; or liability for debt?

Is a form of marriage between slaves permitted or enforced? With what legal forms? Is the tie permanent? What is the status of the children? Is there any estimate of the fertility of slaves compared with free persons? What is the status of female captives? What is thought of the intercourse or marriage of a free man with slave women, or of a free woman with a slave? Has such intercourse, marriage, or the actual birth of a child of either sex any effect on the status of either or both parties? What is the status of children born of such intercourse between a female slave or a male slave and a free spouse? May such children inherit property if the father, or if the mother, were originally free? May they succeed to a position in the tribe? What is done when a free man marries a neighbour's slave, or lives with her? Is payment demanded by the master? What does such payment imply, freedom or change of master? Do children by such marriage revert to the owner of the mother, or do they belong to the husband or lover of the slave woman? Does payment change this? Do slaves form social groups or any kind of associations of their own? If so, are these associations recognized by freemen, or secret?

Manumission. Describe conditions and modes of redemption and manumission; is gradual redemption permitted? If so, how is a slave's sale affected when he has paid an instalment? What is the usual age? Are slaves redeemed into the status of serfs, or into a state of freedom? What are the status and occupations of freedmen and freedwomen; the status of their children; the effect of marriages between freedmen and freedwomen, freedmen and slave-women, etc.? Have freedmen or freedwomen any obligations towards the former master? If so, is that obligation handed down to their children; may it concern the heir of the former master? Collect evidence as to public and religious opinion in relation to the custom and law of slavery.

CHAPTER IV

SOCIAL LIFE OF THE INDIVIDUAL

Daily Routine

A FULL description should be given of the typical daily routine of men and women. The observer should base this on records of the activities of men and women well known to him. An approximate time-table and calendar of village or household life should be constructed, including the hours for rising and retiring, beginning and ending work, preparing and taking meals, driving out, watering, milking, and bringing in domestic animals, or other habitual or seasonal activities, and the time spent in recreation, conversation, or story-telling. The native name for each division of the day should be given. It should be noted whether the habits of the people are orderly, or irregular and desultory. The circumstances of all persons whose routine is markedly different from the majority should be investigated. The difference may be due to definite social status (*v.* Rank), or associated with certain occupations, or it may be a temporal ritual condition (*v.* Ritual and Belief). The scope and toleration of individual idiosyncrasies in the community should be noted.

Habitual Customs and Etiquette
Habits of the household with regard to food, sleep, bodily functions and cleanliness, and all forms of normal behaviour, should be noted. Everyday habits should be correlated to economic and environmental conditions, though they may not be entirely dependent upon them.

Food. It should be noted whether food is taken in common by all members of the household, or whether the sexes or the seniors eat apart, and whether there are set times for meals (*v.* Food, p. 241, and Ritual and Belief, p. 183).

Sleep. In a household where there are one or more married couples, children and adolescents, the arrangement for sleeping accommodation should be observed (*v.* Life Cycle, p. 103).

Fire is required for cooking, heat and light. The location and ownership of the fireplaces are often significant, and it should be noted who is responsible for constructing and looking after them (*v*. Ritual and Belief, p. 183, and Material Culture, p. 240).

Bodily Functions and Cleanliness. It should be noted whether special places are reserved for sanitary purposes, and whether these are near the habitation, in the cultivation, or on uncultivated land; whether they are private or communal, and whether any care is taken of them. Observations should be made under the following headings: Cleanliness (p. 223) (Dirt on the skin is sometimes purposely retained and used ritually). Care of hair, finger- and toe-nails should be observed and difference in treatment by the sexes noted (*v*. Ritual and Belief, p. 182). Beliefs regarding the magic powers of excreta may influence habits of excretion and cleanliness in the house and among strangers (*v*. p. 188).

Neighbouring tribes may differ considerably in their habits of personal cleanliness and as to the disposal of refuse in the homesteads and villages. It should be noted whether such practices are directly influenced by ritual or other beliefs. The presence or scarcity of water is not necessarily a measure of cleanliness; among some peoples where water has to be fetched from a distance it is habitual to wash the hands before preparing or eating food. Native ideas on the subject of cleanliness should be recorded.

Clothing (*v*. Material Culture, p. 234).

Nudity. Habits with regard to nudity vary; some people who are usually clothed or partially clothed do not object to nudity within the house, or sleeping. Note the different attitudes to nudity held by the sexes. Many peoples where males and children of both sexes are habitually nude, adopt a small apron-like garment at puberty.

The normal behaviour in daily intercourse should be observed. Etiquette may be treated under several headings: (1) *Salutations*, noting if they differ according to rank, relationship, age and sex; forms of greeting and farewell; *forms of meeting* between enemies in truce, and strangers. (2) *Forms of address*, ceremonial and official phrases, the use of *titles* and kinship terms, and rules of *precedence*; the observance of *avoidance* as a form of respect (*v*. p. 86). (3) *Rules of hospitality* (*v*. p. 100) between kinsfolk, neighbours, strangers, and enemies. (4) *Rules of politeness*, in the household and between strangers; customs regarding the wearing and discarding of clothing by either sex; of propriety in the association of men and women, children, and old people; rules of conduct and respectful demeanour for the young; behaviour towards Europeans and other foreigners. (5) The standard of *decency* in natural functions, behaviour and conversation, for men, women, and young people and children. It should be noted whether any subjects which may be permitted in general conversation are taboo among people in certain degrees of kinship, and what these degrees are. (6) All etiquette with regard to food and meals, whether in the home or on public and festive occasions, and the drinking of intoxicating liquor; the frequency and toleration (or the reverse) of intoxication (*v.* Material Culture, Artificial Drinks, p. 246). All taboos on food should be noted and the reasons for them investigated (*v.* p. 244). In times of scarcity is priority given to any members of the household according to status, age, or sex?

Rules of Hospitality

Rules of hospitality should be inquired into both concerning special social gatherings and on ordinary occasions. Are club-houses belonging either to the local or social groups (village, clan, or age-class, etc.) used also as guest-houses, or are guests received by the headman or other recognized functionary? Is there any form of permanent or hereditary guest-right between social groups or persons related by blood, clan, or other ties? Is it reciprocal? Can guest-right be suspended or forfeited? If so, for what cause?

Guests will usually have some recognized credentials. They may be traders, pilgrims, or persons who relationship to someone in the local group, clan, or tribe is known. Is the acceptance of hospitality accompanied by any rite, such as blood-brotherhood, or does it set up obligations, and if so of what nature are these?

Are guests given wives or temporarily accommodated with women? If so, is it desired that they should beget children? What payments or services do guests offer to their hosts? Is the guest sacred, and if so, how is this shown? Is his blessing valued, or his curse feared? Does he have to undergo purification before being entertained?

Describe any concrete examples you may be able to observe of the arrival and treatment of guests. The arrival of absolute strangers in a locality would be rare except under European influence, unless they were refugees from some other tribe or community. How would such people be treated? Is any right of asylum recognized? Does a host give protection to refugees? Is he bound by honour to espouse their cause? Would refugees be given an honoured or degraded position in the society into which they were received? Note whether shipwrecked people are killed, and why.

What is the general reaction to the knowledge that strangers have trespassed on hunting or grazing grounds? Can permission be asked, and would it usually be granted? Would any recompense be expected?

Feasts (*v.* p. 185).

Training and Education

Careful field studies of the training of children should be made. General observation shows that infants and young children are usually indulged and treated with great kindness, while correct behaviour is expected of children before they reach the age of puberty. Although formal education may never be given, or only at definite periods or for specific purposes, the training of all individuals is a continuous process resulting in more or less recognized reactions to habitual treatment by parents, other adults, and older children. Investigation should be made as to the responsibility for such training as is given; it may rest with either or both parents or with some other relative. It may change in the varying phases of the child's life.

Segregation of the sexes affects the early training of children; where there is no sex segregation, and women participate in the economic, religious and ceremonial activities it should be noted to what extent children are also passive or active participants.

The phases, infant, toddler, child, may fuse imperceptibly, or the passage from one to another may be emphasized by ritual or customary occupations. The first training that any child receives concerns its food and bodily habits (*v.* Suckling, p. 106); the next stage is mainly concerned with keeping a child out of danger and preventing it from being destructive, while later comes the training in the occupation and traditions of the society in which the child lives. In complicated or "civilized" communities the infant's and child's training is largely a matter of controlling its impulses. It is important to observe how far these motives regulate the behaviour of guardians of the young in simple societies, as well as the child's reactions.

Observations of child behaviour in simple societies are of great value as checks on current psychological theories of child development. Thus theories concerning innate aggression, and aggression as a reaction to frustration, need to be checked by means of direct observation of infants and children in diverse cultures.

It should be noted what and when attempts are made to regulate the infant's feeding habits and habits of micturition, defaecation, and general cleanliness, and at what age such attempts are made; whether infants are allowed to scream or whimper; whether they are left alone, or are habitually carried about; the method should be noted, whether carried on the back, hip, in the arms, etc.; whether any cradle or receptacle is used. The habitual methods of soothing should be noted. Note the effects of the mother's occupational habits on the training of the infant, e.g. if she works in the gardens or plantations away from the homestead, does she take the infant with her? If not, what provision is made for it, and at what age? The use of a *cradle, cradle-board* or of any kind of *swaddling* should be noted (*v*. Deformations, p. 225). If swaddling is employed, the extent, method and duration should be noted. How often are the swaddling bands taken off, what measures are taken for cleanliness, all ritual related to children, should be noted, together with any explanations given by women and men.

The manifestation of fear shown by the child, and its treatment by adults should be noted. In this respect there may be considerable difference between the children of the lower cultures and the civilized. The "civilized" child, guarded against natural accidents, is usually early taught the fear of punishment and sin; in lower cultures, the child, less protected, is seldom punished and is sometimes even considered incapable of sin until he is old enough to be initiated or to learn the customary taboos. Manifestations of anger and affection should be noted, e.g. angry outbursts (tantrums) may be left unchecked by adults or older children. If checked, the method of checking should be noted. It should be discovered whether there is any system of training by reward and punishment, or threats of punishment, and examples should be recorded, whether such rewards and punishments are actually given or merely verbal, and whether supernatural beings such as "bogy man", spirits, or deity are called upon.

Weaning

Very often there is no abrupt process of weaning. On the other hand, weaning may be enforced by abstinence, or by smearing noxious material on the breast; infants are sometimes sent away from their parents to the home of some other relative for weaning. The length of the usual period of lactation should be noted; how soon other foods are given to suckling infants, and the nature of this food. All ritual observed by the parent or the child should be recorded.

Speech and Locomotion

Note should be made whether the child's attempts to walk and talk are guided by adults. The first attempts at locomotion, rolling, crawling, by progressing in a sitting position or on all fours, standing, toddling should be recorded. The beginning of speech should be observed; note should be made whether "baby talk" is encouraged, whether children are taught salutations, how to address elders, and the use of kinship terms, and at what age correct usage is expected.

Sleep

Habits with regard to sleep should be observed from infancy onwards. It should be noted at what age the child sleeps away from its mother, and what accommodation is made for it—a separate place or mat.

Food

After weaning it should be noted whether children are given regular meals or special food, or whether they are allowed to take food whenever they want it; whether they observe special food taboos, or taboos incumbent upon the adults in their family.

Children's Activities

It should be noted how far children are occupied in imitating the activities of the adults in whose charge they are and how far they engage in definite play, to what extent this is organized by themselves—older children or adults. Are there any traditional games that are definitely taught to children? Children's songs and games should be recorded (*v.* Material Culture, Toys). Cooperation, competitiveness and leadership should be observed, as well as expressions of artistic, constructive, and acquisitive tendencies, display, boastfulness, and curiosity. How is the anti-social child treated by his contemporaries and by adults?

It should be noted at what age children are expected to work for the household or the community, and whether there is sex differentiation in this; whether children pick up their knowledge and skill casually or whether special training is given and if so by whom (*v.* Initiation, p. 108). What freedom do children have to leave the homestead and to visit those of neighbours or relatives?

Life Cycle from Conception to Marriage

Conception

Accepted theories regarding conception vary widely. Though it may be recognized that conception follows sexual intercourse, this does not imply that the two events are connected as cause and effect. Among Europeans, in spite of physiological knowledge, children are taught fictions, and the formula that a child is a "gift of God" is a commonplace among adults. Though this does not imply a lack of physiological knowledge, it does indicate a certain standardized attitude towards reproduction and, at second remove, towards sex. Among primitive peoples there is usually an accepted theory to account for conception; it may be ascribed to spirits, often ancestral, or to the agency of sea-foam, the sun, etc. Other beliefs are connected with certain places or foods associated with spirit-children or ancestral spirits, or with dreams concerning these processes, or it may be the man who "finds" the child and passes it on to the woman. It should be discovered how soon birth is expected after the "diagnosis" of conception has been made.

A convention of ignorance of paternity may be associated with certain beliefs and with the practice of immature and impartial coition. Where it exists, inquiries should be made into the accepted causation of conception in animals.

Contraception

Belief in contraception is often held; practices may be purely magical, or involve the use of herbs or drugs (themselves practical or acting by means of magic). What methods are employed? Is the practice socially recognized or considered anti-social? What are the alleged motives for the practice?

Pregnancy

The normal attitude towards pregnancy in married and in unmarried women should be noted. Is it a matter for pride, the outcome expected to follow immediately on married life, to be delayed or to be avoided if possible? Are preferences concerning the sex of the infant felt or expressed, and if so why? For economic or ritual reasons? How soon is pregnancy recognized? Note any special observances, diet, or restrictions for pregnant women, prospective father (Couvade, *see* below), or other relative. How long before the actual birth do these become operative?

Abortion

Is abortion ever practised? If so, what is the reason stated, social, ritual, or economic? What is the method used? Rites accompanying any actual or believed practice of abortion should be noted.

Birth

Where does birth normally take place? In the household of the parents of the parturient woman or that of her husband; in usual sleeping part of the house, or some specially prepared hut, or in the bush? Are any preparations made in anticipation of childbirth? What assistance is provided during labour, by whom is it rendered? By someone in a specific kinship relation, or by an expert? (*v.* Midwifery, p. 202). Note any customs or beliefs regarding the after-birth, umbilical cord, or other matters incidental to child-birth. Is the mother subject to special treatment, rest, diet, restrictions, or purification after child-birth? If so, for how long?

Are there any special customs incumbent upon the husband? Is the husband treated as though it were he, rather than the woman, who is confined (*v.* Couvade)? If so, is this done only for the first child, or for all children? Are such customs considered necessary for the safety of the infant, or what is the alleged reason for them? Is the husband believed to suffer physically during the period of pregnancy? (It is fairly common for English peasants to believe that the husband will suffer from toothache.)

When is the infant first taken out of the house? Is there any ceremonial connected with this? Is the father, or are other members of the family or community, publicly shown the infant? Is the infant given any clothes, ornaments, or charms? Is the first occasion of any action on the part of the infant marked by any ceremony, i.e. cutting a tooth, standing, walking, the first wearing of a garment?

If a woman dies in labour, are any means taken to save the child? Is any special importance attached to the firstborn or the first child of either sex? Are they treated in any different way from subsequent children?

How is the child treated at birth (washed, clothed, given food, etc.)? Is any attempt made to save still-born children? Are there any customs or beliefs about twins or unusual births, or infants born with any abnormality, such as a caul or with teeth, or children whose upper teeth appear before the lower? Are twins or triplets common? Is the mother of twins or triplets treated in any special manner? Twins are usually subject to special treatment, often of a magical nature. One or both may be killed (*v.* Infanticide, below). Occasionally they are honoured. Inquiry should be made into the treatment of twins, of same and of opposite sex. If the mother is unmarried is any difference made in the treatment of birth?

Infanticide may be practised in times of stress, such as a drought or other period when there is great food scarcity or it may be practised normally in certain circumstances. Such circumstances should be investigated; they may be due to deformity or peculiarity in the infant or in its birth, e.g. a monorchid, or a child born with teeth, twins or triplets—or there may be ritual or social reasons given for infanticide. Certain women on account of their social or ritual status are supposed not to have children, e.g. the sisters of the Shilluk Kings. If they do have children, what becomes of them? Are the children of unmarried mothers killed? How is the corpse of an infant killed at birth disposed of?

Suckling. What is the normal duration of suckling? Are there any restrictions upon diet, occupation, or use of objects for the mother, the father, or other relatives, during the lactation period, and for how long must these be observed? Are there any special taboos for either parent? (*v.* Ritual, Continence, p. 122).

Will a woman give the breast to an infant other than her own, either casually, continually, or on ceremonial occasions only? Are infants given the breast only when they cry for it, does the mother feed the child spontaneously, or at regular intervals? Is the child ever purposely denied the breast? Do infants take the nipple, or is it thrust into their mouths? Record any cases of difficult feeding, and the measures taken, i.e. when an infant refuses to suck or when the mother cannot give the milk.

Weaning (*v.* p. 102); *Fostering* (*v.* p. 74); *Adoption* (*v.* p. 73).

Naming. When is an infant first given a name? Does an infant acquire status when it is named, i.e. is it only after it is named that it becomes a member of the community or is thought to possess a "soul"?

Is any personal name kept secret? Does the knowledge of a secret name give power to the possessor of such knowledge over that person?

If genealogies have been recorded, the examination of all the names occurring in one or two families will give the basis for investigating the meaning and the customs concerning them, and a fair sample of the frequency of different types of names.

Sexual Development

Introductory

The sexual development of children is an extremely important aspect of their growth, and exact observation is desirable. It should be noted whether children of either sex are treated preferentially, and at what stage such preferential treatment begins, whether children are taught that certain forms of behaviour are proper to and expected of either sex. It must be emphasized that there is no *a priori* reason for a uniform pattern, least of all for our own standard in this respect, and all varieties should be noted. It should be noted whether infantile and child-hood masturbation and sexual play in either sex are tolerated, encouraged, or checked, and what the adult attitude is to children's curiosity with regard to procreation and birth. Is such knowledge given or withheld from children, are fictions considered correct for children? Is the normal routine such that children observe the coitus of their parents; do they imitate it in their play; is this encouraged, or checked?

It is important to observe the mutual behaviour of children of opposite sexes. Do they tend to take an interest in each other and to form friendships? At what ages are such attachments usual? Is there a period when the reverse process is seen, and children form groups of either sex, showing contempt and antagonism towards the other sex? Is there a definite age at which children of either sex are instructed to wear clothing, and is physical modesty enjoined on them (including modesty concerning the excretory functions)? At what age are children taught the incest taboos and such formal behaviour as is associated with such taboos (*v.* Avoidance, p. 86), and how far do rules regulating the behaviour of related adults affect the play activities of children (*v.* Behaviour between Relatives, p. 84)?

Puberty. The important physiological stage of puberty may mark an equally important stage in social development. Puberty marks the beginning of adolescence. With girls it may be formally recognized at the first menstrual period, or when other physical signs such as the swelling of the breasts, appear. The beliefs concerning menstruation (*v.* Ritual and Belief—ceremonial cleanliness, p. 188) should be investigated, and all special treatment and taboos noted.

For both sexes puberty marks the admission into adult life— this may be granted automatically, or only after ceremonial *initiation* (*v.* Age, p. 68). Many societies stress male initiation more than female. Initiation often follows on a long period of training, hedged about with numerous ceremonies culminating in complete admission to adult status. Formal instruction may be given in tribal lore, sex behaviour, etc. The initiates are usually regarded as candidates, who can only achieve initiation by coming through their ordeals courageously. Psychological as well as physical fortitude is required, as often the ceremonies are designed to be terrifying.

Sexual relations between adolescents should be investigated, both where initiation for either or both sexes is practised and where there is no such custom. It may be that sexual activity between adolescents before initiation is of no social significance. Customary behaviour will be greatly influenced by the accepted attitude towards chastity for either sex; it should be noted if it is enjoined on either sex. If virginity is highly esteemed, the measures taken to preserve it should be investigated as well as the treatment of breaches of the rules. Where pre-marital chastity is not expected, it should be noted whether young people are permitted complete sexual freedom or whether custom lays down definite conventions. In some cultures adolescents are expected to have "sweethearts", but there may be rules: (*a*) as to who these are, (1) persons who are eligible as husbands or wives, (2) persons who are not eligible and with whom marriage would be considered incestuous (*v.* p. 114), (3) persons of a definite caste or social class (*v.* Slavery, p. 195); (*b*) the degree of intimacy allowed may be regulated by custom; (*c*) places of meeting, methods of making assignation, and the degree or absence of secrecy may be regulated by custom. There are sometimes special houses for the adolescents of one sex where members of the opposite sex come after dark. It should be noted whether such connections are definitely regarded as courtship leading up to marriage, or merely as a pastime suitable for adolescents.

It should be noted whether special behaviour is adopted between pre-marital lovers after marriage with another partner. Chastity or the reverse may be enjoined in connection with ritual and the former may have particular ritual value.

Deviations from normal development should be observed as they occur. Unorganized adolescent sex segregation is common in many societies; it should be noted whether it is associated with *homosexual* tendencies in either sex, and whether such tendencies are given free play or are checked. In some societies with a strong military organization or with age-sets, homosexual practices are usual in certain grades before marriage, and are subject to conventional rules. Such temporary associations may not be regarded as detrimental to subsequent normal heterosexual development.

Adult homosexuality may occur in either sex. It should be noted how it is regarded by the sex concerned and by society generally. Does homosexuality crop up sporadically, or is it an accepted practice? Do recognized homosexuals wear the dress and adopt the occupation of the opposite sex? Is its practice connected with shamanism, magic, initiation, or any religious cult? (*v.* Age and Sex, p. 66; and Ritual and Belief, p. 184.)

Frigidity in women may occur as a rare or comparatively common abnormality. Where it is common it may be correlated with the normal sex behaviour pattern as taught to girls; this should be investigated.

Impotence may be rare and unimportant socially; it may, however, be greatly feared. There may be theories to account for its occurrence, and magical or other remedies to combat it. Its incidence in such cases may be high and should be investigated. In societies where old men are able to take young wives impotence may have considerable social significance. It should be ascertained whether the young wives of impotent men may take lovers *sub rosa*, or whether a husband will arrange for paramours, whom he will expect to beget children for him. If so, it should be noted whether such paramours stand in a definite kinship or other relationship to the husband (*v.* p. 111).

Cases of sterility in women should be observed. What is

believed to be its cause, and what means are taken to cure it? How are barren women treated? Is barrenness a ground for divorce? Is it even recognized that males may be sterile, if so, what remedies are resorted to?

Is *bestiality* known? How is it regarded? Is it connected with magic and performed ritually? Do cases of bestiality occur in myths?

Prostitution may be recognized. If so, the status of prostitutes should be investigated. How do women become prostitutes? Of their own free will, or are they captured and kept, or are they attached to some religious cult? What is their status, economic position and their ultimate fate when they cease to carry on their occupation?

The introduction and spread of prostitution may be due to European or other alien influence. Where this is so any changes in the courting and marriage customs should be noted.

Unmarried adults are rare in primitive society. Where they do occur sporadically, they are often considered to be abnormal (homosexual or mentally deficient). However, celibacy—actual or theoretical—may be related to rank, magic, or ritual (*v.* also Widows, p. 117). Records of unmarried adults, male and female, should be made and their case investigated.

Marriage

Marriage is a union between a man and a woman such that children born to the woman are the recognized legitimate offspring of both partners. This definition emphasizes the social and legal aspects of marriage, the emotional aspect must also be observed. It should be noted whether freedom in the selection of a spouse is stressed or whether this is considered unimportant, what special qualities are esteemed in a husband and a wife, whether affection is usually felt between married couples, whether it is permitted to show such affection, and what the ideals of married life are. It may be regarded as a procreative and an economic partnership, and affection may not necessarily be expected, this being considered only due to own parents, brothers and sisters; in some societies the economic ties are closer between brother and sister than between husband and wife. Marriage may be looked upon as a permanent relationship, with divorce and remarriage as the exception, or each union may only be expected to last a few years. The permanence of the marriage tie in life and in "life after death" should be investigated (*v.* Ritual and Belief, p. 178).

Age at Marriage. The age generally accepted for marriage of both sexes should be ascertained. Is it habitual for a man to marry before attaining puberty, on attaining puberty, or is his marriage postponed for any definite period after this? Is the length of this period determined by custom, such as military or other duties, or by economic conditions, or both? If marriage is postponed long after puberty has a man right of access to the wives of any married men? (*v.* Supplementary Unions, p. 120). Similar inquiries should be made concerning the age at marriage of women.

Male as well as female infants may be betrothed or married to an older partner. If a nubile girl is married to an infant bridegroom and goes to live in her "husband's" house, is she expected to remain chaste until he grows up, or is it customary for her to cohabit with some member of the family who stands in a definite relationship to the bridegroom? If an adolescent is married to an infant bride, may he take other wives before the marriage can be consummated, and, if so, is the infant bride the first or chief wife? Who takes charge of the infant bride, where does she live? At what age does actual cohabitation take place? The relative ages of husbands and wives should be investigated. Is it customary for husbands and wives to be about the same age, for one partner to be a few years older than the other or for there to be great discrepancy between the age of spouses? If the last is usual what effects are observable in the community?

Betrothal. The age of betrothal may range from infancy until well after puberty. Among some peoples it is possible for a child to be promised in marriage before birth. Inquiries should be made as to the age at which the child is informed of the fact, and as to whether infant or child betrothals are binding if they should prove distasteful or unsuitable. Betrothals may be arranged by the parents or wider groups of the parties concerned, or may be purely the outcome of courtship, initiated by either sex. In either case the settlement of a marriage-payment (*v.* Bride-wealth, p. 116) may be the custom. What are the conditions of courtship? Are there go-betweens, and, if so, for what reasons are the selected individuals chosen for this office? Is the consent of the following necessary: parents, other specific relatives, clan, or other social group; chief or overlord of either or both parties concerned? What is the usual duration of betrothal? Is the behaviour of a betrothed couple regulated in any way by custom, e.g. is avoidance or any other restriction enjoined or are they permitted to have sexual intercourse, or other privileges? Are ceremonial visits made between the groups of the prospective bride and groom? If so, inquiries should be made as to the nature of these groups, e.g. clansmen, age-fellows, certain relatives, and an account obtained of ceremonies, payments, or exchange of gifts.

Types and Regulations of Marriage.[1] There are several types of marriage. One type only may be legally recognized, or all may be tolerated in the same society. *Monogamy.* Where monogamy is practised it should be ascertained whether this is due to poor economic conditions, scarcity of women, is enjoined by authority, or is the socially accepted norm. *Polygamy* must be distinguished as either polygyny (plural wives) or polyandry (plural husbands). *Polygyny.* Where the plural wives are sisters the term *adelphic* is used. If one of the wives has a position superior to that of the other the polygyny is *disparate.* Polygyny may be permitted to the whole community or only to a privileged few. If the latter, is their privilege due to rank, individual achievement, or wealth? Is there a limit to the number of wives allowed? Does each wife have her own household, hut or room, or do they all live together under the control of the chief wife? Is the husband's attention to his wives regulated by custom? Does each wife have special duties or privileges towards the husband or the upkeep of the joint household? Must the consent of the chief or first wife be obtained before taking other wives? Are the same marriage ceremonies observed in the case of subsequent wives as in those of the first wife, or are they less elaborate? Where polygyny is the custom it should be noted how soon additional wives are acquired after the first marriage, what is the usual number of wives to men of various ages and status, and the relative ages of husband and wife. In some societies chiefs or wealthy men are able to keep up large *harems*. Where this occurs the social structure of the *harem* should if possible be investigated: how the women are procured, whether there is a difference of social status among them, what measures are taken to ensure

---

[1] *Residential Location of Married Couples.* Inquiries should be made as to the residential location, temporary and permanent, of married couples. The definition of marriage as patrilocal when the couple lives in the locality of the husband's parents, and matrilocal when the couple lives in the locality of the wife's parents, is misleading and should be avoided.

their faithfulness (*v.* Eunuchs, p. 140), what are the penalties for adultery (*v.* Adultery, p. 120), as well as the fertility rate of the women? What effect has this institution on the rest of the community?

Polyandry (*v.* The Family, p. 70). Where the husbands are brothers, the term *adelphic* is used. If one of the husbands has a superior position to the others the polyandry is *disparate*. Polyandry may be common, or practised only by certain individuals, or a certain social group as a result of either a shortage of women or economic pressure. Are the husbands necessarily related by ties of clan, consanguinity, or blood-brotherhood? Where a marriage-payment is customary do all the husbands contribute equally or in certain proportions towards it? Are there any regulations as to access to the wife, and, if so, what are they? Does a child of a polyandrous union consider all the spouses of his mother as his fathers, or does one individual claim this function? If so, how is this regulated? Is any ritual performed to establish fatherhood? Are all the husbands responsible for the support of the woman and her offspring, or is the sociological father of her children alone responsible? Where do the husbands usually reside?

Laws regulating Marriage may be laws of *prohibition* or *injunction* or both.

(*a*) *Prohibition*

(1) *Incest* (*v.* Ritual and Belief concerning social structure, p. 191) *is sexual intercourse between individuals related in certain prohibited degrees of kinship*. In every society there are rules prohibiting incestuous unions, both as to sexual intercourse and recognized marriage. The two prohibitions do not necessarily coincide. There is no uniformity as to which degrees are involved in the prohibitions. The rules regulating incest must be investigated in every society by means of the Genealogical Method (*v.* pp. 52–62). The prohibition may be so narrow as to include only one type of parent-child relationship (though this is very rare), or those within the elementary family; or so wide as to include all with whom genealogical or classificatory kinship can be traced. The more usual practice is that unions with certain relatives only are considered incestuous, the relationships being regulated by the type of descent emphasized. In some societies unions with certain persons related by affinity are also considered incestuous.

What penalties for incest fall on (*a*) the individuals concerned; (*b*) the community as a whole? Are such penalties enforced by authority, or are they believed to ensue automatically by the action of a supernatural force? Is there any correlation between the severity of the penalty and the nearness of the blood-tie of the partners in guilt? Should children be born as the result of incestuous unions, how are they treated? Are there any methods, ritual or legal, by which persons who fall within the prohibited degrees and wish to marry can break the relationship and become free to marry?

Myths recording incestuous unions of gods or heroes are frequently found among peoples who have a horror of incest. Native opinions concerning such myths are of interest.

Legalized Incestuous Marriages. In some societies where definite rules against incest exist for the people as a whole, unions within the prohibited degrees are permitted for certain people—chiefs or others of high rank. Where this occurs it should be discovered whether it is the privilege (or duty) of the chief and the heir-apparent or of the whole ruling family. Are all degrees allowed, or is the mother-son or any other particular form of union prohibited? If the mother-son union is not definitely prohibited, does it in practice occur, or is it considered unattractive or undesirable or even unimaginable? Inquiries should be made both among the families in which incestuous unions are allowed and those in which it is forbidden, and the psychological and sociological effects of such unions considered in both groups. What reasons are given for these unions by those who practise them, and how are they regarded by those to whom they are prohibited? Is there more, or less, strict sex segregation in the families where such unions occur than among the ordinary people? What marriage ceremonies are performed? Is bride-wealth paid? Is there anything to show that the early family relationship between a given couple is broken and a new tie formed? Can these couples divorce formally? Are these the only unions allowed in these families (or to such persons)? Do persons or families who marry within the prohibited degrees use relationship terms in the same way as the ordinary population? Do such unions tend to stabilize a physical type (demonstrate by measurements and photographs, if possible), or is it generally believed that they do? Are any special character traits demonstrated by or attributed to those who practise these unions? Is it necessary for the heir to be the child of an incestuous union? For ceremonial incest *see* Ritual Union, p. 122.

(2) *Exogamy is the rule prohibiting marriage within a specified group.* This social group is usually the clan (*v.* p. 89). The application of the rules of exogamy should be investigated in the same way as those relating to incest. The infringement of the one set of rules does not necessarily imply the infringement of the other, though they frequently coincide, e.g. with both patrilineal and matrilineal descent the union of brother and sister is incestuous, and also violates the rule of exogamy. Among peoples with matrilineal descent union between father and daughter would be incestuous, but would not violate the rule of exogamy.

(*b*) *Injunction*

It is frequently found that certain marriages are enjoined. The injunction may vary in force and when less strict the marriage may be described as *preferential*. Where it is incumbent upon certain persons to marry certain other persons this may be regulated by:

(1) Kinship, e.g. marriage with mother's brother's daughter. This is the cross-cousin marriage. There are two kinds of cross-cousin marriage, one in which a man marries the daughter of his mother's brother, another in which he marries the daughter of his father's sister; both kinds may be practised. Sometimes marriage with one is enjoined, while the other is prohibited (*v.* Descent, p. 54).

(2) Classificatory relationship, e.g. marriage with the classificatory mother's brother's daughter. Is this marriage regarded as equivalent to a marriage between close kin or is it resorted to only when no person of the required degree of actual kinship is available? The so-called marriage classes or sections of Australia are groups standing in certain relations of classificatory kinship to one another.

(3) Relationship by marriage, e.g. mother's brother's widow, deceased husband's brother, father's widow, etc. (*v.* Secondary Marriages, p. 117).

Do enjoined or preferential marriages apply to the whole population or only to people of a certain rank, e.g. chiefs, commoners, slaves? Do they refer to the first or principal spouse or to all the spouses? Where marriages are enjoined are the parties concerned able to evade the obligation should the marriage be distasteful to either party? If so, whose privilege is it to do so, and how is this brought about?

(4) *Endogamy is the rule enjoining marriage within a specified social group* (v. Caste, p. 94).

Dowry. This term should be used to refer only to the gifts or payments made by the father or group of the bride to the bridegroom or to the bride herself. Where dowries are given full notes should be made of the types of property accepted as dowry (v. Property, p. 149), whether the dowry is inherited by the bride's daughter or whether it can be left at will. Do all brides receive dowry or only those of rank or special distinction? Frequently where bride-wealth is customary the bride also brings her own household utensils with her to her new home.

Marriage Payment or Service. Bride-price, bride-wealth, marriage payment, are terms used for the goods, gifts or payments transferred by the bridegroom or his family to the bride or her family on the occasion of marriage. The native term for this transaction should be used. The transfer acts as a guarantee of the stability of marriage; it may also be regarded as compensation to the group for relinquishing authority over a member; the bride is never actually purchased as a chattel.

The payment or service may consist of (*a*) livestock; (*b*) objects of daily use, e.g., hoes, fish spears, pots, weapons, especially spears, etc.; (*c*) valuables and objects for ceremonial use; (*d*) recognized currency; (*e*) gifts of food or service; service may take the form of the periodic supply of food and household necessities, or in its more complete forms, the suitor may live in and work for the benefit of the bride's household. Other members of the bridegroom's group besides the bridegroom may be obliged to render service, and this work may be done not only for the bride's parents but also for her relatives. The duration of the period of service should be ascertained. Does it terminate with marriage, or continue for a more or less defined time after it, as till the birth of the first child? Are further services rendered on the birth of each child? Who settles the amount of payment due? Is there a fixed amount, or does it vary with the status and personal condition of the bride? If owing to economic pressure the bride-wealth cannot be paid what will be done? Will payment be deferred and the debt inherited, or, if it is not paid, what is the status of the children?

By whom is the bride-wealth paid? Though usually presented

by or on behalf of the bridegroom, his father, or his mother's brother (according to the mode of descent) it is often collected from other relatives. Where this is so, ascertain exactly who these people are and for what reason they contribute, e.g. relationship, mutual obligations, common membership of a social group. To whom is the bride-wealth paid? Though usually handed over to the bride's father or mother's brother (according to the mode of descent) it frequently has to be distributed by the recipient. Ascertain to whom it is distributed, and for what reason. Is the time of payment fixed by custom? Is the bride-wealth paid in whole before cohabitation takes place? If not, should any definite amount or proportion be paid? Is it paid all at once or in instalments? If the latter, are there definite times for payments? Sometimes further payments are made on the birth of each child. (For further notes on payments, *v.* Divorce and Status, below.) Where a fixed money payment has been introduced to replace the customary bride-wealth, the social changes brought about should be investigated.

Marriage by Exchange. Two men may exchange sisters as wives, or daughters as wives, for themselves, their sons, or brothers. Where such arrangements are customary, inquiries should be made as to whether any bride-wealth or exchange of presents is also made. If this type of marriage exists as well as some other type, inquiries should be made as to whether the ceremonial for both types is the same.

Secondary Marriages. The choice of secondary mates may be free, but certain secondary marriages may be enjoined, and are usually correlated with the laws of inheritance and the status of widows. The most common is that known as the *levirate, by which a man is bound to marry his brother's widow.* This type of marriage may be practised in societies founded on either a patrilineal or matrilineal basis. Often a man may marry his elder but not his younger brother's widow, sometimes in a patrilineal society his father's widow (not his own mother), and sometimes when several widows are left one may go to the sister's son of the deceased. In matrilineal society, marriage with the mother's brother's widow is common. Marriage with the widow of the grandfather also occurs, and an unusual type is with the widow of the mother's sister's son. In all such cases it should be ascertained: (1) whether these secondary wives are obtained without bride-wealth or whether a smaller payment is made, and, if so, whether this is considered under a different category, i.e. as an inheritance fee, if so, to whom is it paid, to another heir of the deceased or to the woman's relatives? (2) Whether the widow has any choice (*a*) within the group of correct heirs, (*b*) can she refuse, and marry someone of a different group—if so, would bride-wealth be paid to the correct heir, and must some special ceremony be performed to free her from her deceased husband or his group? (3) Can the heir waive his claim to the widow? (4) Is any marriage ceremony performed, and, if so, does it differ from that with a first wife? Among some peoples where such allocation of widows is the custom, the unions do not really constitute marriages, for the widow is still accounted wife to the dead man, and her children by her second husband are looked upon as children of the dead man. Such cases appear to be intimately connected with ancestor cult, and inquiries should be made as to sacrificial duties of such widows and children.

Another type of secondary marriage is that in which a man may take a second wife from the same group as his first wife. *The most common is with the wife's sister. This is usually known as the sororate.* The levirate provides a status for a widow, the sororate gives a husband a second wife, either in the lifetime of his first wife, or at her death, or if she should prove barren. Both sororate and levirate may be practised among the same people. Inquiries should be made as to whether sisters must be taken as wives in order of seniority. Other marriages of the same type as that with the wife's sister are those with the wife's brother's widow or the wife's brother's daughter. Where secondary marriages are habitual inquiries should be made as to the bride-wealth and marriage ceremony. The effect of such marriages on the use of kinship terms should be investigated (*v*. Kinship, p. 88).

Concubinage and Cicisbeism. Can a man or woman have a more or less permanent union with one or more women or men without giving them the full status of consort? Under what conditions does this arise, and what is the status of the offspring of such unions? Are any ceremonies observed when such a relationship is entered upon?

Marriage with the Dead. If a person dies unmarried it is sometimes believed to affect his life in the after-world. Hence a marriage ceremony is performed with a person standing in the correct relationship where marriages are enjoined, e.g. among the Todas. Among the Dinka a man must take a wife for any elder brother who has died unmarried before he may take one for himself; the children of such women are counted to the dead man. Marriages with the dead seem to be intimately connected with ancestor cult, the life in an after-world, and with property. The duties of such wives and their children and the ritual it is incumbent on them to perform should be noted and investigated.

Women taken under Hostile Conditions. During war are women taken captive or are they killed? If they are taken captive do they belong to their captors or to the chief? Are they ever taken as wives, and, if so, what is the status of their children? Are the laws of exogamy, where these exist, observed in marriage with women captured in war? (*v.* Totemism, p. 192).

Marriage may be customary between members of two groups more or less hostile to each other, either actually or by tradition. Wherever such a custom is found, all details of the procedure should be recorded.

Among other peoples, where tribes meet for certain ceremonies, it may be recognized that the visiting tribe has a right on the last night of the ceremony to capture and take away with them women belonging to their hosts, e.g. the tribes about Maryborough, Eastern Australia. Such customs should be noted, and special inquiry made whether the women so captured are unmarried or may be already married, and whether the laws of exogamy have to be observed.

Women given in lieu of Blood-money. Where it is customary to give women in lieu of blood-money, inquiries should be made as to how such women are chosen (*v.* p. 147). Are they usually relatives of the homicides? What is their status in the group which has accepted them? Are their children accepted as heirs to the murdered man? Are they allowed to return to their own people after they have borne one or more children?

Marriage to Gods. Are any persons looked upon as the spouse of a deity? On entering the cult do they undergo any ceremony similar to that of marriage? Are such persons expected to remain chaste, or do they have intercourse with a representative of the deity, or become temple prostitutes? Are such persons allowed to have children, and, if so, what is their status?

This form of marriage is used as a subterfuge in some Indian castes to secure succession or inheritance in the matrilineal line if the direct patrilineal line becomes extinct.

Elopement. Elopement may be the recognized form of marriage, or may be resorted to in special circumstances. Where marriage by elopement is recognized are other forms practised, and, if so, is elopement considered equally honourable? *Sporadic Elopement.* Where the necessary consent to a marriage is withheld, or where the man is unable to pay bride-wealth, a couple may elope. If so, are any efforts made to overtake them, and by whom? If they evade capture for a certain length of time are they allowed to return to the village, and recognized as husband and wife? Are any gifts or payments made to the group of the woman, and where bride-wealth is customary are such payments equal in value to the whole, or a certain fixed proportion, of the bride-wealth? What is the status of the children of a couple who have eloped?

Supplementary Unions: Unions between Groups of Persons. Unions between groups of persons have been the subject of so much theoretic discussion that, wherever conditions exist in any way resembling so-called Group Marriage or Sex Communism, careful inquiries as to the exact nature of these unions and the status of the offspring should be made. These conditions appear to exist alongside true marriage and differ from promiscuity or prostitution in that they are regulated by definite rules, e.g. the *Eriam* of Bartle Bay, British New Guinea; the *Pirrauru* of the Dieri, Central Australia; the rights exercised by certain relatives, clansmen or age-group fellows, such as are found among many African peoples, may be regarded as rights of access, rather than group marriage (*v.* Age of Marriage, p. 111). Under this heading the right to exchange wives or lend wives should be investigated (*v.* Hospitality, p. 100).

Adultery. How is this regarded (1) on the part of a wife; (2) on the part of a husband? Is it a punishable offence? If so, are both parties punished? How and by whom is the punishment inflicted? Does the punishment vary if the adulterer is caught in the act, or the fact subsequently discovered? Is it considered ground for a divorce, or is it usually condoned? Is it actually a common practice or a rare occurrence? Should a married woman elope with a lover what action is taken by the husband, and what is the status of any children that may be born while she is living with her lover?

Extra-marital Sexual Relations. Sexual connection outside wedlock does not necessarily constitute adultery (*v.* Rules of Hospitality, p. 100, Supplementary Unions, above). It may be regarded as adultery if the wife is at fault, and not if it is the husband. Among some people a husband may allow or force his wife to cohabit with another man.

Divorce may be defined as the legal dissolution of marriage, regulating the status of the parties concerned and their offspring.

It should not be assumed that divorce implies a previous infringement of law or custom, and necessarily presupposes guilt. The idea of guilt, however, is not absent from all peoples, and note should be made of any indications of such an idea, e.g. if the wife returns to her people, and the bride-wealth is not returned nor any compensation made to the husband, this may indicate guilt on the part of the husband. Are there any specific reasons for the dissolution of the marriage and do these causes operate equally for husband and wife? How is divorce regulated where there is no bride-wealth or dowry? Where there is a bride-wealth, is it returned in whole or in part? Are other valuables which may have been given by either party returned? What arrangements are made for the maintenance of children of a divorced couple, and to which group do they ultimately belong? Is divorce allowed when a woman is pregnant? If so, what arrangements are made for her and her child?

Does divorce change the attitude adopted by a man or woman to the family of his or her previous spouse, i.e. respect, avoidance, etc.?

Marriage Ceremonies

Ceremonies mark the difference between a legal marriage and an irregular union. They may be very simple or may last for a long period. It is often difficult to distinguish between ceremonies of betrothal and marriage, since the former often appear to be but a preliminary stage of the latter.

The consummation of marriage may take place at a definite time in the ceremony, or it may have been performed previously, or be deferred (*v.* Ritual Continence, p. 122). There is usually some action, or group of actions, performed by bride or groom or both which finally establishes the marriage. Sometimes this does not take place until after the birth of the first child. Some actions may symbolize the union of the two people or the two groups, e.g. by eating together, being tied together, exchange of blood, the bride sitting on her husband's knee, etc.

Special clothing, processions, dramatization of fights and

capture, veiling and unveiling, and feasts, may be essentials in marriage ceremony. The respective parts played by bride and bridegroom as well as by the relatives of both should be noted, also who provide and who attend the feasts, and the food, etc., provided for such feasts.

Seasons for Marriage. Do marriage ceremonies tend to take place at any particular time or year, or are certain months or seasons considered either favourable or unpropitious? If so, can such practices be correlated with seasonal occupations and the food supply, or are magico-religious reasons given? Are omens taken to fix the day for marriage ceremonies?

Defloration. This may be unimportant, or, if important, may be part of the ceremonies of initiation (p. 108), or of marriage. As part of the marriage ceremony it should be ascertained whether it is performed naturally or mechanically. By whom is it performed—husband, overlord, priest, stranger, or a person or group of persons belonging to a certain social group of which the husband is a member, e.g., age-set? Does this act set up a definite relationship between the girl and the performer or performers of this ceremony? When and where does it take place? Is it believed to be dangerous, and, if so, are any magical or other protective ceremonies performed? Is it believed to have any magical virtue? Are "signs of virginity" valued and displayed?

Ritual Union and Ritual Abstinence

Consummation of marriage is not always permitted as soon as marriage is recognized. There may be a period of ritual continence. Should this be the case, inquiries should be made as to the period of continence, the means of enforcing it, and the reason given for it. While bride and bridegroom are kept apart, has any other person right of access to the bride? Continence of married persons may be enjoined during pregnancy, lactation, illness, or mourning, or during some important undertaking such as hunting, hostile expeditions, building a new house, etc. Continence during pregnancy and lactation may be incumbent only on the wife; if so, is the husband allowed access to other women, and does he need to take any ritual precautions? Are breaches of such rites considered to harm the offspring? Continence during communal undertakings may be enjoined on all taking part in them, and on their legal partners; breaches of such rules may be believed to harm the whole community. How are such breaches punished?

Medicine men, magicians, rain-makers, sacred chiefs, and smiths, may be obliged to observe special sexual taboos (*v.* Chastity, p. 108).

Ritual union may be performed before certain undertakings.

Incestuous union may be ritually performed by certain persons for special purposes (*v.* Magic, p. 191).

Certain persons may be, for ritual or political reasons, debarred from marriage, e.g. the daughters of the Shilluk Kings. Are such persons permitted to indulge in free love? If so, are precautions taken to prevent conception (*v.* Abortion, p. 105), and if not, how are their children treated?

Among some people, on special occasions, extra-marital sexual intercourse is not only allowed but actually enjoined; it should be ascertained whether these occasions are seasonal festivities connected with the fertility of the land or cattle, or social occasions such as initiation, marriage, or funeral feasts, or connected with some special cult, or to avert calamity. Do married as well as unmarried persons take part in these ceremonies?

Symbols of Coitus. Are any actions performed by a man or woman regarded as equivalent to coitus, e.g. jumping over the legs of a woman? Are such symbolic acts believed to be able to cause conception? Is the performance of such an act between a man and a woman, not his wife, considered to be equivalent to adultery? Where ritual union is practised (*v.* above), can it take this form? Are there any actions which are legally regarded as indicating that there have been sexual relations between a man and a woman, and for which compensation can be demanded, e.g. sitting upon her mat or bed, loosening or touching her belt? Is the act of eating together by a man and woman regarded in this light?

Adult Status for either sex may be recognized immediately after initiation, marriage or on becoming a parent. Any changes in behaviour, dress or ways of living should be noted.

The status of women may be equal throughout life to that of men, inferior or, in some cases it may be superior, but it should be noted that matriliny (*v.* p. 71) is no indication of superior status. The status of widows should be investigated. In some societies this may be honourable, especially if the widow is the mother of children, while in most it is dependent on her remarriage (*v.*

Secondary Marriages and Marriage with the Dead, pp. 117–19). Where *suttee* is practised the widow has no status at all and death is a logical end.

OLD AGE, DEATH, AND DISPOSAL OF THE DEAD

Old Age

(For status of seniors or elders, *v.* p. 85). The status of the aged deserves special study. As a man or woman grows old he or she becomes of diminishing economic value to the community, but this may be counterbalanced by increase in knowledge either of practical affairs or of ritual matters, and old age itself may be honourable. Who is responsible for the sustenance of aged persons? (*v.* Duties of Relatives, Widows, Widowers, p. 117). Is it looked upon as a burden? Are duties performed in gratitude for past assistance, or in fear of the curse of the aged or the revenge of their spirits after death?

Do old people tend to associate together, or to be solitary, perhaps looked after by quite young children? (*v.* Behaviour between Relatives, Grandparents, Grandchildren, p. 87).

What part do old persons take in public life, ceremonies and feasts?

Beliefs concerning Death (*v.* also Ritual and Beliefs, p. 178).

Beliefs concerning the causes of death vary greatly among different peoples. Natural causes alone are often not considered to be sufficient explanation, though they are seldom entirely disregarded. Death may be attributed to:

(*a*) The malice of spirits, ghosts, or magicians. These may use purely supernatural means, e.g. stealing a man's soul, but deaths which seem to the observer as accidental, as being slain by a wild beast, may be believed to be due to the action of some person or spirit who has instructed an otherwise harmless animal to kill.

(*c*) Sins, acts of commission or omission (*v.* Law and Justice, p. 146), either of the person who has died or of someone connected with him. The punishment of death may be automatic (the outcome of some impersonal spiritual force) or it may be inflicted by the spirit or spiritual beings against whom the person sinned; these may use natural or supernatural means (*v.* above).

(*c*) Physical injuries or a physical condition, e.g. old age, unconnected with any supernatural agency.

Ascertain as far as possible what is the native conception of death. Does the native word for death correspond to our own, or does it include also physical conditions, such as severe injury or disease, which are believed to lead up to, or to be in some way connected with the death of the body? Are people while still alive ever treated as though dead, e.g. spoken of as dead and actually buried alive? Are trance conditions confused with death?

Among many peoples there is a tradition that mankind was originally immortal, and that the foolish or malicious action of some creature or individual brought death into the world. Legends should be recorded. For beliefs concerning life after death *v.* Ritual and Belief.

Voluntary Deaths and Suicide

In many societies, kings, chiefs, rainmakers and other sacred persons are not allowed to die a natural death. The ritual method of death should be investigated. They may be walled in, buried alive or voluntarily put to death in some way. It may be customary for them to show resistance, so that though their own death is certain, others will die with them (*v.* Ritual and Belief—Sacrifice, p. 180). Mass or individual ritual suicide may be practised, e.g. immolation beneath the Juggernaut, suttee. The incidence or frequency of suicide should be noted. Statistics should be collected if possible and reports made of any cases known to the investigator or told to him by any member of the community. The reasons given for suicide and the attitude of the community towards attempted suicide should be investigated. Is suicide considered an offence; if so, how is it treated? It should be noted whether disappointed love, shame, or disgrace are considered normal causes of suicide. What is the fate of the spirit of a suicide after death, and how is his body disposed of?

Treatment before the Disposal of the Body

What signs are taken to indicate that a sick person will not recover? Is the sick or aged person allowed to die in the house, or removed outside the house or village, or to any definite place inside or outside the village? Is the sick or aged person tended, abandoned, or buried alive?

The treatment of the corpse and its disposal should be described in detail. While investigating the customs and rites information concerning the working of the social system and insight into religious beliefs will be acquired. In all cases the age, sex and status of the dead should be ascertained; treatment may be different in each category.

When death is due to certain definite causes, the body may be treated in a way different from prescribed custom, e.g. death from violence—in war, murder, or by accident, i.e. by lightning or wild animals, death in childbirth or during pregnancy; suicide, or certain epidemic disease such as smallpox. In all cases the ritual —or lack of it—should be described and the religious beliefs underlying the differences investigated (v. Ritual and Belief, p. 178).

Treatment of Dead during Epidemics. This may differ from the normal method. The difference may be due to practical expediency, or to beliefs concerning the nature of the epidemic.

Treatment of Corpses of Pregnant Women. Does this differ in any way from the normal disposal of women? Is the foetus removed and disposed of separately with or without ritual? What reasons are given for this?

Treatment of Stillborn or Infant Dead. Ascertain the method of disposal whether accompanied by ritual or not, and the reasons alleged for the treatment.

Treatment of Enemy Dead. The bodies of enemy dead may be left to rot or to be eaten by carrion; they may be treated with ceremony, preserved (v. Head Hunting, p. 177), eaten (v. Cannibalism, p. 177), or opportunity may be granted to the enemy to fetch their dead (v. Warfare, p. 141).

Preparation of the Body. Note should be made of the treatment of the dead body; whether it is washed, painted, decorated, or clothed. Whether the external orifices are closed, if so, how, and the reason given for the practice. Who prepares the body—a person in some specific relationship, non-related but belonging to some specific group, or someone paid for the purpose?

Disposal of the Body

Considerable confusion has arisen from the inaccurate use of technical terms. The following definitions are suggested.

Inhumation or Interment is the practice of concealing the body in the ground or in a mound above the general level of the ground. The word *burial* should be used for this method of disposal only.

Cremation is the practice of destroying the body by fire.

Exposure is the practice of laying the body on some exposed spot to be destroyed by the elements or by animals or birds of prey.

Preservation is the practice of preserving the body. This may be done by means of desiccation, fumigation, or the use of preservatives. The term *mummification* has been used as synonymous with preservation, but is best reserved for the embalming technique of Ancient Egypt, or processes resembling it.

Artificial Decomposition is the practice of using special means to hasten the decomposition of the body.

Sometimes two or more of these methods may be employed in disposing of one corpse, e.g. inhumation followed by cremation, or some time after inhumation the bones may be dug up and either exposed or kept in some special way (*v.* below).

Amongst the majority of peoples a definite length of time is allowed to elapse between death and the disposal of the corpse. How long is this? Are there any definite days, or times, or seasons, when the disposal of the dead in the customary manner is prohibited?

Burial. Notes should be made as to: (*a*) Who is responsible for digging the grave; this may be a relative, or a person not related by blood but by marriage, or a person belonging to some specified clan or social group. (*b*) The position of the grave; this may be in a special burial-place reserved for the local group, the clan, or the family, or near the dwelling-place, actually under or at the side of the house, etc. (*c*) The type of grave; shape, use of stones, or any other material. (*d*) The disposal of the body in the grave; it may be wrapped in mats or skins, or put in a coffin. Note the position and orientation of the body; the body may be contracted, or extended or fixed in a sitting or standing position. Its exact orientation should be noted, and reasons for specific orientation elicited.

Cremation. Where does this take place? Is any special timber used for the pyre? Is the fire for the pyre especially made, and, if so, by whom, and by what method? (*v.* Fire, p. 240).

Exposure. On what is the body exposed: a platform built for the purpose, or one used in daily life; a tree; a ledge of rock, etc.? For how long is the body left exposed? Is it visited or avoided by the living, and for what reasons? What is done with the bones when decomposition is complete?

Preservation. It is sometimes difficult to decide whether *exposure* under certain conditions should not be included under this heading, e.g. in a hot dry climate where exposure results in desiccation. The processes used to preserve the body should be carefully noted. Where vegetable matter is used, e.g. the bark of a tree, the species of the plant should be ascertained and, where possible, a specimen of the part used collected.

Are the bodies preserved temporarily in order that they may lie in state or await the season for the disposal of the dead, or is the preservation intended to be permanent? For how long does the method employed preserve the body? Where is the body kept during and after the process: in a hut specially built for it; in the hut of a relative or in a place for the bodies of all the deceased's social group?

Artificial Decomposition. This may be brought about in many ways. The flesh may be scraped or washed off the skeleton; decomposition may be hastened by heat; the body may be put in a termites' nest. What is done with the bones?

Disposal in Water. The corpse may be taken out to sea and sunk, or laid on the shore or reef to be taken off by sharks. It should be noted whether any measures are taken to prevent the body from being washed ashore or devoured by fishes. Are certain parts of the sea or river shore reserved for water cemeteries? Less common is the practice of placing the deceased in a canoe and sending it out to sea or down stream.

Encasing the Body. Sometimes corpses are placed in jars, e.g. Borneo; coffins or hollow effigies, e.g. Solomon Islands; which are not buried but usually sealed. When this is done where are such receptacles kept?

Exhumation and Secondary Disposal. Where inhumation is practised the bones are sometimes exhumed after a time and placed elsewhere. This may be part of the regular mode of disposal, or due to the necessity of awaiting the correct funeral season, or to the restlessness of the spirit, etc. How long is the interval between the first disposal and the exhumation? Is this fixed by custom or do the spirits of the dead intimate to their descendants by means of dreams or in other ways, such as causing illness or misfortune, that exhumation should take place? Is it done at any definite season for all those who have died within a certain period, or is there a separate rite for each body? Where are the exhumed bones placed: in the bush (reburied or thrown away), thrown into water, or placed in an ossuary? Is this secondary disposal final? Are the skulls or any of the bones kept by members of the deceased's clan, kindred, etc., and, if so, what bones are so kept and where? Are exhumed bones treated in any way, e.g. painted, or used for any religious, magical or other purposes? When the bones are exhumed what is done with any grave goods which have not perished? After cremation, exposure, etc., the remains may be removed from one place to another.

Grave Goods. Are any objects buried in or placed on the grave, or ritually burned, and if so, what are they? Do they differ according to the sex, social status, etc., of the deceased? Are they objects which belonged to the deceased in life or which were given to him at death, or made specially for burial? If so, by whom?

When the objects are valuable are they buried with the body or are measures taken to preserve them for the living, e.g. by burying imitations or symbols of them, or by laying them on the grave and removing them later? Are the objects buried or placed on the grave for the use of the deceased in the Other World, or merely to honour him? If the former, are they treated in any way, e.g. broken to release the spiritual part of him? Is food or drink placed near or in the grave and is it removed from time to time? If so, for how long is this done? Has it any relation to the time taken for the body to decompose or for the spirit to reach the Other World?

Should there be communal burial-places, or ossuaries, investigation into the choice of sites and their maintenance should be made. The attitude of the population towards them should be discovered; they may be frequently visited with or without offerings, or feared and avoided.

Ceremonies may be performed for the dying. The ritual treatment of the dead usually includes far more than the actual funeral ceremony, usually beginning at death and continuing at intervals for many years.

All details of the method of proclaiming the death (and sometimes, for chiefs and important persons, the means of keeping it secret for a period), the calling of mourners to attend the ceremony, should be ascertained. The ceremonial may include sacrifice, making an effigy, wailing, dancing, miming, feasting, speeches glorifying the dead or his clan or social group, and the assumption of mourning. There may be differences in procedure according to the rank, sex, age, and social position of the deceased. Actions may be performed with the professed object of driving away the spirit, or confusing it so that it shall not find its way back to its old abode, or means may be taken to induce it to return to its relatives and to protect them (*v.* Shrines, p. 183); both may be performed at different periods.

Sacrifices. These may include objects destroyed or put in the grave, animals and food consumed at feasts (it should be noted whether any special parts are preserved on the grave, or in the house of the chief mourners), or *human sacrifices*—slaves, widows (suttee), or other persons buried alive, or killed and buried with the dead—self-immolation or mutilation during the funeral ceremony. The accepted reasons for sacrifice should be investigated; it may be a means of appeasement of the anger of the deceased, provision for his requirements in the after world, or an expression of excessive grief.

Mourning is a ritual carried out by persons standing in some specific relationship to the dead. What is this relationship? It may be political, i.e. all subjects may mourn for a chief or king, or it may be regulated by membership of a social group or by kinship or affinity. Are any special taboos incumbent upon mourners? Does the type of ritual and its duration vary in proportion to the degree of relationship to the dead? What is the nature of the mourning—special clothes, ornaments (or the taboo on wearing ornaments), smearing the skin with ashes or paint, special modes of hair-dressing, scarification or mutilation? Are relics of the dead—bones or teeth—worn by mourners? Are there any ceremonies and feasts when mourning is finally discarded? Are there certain persons for whom mourning is not observed, and what are the reasons for this?

Shrines and Relics, Cenotaphs or Memorials. A shrine is an object or structure reserved for, or sacred to, a spirit; the spirit may be temporarily or permanently immanent in the shrine. The grave itself may be marked in some way and treated as a shrine, offerings may be made and ceremonial held there. Shrines may contain skulls or bones. (For household shrines, with or without relics, *see* Ritual and Belief, p. 183.)

Relics. Where decomposition or exhumation is practised, bones and teeth may be preserved and become cult objects of importance (*v.* above, and Ritual and Belief, p. 178). They may be kept in special places or in the houses of relatives. Special importance may be paid to the skull, teeth, and jaw bones; they may be treated with oil, paint, or be moulded with clay to become *quasi* life-like representations to the dead. They may be treated with considerable ceremony, and may even be ceremonially destroyed. (For treatment of enemy skull, *see* Ritual and Belief, p. 183, Head Hunting, p. 177.)

Memorials, anthropomorphic or otherwise, may be put up customarily for all dead or only for important dead; these may be treated as shrines, or may be merely secular monuments. Cenotaph shrines may be erected for clan, tribal, or mythical heroes, and become important centres of cult (*v.* Ritual and Belief, p. 183).

It is frequently believed that contact with a dead body causes contamination, ritual uncleanness, or unholiness (*v.* Religion, p. 186). It should be discovered whether the whole population is liable to this disability, or whether only certain persons are liable because of kinship or other ties to the dead or because of their occupation or rank. What are believed to be the consequences of this condition, and how is it counteracted? Ritual purification may be believed to be effectual; methods should be described. Among a few peoples, e.g. the Veddhas of Ceylon, normal death, and among others, death under certain specified conditions, is considered so dangerous that the dead are left where they died and the place deserted. Are objects that have been in contact with the corpse, or property belonging to the dead, subject to contamination? If so, what is done with them?

CHAPTER V

POLITICAL ORGANIZATION

Political Systems

Introduction

IN ITS widest sense the political organization of a people embraces, on the one hand, the whole complex of institutions by which law and order are maintained in the society, and, on the other, all the institutions by which the integrity of the group is maintained in relation to neighbouring communities of a similar kind and protected against attack from without. Thus political organization includes the legal institutions by which the juridical rights of every member of the society are safeguarded and his juridical obligations enforced (*v.* Law, p. 146), the organization of local, i.e. village, town, tribal subdivision, government, and the system of tribal or national or state government. It embraces also the military or other organization by means of which offensive or defensive action is taken against enemies who threaten the unity, security or independence of the society either from outside or from within.

Political organization in primitive societies varies widely. At one extreme are the simplest hunting and collecting cultures amongst which the largest cohesive group may number only a few score persons. In these societies there are no explicit judicial institutions (such as courts) or administrative leaders (such as chiefs) or military machinery (such as warrior regiments), and very many activities and relationships that we should consider as falling into the political organization are subject primarily to ritual regulation. Kinship bonds and ritual activities largely fulfil functions which in more complex societies are associated with explicit political offices, machinery, and rules. At the other end of the scale, where the nation or tribe may number up to a million or more people, specialized political institutions such as regular courts of law, chiefs and councils exercising legislative, administrative and executive functions, a well-defined military organization, and even a state religion may be found.

Between these extremes there are peoples with considerable populations both agricultural and pastoral occupying large territories, who regard themselves as units although they have no centralized organization or supreme chief. Among some of these even local, clan or lineage chieftainship or councils may not be organized. The basis of unity and the institutions by which law and order are maintained and defensive action against outside enemies organized, require careful investigation in these cases.

Allowing for the great variety of political institutions, certain main factors can be isolated for the guidance of the investigator. These are summarized here:

Territorial Foundations

Among sedentary peoples the widest political group is invariably territorial, and membership is based on local birth or lengthy residence. Individuals or groups of alien origin are absorbed into the political system of the host group, sharing the rights and duties normally devolving on members by birth. Formal incorporation by a well-defined legal step, as is usual in modern societies, has not been observed among simpler peoples. But the absorption of alien groups is often accompanied by a reinterpretation of their origin, the alien group assuming the position of a segment, e.g. a clan or a subordinate class, of the host society. Alien sections may also have intermediate status, retaining a partial autonomy, e.g., their own headmen, courts, customary law. Or finally they may remain strangers, deprived of all or some civic rights in the country of adoption, and still subject to duties, e.g. payment of tribute, towards the country of origin.

In large territorial groups the political organization may entail subdivision into villages, townships and districts, and into central and local administration. Large towns are similarly subdivided into wards or quarters. Local and central administration may be on parallel lines or differ in principle. Where they are on parallel lines the local group retains a limited autonomy, specially in the sphere of law, in such matters as the collection and payment of tax and tribute, and the appointment of political functionaries. Sharp differences between local and central administration are usually bound up with a State structure (*v.* below).

Among nomadic peoples political membership is mainly determined by descent and not by occupation of a particular territory. But there may be a permanent association of the group with a well-defined territory, and the corporate rights to that territory, e.g. grazing, watering, or hunting rights (*v.* p. 65) may be collectively defended against trespassers or rivals. In some nomadic societies there is a division into ruling and serf classes. This often reflects ethnic and economic differentiation, due to an immigrant pastoral people dominating an indigenous agricultural community. Sometimes, however, the two groups live together on terms of equality, the original inhabitants being regarded as the true owners of the soil, often being vested with special religious powers over the land.

Ethnic Foundations

In most primitive societies the widest recognized political group is the *tribe* (v. p. 66). The respective boundaries, however, of the tribe and the political group must be investigated, for often a tribe or group of tribes of common culture is politically divided, or a political unit embraces or cuts across several tribes. The intertribal political unit may consist of a confederacy of more or less equal and autonomous partners not under a superior central political and legal authority and uniting only for specific purposes such as defence against outside attack or periodical ritual celebrations (v. System of Government, below); or it may arise from a mingling of peoples under a single rule with or without social stratification (v. Class, Caste, p. 93). *Where there is a well-developed central authority, exercising final legal, administrative and military power over a group of people occupying a clearly defined territory whether or not they are of uniform culture and of homogeneous ethnic origin we may speak of a state and of the people as a nation.*

It is necessary in this connection to inquire how the people of a state describe themselves. The name by which they call themselves may express the distinction between tribal origin and political unity; more often it is still a tribal name, attaching to one group which has gained ascendancy over others and is regarded as the representative section.

Where the *state* includes groups of diverse tribal origin supreme political authority may be vested in a distinct ruling body. This body is often recruited from an aristocracy or ruling class, which may be identical with one of the component ethnic groups and govern by right of conquest or in virtue of having been the leading element in building up the state. The historic basis of such a system must be investigated.

In a nation of heterogeneous origins the machinery by which political cohesion is maintained will rest in various degrees on the armed strength or on the prestige of the ruling group, on its ability to protect weaker dependent groups or individuals, and on various institutions (religious, legal and moral) fostering political loyalty. It is important to examine the processes of assimilation, in language and customs, by which ethnic diversity is overcome, and to record all myths, legends, religious ceremonies, and moral concepts which bear on the state, elaborating its putative origin, giving expression to its unity, and providing sanctions (*q.v.*) for its authority.

Inter-group Relations

In simpler societies the political unit is normally the widest effective social group. Within this group the customary rights and obligations of members are enforceable through the medium of a regular system of settling disputes and correcting wrongs. Any non-member of this group is a potential enemy, not protected by established sanctions. But this strict limitation is often set aside, enabling regular peaceful relationships with other groups to be maintained. These relationships may be:

(*a*) Non-political, e.g. trade or intermarriage. It must be investigated how these relationships are made possible in the absence of a common body of customary law.

(*b*) There are also regulated interrelationships in the political field. These take the form of pacts and treaties between political groups, regulating warfare, reconciliation after feuds, and the exchange or ransom of prisoners (*v.* below, Warfare). Through such pacts unrestricted vengeance, such as often obtains between independent political groups, may be replaced by blood-money (*v.* Law) or other obligations of redress, and the individual acting in disregard of the pact would forfeit the support of his group and be subject to legal or ritual penalties, e.g. outlawry (*v.* Law). In some societies inviolable go-betweens are entrusted with inter-group negotiations. Among nomadic peoples periodical meetings occur, sometimes taking the forms of religious ceremonies, at which outstanding disputes are settled (over blood-money, boundaries, grazing or watering rights). The custom of blood brotherhood is a common device by which people are enabled to visit or traverse the territory of potentially hostile neighbours. Incidence of blood brotherhood and the ritual associated with it should be described in detail. Kinship ties that run across tribal or clan boundaries may fulfil the same purpose (*v.* Kinship).

(*c*) In certain societies a common religion creates a community wider than the regular political groups. This is found both amongst very primitive peoples such as the Australian aborigines and amongst more developed peoples, e.g. where Islam has penetrated.

System of Government

Political units are clearly definable where there is a centralized organization (*v.* above, Ethnic Foundations), headed by a chief or king. In simpler societies where no centralized government exists there may be a council of elders (*v.* below) guiding collective action. Or there may be no supreme tribal authority; and the political unity of the tribe emerges only in co-ordinated or collective action, by the more or less independent subdivisions of the group each under its own elders on special ceremonial occasions or in opposition to enemies. The composition of the group acting in co-ordination and the initiative behind the action must be examined, so as to show the extent to which political activity is specialized or rests on the solidarity of social units, such as kin groups, leadership of which is vested in the most senior persons.

Nowadays many primitive societies, even though originally without chiefs, have chiefs appointed by the Colonial Government partly for reasons of administrative convenience and partly owing to the once widespread misconception that all primitive peoples have chiefs or headmen. Such superimposed chieftainship will be revealed, not only by historical investigation, but often in the weakness or limitations of authority. The results will be similar where government chieftainship is the successor of some other more restricted or transient form of leadership.

Chieftainship

Political leadership under traditional conditions shows greatly varying forms. Its scope may be narrow and specialized, or very comprehensive, extending also over non-political fields of communal action. Some societies have different heads for war and peace; or there may be, in addition to the political head, special functionaries responsible for hunting, for communal farmwork, for ritual of public importance, or for the supervision of occupational groups, e.g. craft guilds (*v.* Experts, Ritual and Belief and Knowledge and Tradition, pp. 181, 193, 195); in other societies all these responsibilities appear combined, in varying manner, with political leadership.

Leadership in war depends upon the military organization of the group (*v.* War, p. 141). It is often a temporary office, materializing when needed. It may imply a formal appointment, or only the fortuitous emergence of an individual commanding a spontaneous following. Such command may rest on membership of a large kin group, which for reasons of kinship loyalty would lend its support. Or it may rest on wealth, on tested personal ability, on the reputed possession of supernatural powers, or on descent from men who previously occupied a similar position. Where age-sets (*v.* p. 68) are found, leadership in war may be vested in senior age-sets. In a developed political organization, it may be vested in the holders of special court and state offices.

Permanent political leadership—which may properly be termed chieftainship—is hereditary or elective. There may be formal election by an assembly, or informal selection by men of status and responsibility, i.e. the council of elders (*v.* below). Some societies have special "electors" or "king-makers", invested with priestly authority. The nature of such electors should be investigated. Note whether their office is hereditary, or how they are appointed, they may belong to a lineage with some historical or mythological connection with the chiefly line. Hereditary and elective or selective principles are often combined, where there is no law of primogeniture or ultimogeniture and there may be a number of possible heirs. The manner by which the successor is chosen from among the potential heirs must be investigated: it may be a question of personal qualifications and popularity or believed supernatural powers. There may be formal or real fighting among the heirs; an interregnum either organized or amounting to civil war may be normally expected after the death of a supreme chief before the installation of his successor. Genealogies should be taken to check the accuracy of general statements obtained from informants on the rules of hereditary succession. Inquiries must also be made as to whether succession to chiefship corresponds to the general system of inheritance in the society or is governed by special rules which must be investigated. Related questions are—the admittance of female successors, and the expedients resorted to where there are no legally eligible successors or these are very young, very old, or infirm.

In tribes organized in descent groups hereditary chieftainship may be vested in a kinship group such as a lineage of a particular clan (*v.* p. 90). In royal lineage or dynasties there is often a marked tendency to split, the claims of rival sections may need adjustment or lead to civil war and the formation of a new dynasty.

Territorial subdivision leads to a hierarchy of chiefs, whose supreme head we may call *king* or *paramount chief* and whose position may be based on a myth of descent from a tribal hero or a god, or the king himself may be considered divine (*v.* below). Where the stratified society is also ethnically heterogeneous, the dynastic claim may be by right of conquest and vested in an alien ruling class. In these complex societies kingship often differs in nature from the chieftainship obtaining in the subordinate communities, and local chieftainship requires confirmation by the king or paramount chief.

His office confers upon the chief not only prerogatives, but also more or less clearly defined obligations towards the community.

(*a*) The responsibility of the chief towards the community may be essentially of a religious or magical nature. He may be credited with an especially intimate association with the gods or the ancestor spirits, which he must utilize in ceremonies or through ritual obligations for the benefit of his people. He may have to observe special rules in his daily life, in his association with other people, his meals, his marriage and sex life, and in connection with the sacred insignia of office. Rituals surround his installation and his death (*v.* Ritual and Belief, Voluntary Death, p. 181). The attitude of the community in the event of calamities which the chief is believed to be capable of warding off, or when his supernatural powers appear to be waning should be recorded. Many of these features appear without their sacred connotation in secular chieftainship, i.e. court etiquette, as traditional regalia, etc. Some connection with magical and religious powers is almost universal in chieftainship, but religious and secular leadership may be divided between two persons occupying corresponding positions in their respective fields, or the two offices may be held in the same descent group by classificatory brothers. Investigation should be made into the role, status, and ritual of the queen mothers, and the king's sisters.

(*b*) Chiefs usually have economic prerogatives such as monopoly rights to fruit-bearing trees, or to certain kinds of game, or the right to command communal labour. The chief may be entitled to certain regular gifts, pecuniary or in kind, voluntary gifts shading over into compulsory dues (tithe, tribute, tax). His title to land and his right to dispose of the communal land resources must be specially examined. The wealth which the chief may derive from his office may be personal or vested in his office. Can it be accumulated? And is it such as to outlast his life or tenure of office and so benefit his descendants? Vague references to the "democratic" character of chieftainship should be avoided, or meaning must be conveyed through statements of detailed observation. All the economic obligations of the chief must be recorded—such as duties of hospitality, of assisting needy subjects, expenses on the occasions of communal ceremonies or feasts, or other ritual expenses, or the organization of communal labour for the benefit of the community. Important seasonal activities, e.g. clearing the bush, planting, harvesting, etc., are often initiated or arranged by the chief, Other privileges of chiefs, such as preferential rights to wives and women who may be taken into the royal household and the corresponding obligations should be investigated.

(*c*) The chief in the proper sense of the word is almost invariably also the supreme judge of the community, though in Islamic communities men learned in Koranic law may act as professional judges. Thus the chief is appealed to as arbiter or judge in disputes and offences. His person is often protected by special laws, and attacks upon him may be grave offences. The chief may be irremovable; in this case it is essential to find out how misrule is countered. Or there may be recognized means of deposing a chief—by common consent, by some customary procedure, or by rebellion. Here the usages in the event of an interregnum must be explored, and the distinction between the chief's office (which may be sacred or semi-sacred) and his person presents an important problem.

Councils and Officials

Chiefs are usually assisted by councils. It must be shown how they perform their double role of agents of the chief and spokesmen of the community, and how far and by what means they may control the chief's actions.

The constitution of the council varies in formality. Membership may be hereditary or by appointment. In the latter case the selection may be limited to members of a descent group, or based on personal qualities. Age and seniority are nearly always important qualifications, and here we may properly speak of councils of *elders*. Often their office carries with it a title or prestige and sometimes economic privileges. In complex societies their duties may be specialized, approaching to a bureaucracy of public officials, civil and military dignitaries, officers of the royal household, etc. The hierarchy of offices may go hand in hand with a graded system of ranks and varying economic prerogatives (referring to land or a share in the State revenue). Slaves often hold responsible offices which involve close personal loyalty to the king.

Where eunuchs are employed their office and the possibility of their attainment to political power should be investigated.

Where the political unit is territorially subdivided the political officials may act as governors of districts or towns. Alternatively, the local chiefs may represent their communities on the central government. Fiscal and judicial functions may be similarly devolved; certain offences being dealt with by the local chiefs, while others are the prerogative of the king and his council.

In addition to councils and officials there may be executive agencies of a different order, either autonomous and merely utilized for the purpose of government or especially designed for such purpose. Examples of the former are age-sets (*v.* p. 68) or secret societies (*v.* p. 194); of the latter, when these constitute bodies corresponding to a police force or organized army. In societies without a well-developed form of chieftainship or other central organs of government, *ad hoc* councils of clan heads or village elders may meet for judicial purposes and for the regulation of community life, e.g. in the distribution of farmland.

Military Organization

In simpler societies all able-bodied men may join on occasion in warlike expeditions. Here it must be examined whether the initiative lies with the political authority or with individuals who can be certain of the approval and aid of the community at large, e.g. where feud or self-help is a recognized procedure.

In many societies one or more age-grades or sets become the military executive of the community. These military duties may be incidental, or the young men of the selected age-set may be specially trained for the task and subjected to strict discipline. They then often live apart, under their own leaders, and must refrain from such conduct as is considered harmful, e.g. sexual intercourse, or the eating of certain kinds of food. Entry into the military organization may be by an initiation ceremony (*q.v.*).

In stratified societies all individuals holding office or rank may have military duties, implying either personal service or the contribution of levies from their households or the groups over which they exercise authority. Chiefs, sub-chiefs or members of certain ranks each contribute a body of armed men in case of war. The soldier is rewarded with a share in the booty, sometimes with grants of land, or with titles and other honours. Thus the armies include many slaves, serfs, or landless strangers, who join for the promised gain. Where military service is bound up with the tenure of land or with ranks we may find analogies with a feudal system.

Warfare in Relation to Political Organization (see also p. 152)

Among the simplest societies warfare is often limited to sporadic conflicts between contiguous groups. In State organizations wars may be extended over several adjacent territories, and attacks made on territorially remote groups. But wars mostly occur between groups which normally maintain close contacts—of trade, intermarriage, or intermigration. This combination of intermittent hostility with contacts and even kinship relations is by no means rare, and the complex adjustment of relationships involved is of great interest. One of the consequences of this form of warfare is the ever-present danger of killing a kinsman who belongs to the enemy group, and so committing, not a legitimate act of war, but a crime or sin. It is important to find out what precautions are taken to avoid killing kinsmen, and what ritual or other measures are taken to expiate the offence, if such a person is killed. The hostility between neighbouring groups is often traditional and deeply ingrained in the social organization. Legendary or mythical explanations may be given for it.

In some societies raids into neighbouring territories have the aim of capturing livestock or slaves or of achieving revenge. Sometimes magical or prestige motives play a part, as where the aim is to obtain enemies' heads (*v.* Headhunting, p. 143), or the capture of some prized or ceremonial object. Among agricultural peoples raids on crops or food stores appear to be absent, though the capture and destruction of crops are sometimes entailed in retaliatory attacks. Retaliation by revenge or blood feud on a collective scale appears as a typical relationship between autonomous political groups, unless outlawed by special pacts (*v.* below).

With an expanding State organization, wars of conquest occur, aiming at bringing one territory under the domination of the other, making it tributary, or capturing its wealth and resources (pastures, farmland, trade). There are many instances of invasion and the settlement of the victors on the land of the vanquished in the history of tribes which became states through establishing themselves as the ruling groups in invaded countries.

A state of permanent and unrestricted warfare is rare, and would indeed lead to mutual extermination. Thus wars are in various ways regulated by accepted rules of fighting and by formalities concerned with the opening of hostilities and the conclusion of peace. Lasting pacts proscribing warfare exist between many primitive societies; they sometimes develop into treaties of mutual assistance, and always extend the scope of law beyond the boundaries of the single community (*v.* above—Inter-group Relations).

Conduct of War

From the above it will be seen that the causes leading to war and the types of warfare vary with the political organization of the people concerned. This will influence such matters as the organization, equipment and provisioning of the armed forces both in times of peace and during a campaign. The following notes will be of help in recording the conduct of war. As it will scarcely be possible for any investigator to-day to see native warfare in action, he will have to make his records from a series of narrations of past wars from reliable informants. All practical and ceremonial preparations for war and the absence or presence and type of declaration or preliminary negotiations should be recorded. The part played by special war chiefs or ritual experts should be noted. Are there challenges and single combat? Are calculated insults performed? Is enemy territory burned, devastated or occupied? Enemy property destroyed or captured? If the latter, is the loot the prize of the captor or dealt with in some organized manner? Are prisoners taken? If so, how are they treated? What is the fate of enemy women and children? Do women take part in active warfare or have duties with regard to provisioning or defence assigned to them? How are sacred objects of the enemy treated? Are they destroyed or preserved and honoured by the captors? All ritual or taboo observed by the warriors or their relatives, as well as all practical and ritual acts concerning the cessation of hostilities and peace-making should be noted. Are there ceremonial methods of greeting the return of warriors?

The dress and badges of warriors should be noted. Are trophies or emblems of success in war worn during warfare, ordinarily or ceremonially? (*v.* below, Head-hunting). Are any signs of honour given to a warrior who has slain an enemy? If so, must this be in battle or can a warrior claim homicidal honours if he kills a captive or a woman?

The tactics of warfare will be influenced by the types of weapons (*v.* p. 259) in use and the presence or absence of horse or other riding or draught animals. All these should be described.

In some cultures war may be a way of settling blood feuds. The enemy dead may be counted and if this count tallies with the number slain by the enemy in a previous war, the debt is considered paid, and hostilities cease. Such warfare is associated with a state of intermittent hostility between neighbouring tribes.

Head-hunting

The practice of cutting off parts of the slain enemy's corpse, particularly the head, is a widespread feature of savage warfare, and persisted in Europe well into modern times. Besides the value of the head, etc., as a trophy, there are a number of ritual or psychological motives for such behaviour (*v.* Head-hunting, Ritual and Belief, p. 117). After the head, the hands and feet are the most usual parts to be abscinded, in some cases perhaps merely to prevent pursuit by the ghost, and in others in order to gain specific status, e.g. the right to wear scarlet gauntlets by a Naga warrior who has brought home a pair of enemy hands. The head (or scalp) when brought back is commonly treated as a source of fertility. It may be mounted high up on a stake to make the crops grow high, exposed on a boulder or a stone for the benefit of the village, or hung in a clubhouse or dwelling-place. Often a head so taken will be hung on the bows of a canoe or, e.g. in Assam, of a canoe-gong, as a necessary part of its inauguration for public use. Finally the trophy may be hung on a pole in the village and used as a popinjay or buried face downwards in the earth. Measures are commonly taken to placate the ghost and induce it to remain in the village of its captors. Special treatment is often accorded to the lower jaw, while heads taken may be divided between those who took part in the raid or, as by the Jivaro of the Amazon Valley, subjected to the removal of the skull and the shrinking of the flesh by means of hot sand and red-hot stones to create a vivid representation of the features of the late possessor. Head-hunting is often accompanied by some form of ritual cannibalism and by some such fertility ritual as pelting with grain. The Angami Naga (unlike their neighbours) bury the head, and that face downwards. The Kachin of Burma are reputed to bring back the head merely as evidence of success, while the Kuki-Chin tribes do so in order that the ghost may attend a dead chief as a slave.

Secret societies and age grades (*v.* Age, p. 68) may have considerable influence in the organization of public life. Inquiries should be made as to the functions of secret societies, their numerical strength, their status and relations to other authorities. Their power may be accepted and employed for the punishment of crime or other political purposes (*v.* Ritual and Belief, p. 194).

LAW AND JUSTICE
INTRODUCTION

In all societies the relations between the members of the community are regulated by a body of observances, traditions, rules and accepted religious and moral standards. Observe as fully as possible all the forms of conduct pertaining to such aspects of communal life as the personal relations between kinsmen, clansmen, and members of the community; the status of husband and wife and their respective families; economic relations; regard for human life; personal honour; institutions such as rank, chieftainship, marriage, property inheritance (*q.v.*), religious observances, etc. If anything happens to excite general interest, say a family quarrel, a lawsuit, or a breach of the marriage regulations, note what is said, what actions are praised, blamed or penalized, and what public opinion seems to require. Discriminate between those manners and customs of which rigid observance is required, and those of which breaches are tolerated. To what extent does this toleration depend upon the preservation of a "decent secrecy"?

Ascertain how standards of behaviour are impressed upon and taught to the members of the community (*v.* Social Life of the

Individual, p. 98); whether they are codified or merely inherent in the culture as manners and customs. Are any special persons held to be repositories of, or specially skilled in, knowledge of the law, etc.? The connection of the different norms with religion, myth, cult, organization of secret societies, etc., should be investigated.

Whenever possible note how rules come into being, whether they are all customary rules of behaviour which have grown up within the community, or are sometimes specifically declared by some influential individual or body and enforced by authority. The power of outstanding individuals to introduce changes in fashion and custom should be examined.

Sanctions regulating the norms of social conduct may be positive or negative. In the former, observance of the social standards is approved by the community, the individual is rewarded, held in respect, etc.; in the latter, their infringement is threatened with unpleasant consequences. Sometimes compliance with the standards is secured through informal social pressure, e.g. by the threat of ridicule, contempt, scorn, or ostracism, usually very effective forms of punishment. The observance of other norms is secured by supernatural sanctions, any violation being followed automatically by evil results, without any overt interference on the part of the community. The breaking of taboo, e.g. often renders the offender "unclean", and may result in disease or even death, or a punishment may be inflicted by some supernatural power or spirit. Other norms are sanctioned by the organized reaction of the community, acting as a whole or through its authorities, or certain groups, or individuals. Thus the magician will use his power to bring about illness or death of the culprit, or a regulated fight may ensue, or the culprit may have to undergo an ordeal or trial. Occasionally a group of people will organize an armed party against the offender on their own account, but with the approval of the community; or again the culprit may be tried and punished by the judicial authorities of the community. Such sanctions may be referred to as organized sanctions as opposed to the ritual and unorganized sanctions which do not involve the active participation of the community through recognized agencies. Careful investigation of the different forms of sanction should be made noting the basis on which they rest—beliefs, moral precepts, actual institutions. In the case of ritual sanctions, note what types of action involve their operation; what happens to the offender—is the working of the sanction inevitable; can he be purified in some way, e.g. by expiation, sacrifice, confession, or other ritual? What attitude is adopted towards him by the community—is he avoided and cut off from social intercourse or driven out of the society, etc., and do the consequences of his action affect his relatives or other members of his household, local group, etc., as well as himself? (*v.* also Taboo, p. 185).

Justice
In the strict sense of the term, law exists in a society only if there are constituted legal tribunals (*v.* Political Systems, p. 132) vested with the power to enforce their judgements by means of organized sanctions. Such tribunals have established procedures for bringing offenders to justice. Many primitive societies, however, lack such explicit legal machinery and procedures. Provision does, nevertheless, often exist for action to be taken by the community as a whole possibly under *ad hoc* leadership of elders or ritual experts, against grave breaches of customary codes of conduct or belief, e.g. in cases of incest or witchcraft. In most societies also institutions exist which enable a wronged individual or group to bring a cause before the constituted legal authority or before an informal tribunal of arbitrators. Legal procedure in the strict sense is concerned almost entirely with the breach of these norms which involve organized sanctions. Note what actions involve such sanctions, and especially what happens in the case of homicide and bodily injury; incest, adultery, seduction, rape, and breach of the laws of exogamy; theft, the killing of other people's animals, damage to their property, etc., slander, disturbance of the peace, revolt against communal authority, witchcraft and black magic. Inquire into the responsibility of the culprit for wrongs committed voluntarily or involuntarily, accidentally, through carelessness, etc. Is any distinction made in the penalty according to the motive of the culprit, or are only the consequences of his act considered? How far are slaves, women, children, idiots and animals regarded as responsible for their actions, and if they are not held responsible does anyone else bear their responsibility? To what extent are the relatives of the man, his clansmen, members of the same association, age-set, or other social group, involved in joint responsibility for his actions? Is any distinction made according to

whether the wrong affects a stranger or a member of the same community, or to the status, age or sex of the victim or culprit?

Where there is a central authority to which cases are brought for trial the composition and powers of such authority should be investigated as well as its territorial sphere. It should be noted whether all cases are brought before the authority or whether, in some types of wrong, the victim or his group take direct action against the culprit or his group.

Where the aggrieved party takes direct action, even where there is a central authority, is any form of blood-revenge practised? If so, is it employed for deeds other than homicide? Whose duty is it to take blood-revenge—close relatives only, clansmen or members of other social group, masters for their serfs? Who takes vengeance for a woman—her husband or her own family? What happens when culprit and victim belong to the same family, clan or other social group? How long does the blood-feud last: is it satisfied by the killing of the culprit or a member of his social group, or does homicide give rise to a continuous vendetta? How is this ultimately ended—by any forms of ritual or by intermarriage, and with what ceremonies? Is blood-vengeance ever commuted to payment for the injury, i.e. blood-money? If so, is the aggrieved party at liberty to refuse such compensation? Does the payment depend on agreement, or is there a traditional tariff defining the payment to be made for all possible injuries? Does the amount vary with the age, sex, or status of the victim or culprit? Must compensation be paid in any special form, e.g. cattle, garden produce, women, slaves, etc.? What people are expected to contribute to, and what people can claim a share of, the blood-money? Who regulates the payment and what happens if it is not made?

Where organized trials are held, notice the composition of the court, when and where it is held. Who initiates the prosecution—the plaintiff or the community through its agencies of government? Where both occur, for what types of action are they respectively employed? How are culprits detected, arrested, summoned, and brought before the court; how are the proceedings conducted; what proofs of innocence or guilt are demanded; what regard is paid to evidence? Notice the employment of oaths, ordeals, counsel, etc. How is judgement arrived at, and how pronounced, and are there any rights of appeal? If so, to whom? What kinds of punishment prevail, e.g. death, mutilation, chastisement, outlawry and banishment, slavery, confiscation of property, fines; and to what crimes are they respectively applied? Where fines are imposed, what objects do they consist of? Who contribute towards them and who benefit from them, and in what proportion? Does the court share in the fines or claim fees? If the latter, who pays them? How is the judgement of the court enforced? Where there are subordinate courts, what kinds of cases do they deal with, and to what extent are their judgements binding or subject to appeal and revision by a higher court? Are there any ways in which disputants can settle their conflicts without reference to a tribunal, e.g. by regulated combat, payment of compensation, licensed plundering, etc.? Endeavour always to distinguish between major and minor offences. Notice especially what are regarded as offences against the community as a whole and what against the individual, and indicate any differences in their respective treatment, sanctions, etc. Attend all the trials you can, and record all details of cases that come up for trial. Note any rights of asylum for fugitive criminals, slaves, etc. What persons or places have power of sanctuary? Does the refuge protect against the agents of the law or only against private foes?

Where there is no central authority, with an organized system of justice, there may be persons or groups of persons whose authority may be accepted in the community. Such persons may be heads of a clan, lineage, or family, elders in the local community, officiants of religious cults, persons with magic or esoteric knowledge, or even a person recognized for his outstanding character, independent of recognized status. In all such cases the sanction for their authority should be investigated (*v.* Chiefs—Councils and Officials, pp. 128–40). It should be noted both in the cases of organized and unorganized systems of justice whether women as well as men hold positions of authority, and if so, what are the sanctions for their authority.

Property
(*v.* Economics, Law, Land Tenure, Inheritance)

The concepts of property and ownership are closely linked. Ownership is best defined as the sum total of rights which various persons or groups of persons have over things; the things thus owned are property. Property thus defined in terms of the relations of persons to things is a distinct concept from that of *capital* in which things appear as the basis of production and in which the emphasis is on the use to which things are put rather than on the rules by which they are controlled (*v.* Economics).

Ideas concerning property may vary, not only in different societies, but even within a single society according to the nature of the property and the type of ownership right involved. A full understanding of these ideas calls for careful analysis; simply to label a system as "individualistic" or "communistic" is never adequate and often misleading. Language alone is no adequate guide. Verbal claims of individuals cannot necessarily be accepted at their face value. Accepted equivalents of *my* and *your* may not indicate ownership in our sense; the meaning of such terms may vary with context.

Examine the types of property involved, the types of rights that can be exercised over such property and the rules governing the exercise of such rights, and finally the types of individuals and groups that may exercise these rights. The following list of broad categories indicates the general types of property that need consideration; actual details will vary in different societies (*v.* Material Culture).

Types of Property
(*a*) *Real Estate*, e.g. land, roads, bridges, wells, buildings, trees. Improvements of land, plantations, and even forest trees may be deemed property quite distinct from the land on which they are superimposed (*v.* Land Tenure). Are there buildings owned and maintained by associations other than ordinary household groups?

(*b*) *Household and Occupational Equipment*, e.g. general household furnishings, implements of husbandry, tools, weapons, the equipment of specialists. The ownership of implements and equipment links with the distribution of labour between the sexes, age groups, etc. Understanding of the control and inheritance of household equipment requires an understanding of kinship structure (*v.* p. 75); in contrast the ownership of specialists' equipment may be restricted to craft associations. Spears may be items of bride-wealth or may have ritual value. Is *skill* in use or manufacture treated as an item of property?

(*c*) *Personal Effects*, e.g. clothing, ornaments, currency. How were such objects obtained? Who will inherit? Is jewellery used as a cash reserve in times of emergency? Has the individual holder unrestricted powers of disposal? Do such objects include heirlooms and items of bride-price?

(*d*) *Ritual Objects.* In most societies, including our own, objects are found which are valued on account of their ritual and traditional associations rather than because of their intrinsic economic worth. Such objects may be the heirlooms of social, political or religious groups—and as such inalienable. In other instances such objects circulate by means of traditionally sanctioned gift exchanges. Since the values involved can seldom be fairly expressed in terms of any common medium of exchange these objects must be distinguished from currency; indeed the substitution of cash for ritual objects in some traditionally approved transaction such as bride-price may have important economic and social significance. What are these objects and what is the nature of their ritual value?

(*e*) *Stocks* (including livestock, foodstuffs, raw materials, standing crops) (*v.* Livestock, Agricultural Life, Mining and Quarrying, Metal Working, Pottery, Salt). In primitive societies accumulated reserves of wealth are frequently invested in stocks of livestock and material rather than land or treasure. The display value of such property is often important; quantity may be more important than quality; with livestock, external markings alone may determine the value. To assess property values list stock in terms of native categories as well as those of Europeans. The sex and reproduction rates of animals are important. What rules govern the ownership of animal offspring? Cattle are often items of bride-price? Do the affinal kin retain any rights in such animals? What happens in the event of divorce? Are animals or goods pledged in advance of payment? Can such pledged debts be inherited? For what purpose are animals kept? For meat? Dairy produce? Wool? Manure? Prestige only? Is killing ritually prescribed? Who gets a share of the meat?

(*f*) *Rights to Economic Utilization* (*v.* Land Tenure). The rights of use (*usufruct*) must be considered not only in the case of simple land but in any form of fixed capital or real estate, e.g. the right to use a millwheel, the right to draw water from another's channel.

(*g*) *Rights in Persons and Human Services.* Property rights in another's labour may take various forms ranging from the crudest slavery (*v.* p. 95) to the mutual rights of husband and wife. Tribute to political superiors may be paid in labour; bride-price may take this form; debts and mortgage transactions may be settled by voluntary serfdom (*pawn*). Is labour service a recognized alternative to other forms of payment as a matter of choice?

(*h*) *Other Forms of Incorporeal Property.* Titles, songs, names, spells, skills, etc., may be matters of precise ownership and the subject of purchase, sale, inheritance, gift, etc.

It must be stressed that the above categories are by no means mutually exclusive. In some societies cattle may be best considered as ritual objects; other ritual objects may appear at least temporarily as the personal effects of individuals. Make an inventory under the above main heads of all the types of property to be found in the community. Select also a sample of small localities (homestead, nomadic camp), for detailed examination in quantitative terms (*v.* Sampling). If distributions of types and quantities are recorded on maps significant variations may be revealed.

The types of individuals and groups that may exercise rights over property will emerge from an analysis of the social organization. In general all social groupings may be expected to have property rights of one sort or another. The original inventory of property suggested above records the distribution of property in terms of locality; consider this same inventory of material in relation to the persons and groups who exercise control. For example what rights are exercised by men; women; children; the simple family; the joint household; the bilateral kin group; the lineage; the clan; the village group or ward; age-sets; secret societies; political associations; craft guilds; religious sects; ritual, magical and religious leaders; kinship elders; political chiefs.

The types of rights exercised vary according to the type of property and the type of controlling individual or group. There will be economic, religious, legal and political aspects of such rights. The following types of rights must be distinguished: (*a*) rights of use; (*b*) right to control the use or disposal of property by others; (*c*) rights of disposal; (*d*) rights to derive an income or other benefits from the use of property by others; (*e*) rights to be described as the titular owner of property without further benefits. Several such types of right may be simultaneously exercised over the same piece of property by different persons or groups. For example a householder may have the exclusive use of certain cattle but may be unable to dispose of those cattle without consulting a wider group of kinsfolk; he may be able to dispose of the cattle without consulting his political chief, yet if he kills an animal he may still have to give a share of the meat to that chief.

Make a detailed inquiry into the origin, use and title of all items found in a particular small locality such as a single homestead. Observe whether verbal claims to ownership are confirmed by the practice of actual use. Determine of each object: Where was it made? Who made it? How was it acquired? Where is it kept? When is it used? Why? By whom? With some types of property an even wider inquiry will be necessary to cover all the types of right that may be involved. For example, with a fishing canoe: What is the composition of the crew? How are they selected? How are they associated with one another? How are they rewarded for their labour? Who decides how and when the boat shall be used? Who shares the proceeds of the catch? Who initiates repairs?

What is the legal procedure for settling disputes relating to property? How does such procedure work out in practice?

Inheritance

Distinguish between *succession to social rights and political office, and inheritance of rights over land and property. While strictly speaking inheritance comprises only the rules governing the transmission of property from a dead person to his heirs, the transfers of property during life (gifts inter vivos, dowry, bride-price, etc.), necessarily affect inheritance.* An inheritance rule such as *ultimogeniture* (inheritance by the youngest son) may even be based on a presumption that other "heirs" have already received their share during the lifetime of the deceased.

The various rules of inheritance reflect two conflicting tendencies; firstly the obligations of an individual towards his near kin, and secondly the desire of an individual to exercise choice in the disposal of his property. The majority of traditional inheritance rules emphasize the first of these tendencies; only with inheritance by *will* is the second fully satisfied, though a system which permits *adoption* may be regarded as a compromise. In primitive societies inheritance by *will* is rare, while the varieties of obligation towards kin are very great; a thorough understanding of the kinship structure (v. pp. 70–90) is therefore fundamental to the understanding of any system of inheritance. It is also essential to appreciate the nature and the limitation of the rights which living individuals have in different forms of property (*v.* Types of Property, above; Land Tenure, below); by no means all such rights are transmissible to an heir.

Record the native rationalization of what the system of inheritance is supposed to be but collect also concrete examples of inheritance situations. The rules may differ according to the type of property (*v.* p. 149) and the sex of the heir. Has a woman equal rights to a man in land? Do certain objects pass only to persons of a particular sex?

Craft and esoteric property may pass only to initiates outside the kinship structure. Are property rights in persons inherited, e.g. the *levirate* (inheritance of a man's widows by his brothers?) (*v.* Marriage, p. 117). Are obligations and debts inherited? Rules governing the inheritance of land are especially important in their social and economic effects in the form of local group fission and the fragmentation of holdings. Rules in which the bulk of the property passes to a single heir, e.g. *primogeniture*, clearly induce less fragmentation than others where several heirs, both male and female, share the property in fixed proportions, e.g. Roman Law *legitim* shares, Islamic law, etc. Supplementary customs may counterbalance this trend towards fragmentation of holdings. Consider cousin marriage from this angle (*v.* p. 76). Where there are multiple heirs do they all receive the same kind of property? It may be that certain types of property are inherited in the male line, and others in the faemle line. Does a woman's property pass to her daughters only? While individual rights are commonly inherited in the direct line of descent, father to son, or mother's brother to sister's son (*v.* p. 85), there are other cases where the heir is the eldest surviving member of an extensive group of cognates, for example, a man's brother will inherit before his sons. Note whether this is due to foreign influence.

Where inheritance by will is recognized, are the powers of disposal vested in the testator restricted? Can he disinherit a "legitimate" heir?

What happens in a case of dispute? In general what are the legal and moral sanctions which support the conventional rules? Are changes taking place under present conditions? What is the direction of these changes and how do they react upon the property and land tenure rights of the individual? In traditionally matrilineal societies, note any tendency for the inheritance rights of a man's children to improve at the expense of those of his sister's children.

Land Tenure (*v.* Economics, Property, Inheritance)

Land tenure may be best understood in terms of the *rights* of persons and groups in land (*v.* p. 168). The most fundamental of these are the right to use land in various ways, and the right to take a share of the produce of land, either directly or as rent, without contributing labour. Other rights are: the right to transfer holdings, the right to alienate land (as in sale or gift), the right to grant rights of use to others; there may also be vague titular rights, which, though carrying a certain prestige value, do not under normal circumstances imply any material benefit for the owner. For purposes of analysis we need to know the types of land over which rights can be held, the nature of the rights, and the persons and groups holding these rights.

As regards types of land, it is not sufficient to record and classify merely the primary economic forms, agricultural land, pasture, forest, etc. Subsidiary uses of land must also be considered. For an agricultural community there is not only land that has been cultivated in the past, but land which might be cultivated in the future; there may be fuel and thatch reserves, building land, water supply, fishing reserves, sacred areas. The rules of ownership that apply to one type of land do not necessarily apply to another. Moreover a particular piece of land may be used in different ways on different occasions. What is fallow to the agriculturist may be pasture to the herdsman; standing trees may not belong to the man who cultivates the land on which they stand; chiefs may be entitled to perquisites of particular kinds of produce regardless of where they occur; in short different persons may have different rights in the same land without a conflict of interests.

For the types of individuals and groups to be considered, *v.* Types of Property, above.

Study the manner in which the land is subdivided by the people themselves. There is great variation in the degree to which individual holdings are defined, in some cases these variations may be correlated with the technique by which the land is used. There is also great variation in the manner in which individual holdings are owned. There may be permanent ownership or temporary ownership by individuals and by groups.

Where land is used for hunting, collecting, and grazing, sharply demarcated boundaries to individual holdings are often impracticable. A whole group may claim collective rights over a somewhat loosely defined area (*v.* Social Structure—territorial arrangement, horde, p. 65).

Where agriculture is practised holdings tend to be more precisely defined where the cultivation is intensive. If land is very plentiful the rights granted may be simply to cultivate land, without particular regard to area or place. Where land is worked in rotation the boundaries of holdings sometimes, but not always, persist through the fallow period. How far do the rights of individual owners also persist? Is there any system of reallocation to those who are short of land? Who has this power? Has the individual who is short of land any *right* to the temporary use of the surplus holdings of another? How does such usage affect the long-term title to the land? Where techniques permit the land to be maintained in continuous cultivation it is possible for the same individuals to be cultivating the same land year after year; yet even here tenure may fall short of absolute individual ownership. Frequently there is restriction against alienation, especially by outright sale. It may be more accurate to speak of ownership by a group of kindred than of ownership by individuals.

With these possibilities in mind obtain data on the size of holdings and the variations of size according to the type of land held. Make maps and sketch plans to record these holdings (*v.* Sampling). How are boundaries defined in native terminology? In writing? By boundary stones? Through the memory of village elders? In court records? Are fees charged for such recordings? If so, do you consider the records to be up to date in accordance with observed facts? Are titles thus recorded indefeasible in a court of law? Are they in fact a source of dispute? How are disputes settled? (*v.* p. 135). Investigate all methods of transferring right in land, e.g. sale, pledging, mortgage, gift, loan, lease, etc. Are there religious links between individuals, or lineages, and particular pieces of land, e.g. are the bones of ancestors supposed to be associated with particular localities? Claims to land are likely to be backed by myths, traditions and history. How precise are these traditions? Are they mutually consistent? Are they a source of dispute? In different contexts informants may speak of household land, lineage land, clan land, village land, chief's land, etc. Do these concepts overlap? What do they signify in terms of rights? The commonest forms of title derive either from ancestral right (*v.* Clan Organization), often backed by mythical charter, or from fiefs allocated by some political superior (*v.* Political Structure), or from a combination of the two. Where there is a chief he may receive rent or tribute in recognition of his title; the chief in turn may have reciprocal obligations of a ritual, legal, or military kind (*v.* Chieftainship, p. 136).

Where there is a hierarchy of chiefs, the paramount chief, the subordinate chief, the village headman, and the individual tenant may all have distinct though mutually consistent rights respecting a single piece of land. There may be individuals, lineages, or groups who are called "fathers or owners of the land". Actually they may have no rights over the land and never cultivate it, but it may be necessary for them to initiate or organize such operations as burning bush, clearing forest land, etc. They may be responsible for various ceremonies concerned with fertility of the land and for all land magic. Their activities, status, duties and rewards should be investigated. It may be found that these people are, or are believed to be, descendants of aboriginal inhabitants, occupying the land before the incoming of an agricultural people.

Rent. Record details, both theory and practice, of all forms of tribute and rent. Are dues paid for the right to utilize land? To whom are they paid? What proportion of the total yield does this represent? Are the dues reckoned in the form of a fixed annual rental, a proportion of the gross yield, a levy on labour services, a token quit rent in some traditionally approved form? What is the native explanation of this payment? By what sanction is it enforced? What reciprocal services, if any, are received in return for such rent or tribute? If the land is leased or mortgaged by the original tenant, who then pays the tribute to the overlord?

Changing Forms of Tenure. Under conditions of social change land tenure provides some of the most difficult of all administrative problems. The following changes may be taking place: new classes of persons may acquire rights at the expense of the traditional rights of others; land values may increase sharply and encourage holders of unimportant rights to exaggerate their claims; money payments may replace other forms of traditional payment in all forms of rent transaction; traditional objections to alienation may be circumvented by devices such as perpetual mortgage; the decisions of European courts may serve to confirm and perpetuate the rights which under the traditional system were provisional or temporary; customary law may be found to apply to one piece of land and a conflicting government law to another. Under such conditions there is a general tendency for native practice to circumvent rules laid down by the administration and this in turn leads to prevarication and suspicion as a normal response to all official inquiries relating to the tenure of land. Special care is therefore necessary to distinguish current practice from tradition on the one hand, and government approved behaviour on the other.

CHAPTER VI

ECONOMICS

Introduction

Economics may be briefly described as the study of that broad aspect of human activity concerned with the utilization of resources and the organization whereby they are brought into relation with human wants. It has sometimes been thought that the simpler societies have no economic organization worthy of the name. But the study of any society, however simple, will show that the resources of a people are handled in a systematic way with regard to means and ends. And this economic organization is fundamental to their life, being linked with their social structure, their system of government, their technology, their ritual institutions.

The general ideas of ordinary economic analysis have been found applicable even to very primitive communities. But to apply a technique of analysis which has been developed against a background of Western institutions to primitive institutions is not easy. Money prices often do not exist, and so a simple measure of economic relationships is lacking. Terms such as "capital", "wages", "rent", "saving" may have to be avoided or used in a different sense from that in which the modern economist ordinarily uses them, since the distinctions they involve may not be applicable to unspecialized "primitive" situations. Other difficulties are mentioned below. The study of economics in simple communities should properly speaking be a job for economists. But so far few economists have tackled it, and most of the investigation has perforce been done by anthropologists. If the investigator has been trained in ordinary economic theory, but has no anthropological training, he should remember continually the importance of the social setting of the economic institutions he is studying, otherwise he will not grasp the value system upon which the economic organization depends. If he has had no economic training, he is recommended to study the fundamental principles expounded in one of the recognized economic textbooks.

The economist who studies civilized communities can assume

for his own purposes and those of his readers a common knowledge of the technical conditions of industry, the processes of exchange, the system of property holding, and the whole social background of the economic system. Since these assumptions cannot be made for most primitive communities, the anthropologist must give some account of these features (*see* sections Social Structure), in order to show how far the economic organization depends upon them. The suggestions given in this section assume then that the investigator is also studying the other aspects of the social life of the community.

But economics must not be confused with technical activities. A description of the arts and crafts of a people and of the way in which things are made is interesting, especially when related to problems of invention and the adaptation of processes, or to native ritual, myths, and social organization. But from the point of view of an economic study these technical methods may be taken as "given factors" in the organization of production. What is important here is the effects of different techniques on time and efficiency of production per unit output, division and organization of labour. With regard to pottery, who produces the clay? Is the time put in on pot-making an alternative to other pursuits or taken from leisure? Are the pots exchanged, and for what? What relation do earnings from pot-making bear to a person's total income? . . . . and allied information.

The purpose of economic inquiry in a primitive community is rather to study the volume and nature of production, to ascertain what may be called the real incomes of individuals, to measure the amount of resources, including labour, needed to get these incomes, and hence to give some significance to such a concept as a local "standard of living". Fully statistical material such as the economist is accustomed to deal with is usually unobtainable in a primitive community where there are no prices, and even for most peasant communities where money is used for a limited range of transactions only. But quantitative information may be obtained in the form of maps of land tenure, showing areas utilized by individuals, with the yields of crops in agriculture; or the stock carried in pastoralism; analysis of the amounts of labour utilized in different undertakings; the type and quantity of goods owned and utilized by households. Within certain limits such material can be subjected to statistical analysis and give valuable information about the economic position of groups and individuals. Where money is used and goods and services can be priced in such terms, the collection of quantitative data is of course much easier. But it must be remembered that there will still be items not measured in money to be taken into account.

The economic organization of a people can still be conveniently treated under the conventional heads of Production, Distribution, Exchange, and Consumption.

Production

Under this head is considered the organization by which resources are converted into goods or services which satisfy the people's wants. Any expenditure of effort to this end, such as the transport of materials from one place to another, comes into this category.

Side by side with the study of production the student can usefully make a brief general examination of the resources available to the people. What is their relative scarcity? What do the people know of their properties and uses? What are the technical processes at command for using them (*v.* Environment, Knowledge and Tradition, Technology).

The problem of how far technical progress is possible in existing conditions is important. Reference should be made therefore to those potential resources, such as forest timber, unutilized land, or water supplies available for irrigation, of which the people are ignorant or have not the technical means for turning to account at present.

The type of economy may be roughly classified according to the major productive technique used, as follows: *collecting* of fruits, seeds, roots, etc., *hunting* and *fishing*, *agriculture*, *pastoral* care of domestic animals, *industry*, or *trade*. Too much emphasis, however, should not be laid upon the classification in terms of these techniques. In the first place they are seldom found to be mutually exclusive. Pastoralists often practise some agriculture; agriculturists collect forest fruits, or hunt and fish; they may also have herds of livestock, or now frequently spend time in wage labour at mines or plantations. In each case the returns obtained from every type of technique in use should be compared in order to find out what relative income or other particular economic advantage they give. In the second place the economic organization of two peoples having the same technique of production may be very different—if only because of differences in their social structure, such as a matrilineal or patrilineal system, existence or lack of a central authority, etc. The analysis of productive technique is important in showing its effects upon the volume of production and upon the amounts of labour involved and its organization. It is important also to consider how far the technique of production is standardized and how far it is flexible. The reasons for adherence to a traditional technique or for change to a new one should be analysed. This is of special weight in conditions of changing social life due to external influences. The relative cost of the preferred technique in materials and labour should also be examined.

Sometimes different technical methods are practised side by side in the same community, allowing of a more efficient utilization of resources than would be possible by one alone. Some tribes for example use different agricultural methods according to the type of soil, and the season of the year. In comparing the economic systems of two adjacent communities the effect of a greater range of technical method upon crop output and more uniform production should be examined. The effects of changes of technique, as for instance the introduction of Western tools, should be examined.

It is worth noting that the natives may themselves have a rationalized explanation of their attachment to one form of economy rather than another. For example, the Masai pastoralist scorns agriculture and gives as a reason that it is wrong to scratch the body of Mother Earth. By feeding the cattle with grass, he alleges, the earth already gives man milk as food, and she should not be forced to contribute twice over. Descriptions should cover not only the type but also the aim of the productive activities of the people—how far the food produced is devoted to subsistence, to fulfilling kinship and other social obligations, to offerings to spirits; how far cattle raised are used for food, for sale, for marriage payments, or bride-wealth, for compensation for injury, for sacrifice. A rough estimate of the proportions of each item normally allotted to such ends should be given. Consideration should also be given to the size of the population and to its distribution in terms of sex and age groups (*v.* Demography).

The scale of economic undertakings, and the manner in which goods are disposed of, may depend to a considerable extent upon these factors. The distribution of the population in relation to the location of the available resources should also be noted on a sketch map (For ownership of resources, *v.* Capital.) In order to realize the demands which the economic process makes upon the community, the time spent by individuals and by groups of people in the production of goods should be estimated. Where possible, such estimates should be based upon a set of systematic observations of the occupations of individuals from day to day in various co-operative activities, at different periods of the year. It is valuable to construct a *calendar of work* for the community as a whole. This calendar, which can be based upon the records of the observer's diary, should be of a synoptic kind. But it should include fluctuations in work due to seasonal variations of occupation, unfavourable or unlucky days for work, weather conditions, customary rest periods, festivals, funeral and other ceremonies.

Care should be taken not to prejudice results by thinking too exclusively in the economic terms of our own commercial organization. Specialized economic relationships such as exist in our own society between buyer and seller, or between employer and employed, are frequently not found in primitive communities. For them the economic unit is also a social unit, and economic relationships are at the same time social relationships. In many societies kinship bonds form the basis for most economic relationships. For instance in the Trobriands a man's income is largely dependent upon the customary obligation of his wife's brothers to support him. Again in Tikopia a great deal of cooking is performed by men as a service to their wives' families, and they receive specific reward for this only on certain formal occasions. Those types of service which are given because of social ties as well as those given primarily for an immediate or direct economic reward should therefore be listed.

In every primitive community the economic life is more than an individual search for food; even in a simple hunting group there is some collective action. The organization of production involves the division of labour, and the integration of each person's contribution into the common task.

Division of labour usually occurs on the basis of sex, age, and skill, and also on the basis of rank where there is social stratification of this kind. Sometimes there is an exclusive division of whole occupations, as when men tend cattle and women perform the agricultural operations. Sometimes the sexes take separate tasks within an occupation, as when men break up the ground and plant the crops, but women do all the weeding. Such division of tasks should be noted. It is often reinforced by taboos or other ritual sanctions, the economic effect of which should also be examined. In the different types of co-operative tasks, is the membership of the group always the same or does it vary from time to time? What factors are responsible for such association, e.g. kinship, availability of workers, possession of requisite skill, etc.? Are any specific tasks associated with the different social groups, such as the family, the household, and the clan?

Certain crafts, e.g. metal-working, building, etc., may be carried out or directed by skilled *specialists*. Do particular villages or households specialize in particular types of craft and ritual? The social and economic reasons for such specialization should be examined. How far is it hereditary? What are the effects upon the income of the individuals and groups concerned? How far is there continuous specialization? In many communities comparative proficiency in crafts allows some people to devote more time than others, but they are not so engaged the whole of their time. In this regard the extent of the demand for the products of such specialized labour should be investigated, and the levels of skill that are reached. How is such skill transmitted? List any educative mechanisms, e.g. mimetic play, observation of elders, deliberate instruction. Is there any system of apprenticeship?

Ideology of Production

In addition to the social ideas that lead to a division of labour as between particular individuals or groups there may also be certain attitudes applicable to the group as a whole which have an effect upon production. Moral values, expressed in terms of praise or blame, may determine to some extent the enthusiasm and efficiency with which the individual approaches his task. Ritual values may lead also to distinction between different types of work, making some occupations dignified and desirable and others the reverse, e.g. iron-working in parts of Africa, or leather-working among Hindus. Proverbs and sayings may sometimes give an indication of the moral ideas underlying native attitudes. In some cases success or skill is rewarded by privilege, as when a successful hunter or fisherman acquires the right to wear a special type of ornament, and this may provide an additional incentive towards efficiency. In specific cases stimuli to work may be provided by more mechanical means such as drugs or rhythmic singing and their effects upon production should be noted. In a more general sense also the rhythm of work is an important factor. List the various incentives to industry. Describe the attitude of the people to work of different kinds.

Quantitative Material on Labour

Attention should also be given to the quantitative aspects of the labour situation. What amount of labour is required for particular tasks? Try to estimate the supply of labour available; and count how many labourers actually assemble. The reasons which lead them to come and others to abstain should be analysed, e.g. what the technical requirements of the work are; the rate of reward they will get; the kinship or other social bonds to the person for whom the task is performed; and any direct benefits they will derive from participation in the work, either then or later?

Organization of Work

Few forms of production proceed haphazard at the discretion of individuals. Frequently there is some individual, often the chief or village headman, who, without necessarily doing any manual work himself, is responsible for initiating enterprise and, directly or indirectly, guiding labour into particular channels. Note the social status of this individual and the sanctions through which he enforces his will. Make a careful assessment of any economic reward that he may receive for these directional services to the community. It is important also to get an idea of the basis of the planning that results either from the action of this organizer or cumulatively from the foresight of various individuals. For instance, a scheme of production that leads to a special glut of food at one particular season of the year may be a regular feature dictated by the requirements of specific ceremonial. On the other hand, it may be a special feature resulting from some economic enterprise on the part of a chief, or it may arise from a need to store food over one particular season of the year. In some communities production is so organized as to make provision for calamities of nature such as flood and drought; data which give an apparent suggestion of "over-production" may have to be examined in this light. In any consideration of organized production it is necessary also to find out in whose interest the planning is primarily carried out. Is it only for the benefit of particular groups or individuals, or for the community as a whole? (*v*. Consumption).

Differential privileges may be important in the economic sense. Individuals such as chiefs and magicians who possess the greatest economic influence may do little or no manual labour. Distinctions of rank and social class frequently have their counterpart, if not their basis, in the economic field. But the investigator should not assume too readily that the individual with the greatest economic advantage is in a position to "exploit" his less fortunate fellows. While freed from the obligation to carry out any manual labour himself, the chief may nevertheless contribute valuable services to the community in other ways (as economic organizer, as "banker" in times of crisis, or as legal, political and religious organizer). Examine in detail the sources from which the privileged individual does in fact derive his "income", and examine also how this wealth is disposed of. Consider also the attitude of the rest of the community towards this position of privilege, since grudging acquiescence may affect the efficiency of the system. A tabulated analysis of the individual's various sources of income and his outgoings will as a rule prove very helpful here.

In nearly all communities there is a very close association between production and ritual. Any analysis of the latter would be incomplete which did not take note of its actual economic effects in terms of production, exchange and consumption of resources. Such ritual may be either of the productive type of agricultural magic, the magic of craft work, or that used to promote the welfare of cattle; or the protective type occurring at marriages and funerals. More concretely, give an analysis of any stimuli to work given by the need to perform ritual. What are its effects, not only upon the output of goods but also in promoting interest in the work? If possible make an estimate of how much time is occupied by ritual which might otherwise be spent upon the production of goods, or in leisure. What is the amount of goods diverted from ordinary consumption to meet the demands of ritual? Observe whether the particular time of year at which ritual is performed has any important economic consequences through its effect on the availability of labour and hence on production.

Capital

The term "capital" is sometimes held to include *all goods in stock*, whether these are later consumed directly, held as valuables, or used to further the process of production.

The productive function of capital is, however, the most important. The investigator should try to find out what proportion of the goods held in store by individuals and groups is used to make possible the production of other goods. Tools are utilized directly, and other wealth, such as food, cloth, beads, cowry, i.e. shells, cattle or pigs, may be employed for the maintenance or payment of labourers or to obtain the means for their payments. An analysis should be made of how far each person in a community provides the tools for his work himself, obtains them from a family stock, or borrows them from others. In the latter case it should be observed whether he makes a return for the loan; inequalities in the possession of stocks of goods between individuals and between groups should be noted. How far are such inequalities levelled out in practice by conventions limiting the rights of ownership, claims upon kinsfolk, or upon the chief, etc.?

Estimate as far as possible the types and amounts of the items that comprise the capital equipment of the community. To what degree is capital maintained by repair or replacement when required? From the point of view of the volume of production, the income of individuals and their standard of living, it is important to estimate whether the stocks of capital are being increased over a period and in what proportion. Such an estimate may be very difficult to obtain. The investigator can, however, get some idea of whether the people are content merely to replace their existing stocks as these depreciate and so maintain a kind of static economy, or whether they attempt to increase their wealth by greater accumulation and so expand their productive system. In some societies such expansion by certain sections of the population is frowned upon by others as tending to upset the balance of prestige and power. Are instances of this to be observed? In pastoral communities and in many agricultural communities livestock, especially cattle, are the most important capital item. Associated with their mode of natural increase (by contrast with inanimate capital such as pots or spears) there may be special modes of inheritance, or of loan and interest-taking. Cattle in particular may be so symbolic of social relationships or so much an object of sentiment that their owners may be most reluctant to part with them for money. They often pass from one family to another as part of bride-price, or in compensation for bodily hurt or as blood-money. Concrete examples of the value of cattle and/or other livestock should be obtained (in goods and in money if possible). The occasions on which livestock changes hands should be listed and described, with indication of the nature and amount of goods given in return.

The function of holding capital for investment is not usually marked in a primitive society, but it may occur. Note should be taken of how far stocks of goods are kept for lending to others. What kind of security for loans is demanded? Are such loans made more frequently to kinsfolk than to others? What interest is demanded? Is this rate always paid? Are different rates charged to outsiders from those to members of the community? The mobility of capital, that is, the possibility of certain objects being put to a number of alternative uses, is important. Indebtedness is apt to be one of the great drains upon the income and capital accumulation of peasant families. Distinguish between indebtedness within and outside the social groups such as the household, kin group and local community. Ascertain whether debts are inherited or wiped out with death, and what penalties are suffered by insolvent debtors, e.g. pawning or sale into slavery. Find out if possible the degree of indebtedness of a sample of individuals or families, how they meet their liabilities, and what are their economic prospects. Can it be ascertained what proportion of a person's income or labour goes to pay his debts or, conversely, is derived from lending stocks of goods at interest?

Care must be exercised in the use of the term "ownership" as the concept is one that differs widely between various communities. In defining native usage it is best to quote concrete cases showing just how the rights of various individuals and groups affect the situation in practice (*v.* Property, p. 148). It is especially important to find out what reward the "owner" receives for the use of his property. This applies not only to ownership of movable goods such as boats and implements, but also to the system of land tenure. In economic analysis it is, however, not the rules of ownership in themselves that are to be considered (*v.* Law, Property, and Land Tenure), but the effect of these rules on the system of production and distribution. It is necessary then to consider how far the system of land tenure allows an individual to expand his resources when he wishes. Can he, for instance, plant crops on the land of other people with or without permission? Again, what does an individual's claim to land actually yield him in terms of real income of goods and services? And how far does it give him security of tenure?

It should be noted that ownership may cover certain rights over the labour of individuals. In circumstances of slavery or forced labour human beings are themselves a form of capital. The inequalities as between different individuals in the possession of capital goods should be examined with special care, as such factors are of great significance in the system of distribution (*v.* Slavery, p. 95).

Distribution

By this is meant the rewards received by the various factors or agents in the productive process. These rewards may be obtained either by the division of a joint product (such as a catch of fish) or by remuneration out of other stocks of goods when the product (such as a canoe or a house) is not divisible. Economists ordinarily classify distribution under the heads of earnings of labour (wages), of capital (interest), of land and raw materials (rent), and of business management (profits).

As a rule, however, no such clear-cut division can be applied in a primitive economic system, but attempts can be made to determine for the economy under consideration what fraction of the goods or money distributed goes to each kind of agent or factor of production. How far is reward calculated in proportion to the service rendered? The best method of analysis is to take a number of ordinary activities (such as fishing, house-building, etc.), and examine how the various individuals involved are rewarded or "paid". Include here not only the ordinary workers, but also the owners of raw materials and tools, the specialists in ritual and technique, and the organizer of the work. The investigator should be careful to describe such transactions as "payment" only after he has satisfied himself that what is given is intended as an equivalent for the work or the materials, etc. The principles of distribution often rest upon a social rather than an economic basis. Rewards are often conventional in the amount and kind of goods handed over. Study different transactions carefully to see what is the variation, if any, according to the type of work or other service rewarded, the time spent on it, or the type of person who is being rewarded. In many cases it is the *service* of attending at the work or the *status of the worker* rather than the *amount of work performed* that determines the amount of payment. Again, make inquiry to see if the "payment" is not reciprocated in its turn.

Just as the distinction between employer and employed may not emerge clearly in a co-operative activity where people assemble because of their kinship ties, so also there may be no clearly defined system of payment from one party to the other. Reward is often given to the various factors of production by means of a feast. To this the labourers themselves, for instance, may even contribute a portion of the food. In all these different types of payment, do the satisfactions of the workers vary? For instance, would they prefer money rather than food? (Sometimes they prefer food, as when they are given grain for helping with the harvest, because they get more of it than they could buy with the money.) How far in modern conditions are money payments tending to replace payments in kind?

Exchange

In modern civilized communities, distribution is largely effected through money which thereby becomes a means of giving a numerical value (price) to items of wealth. In some primitive communities there are established currencies which fulfil some of the functions of money. These should be described and indications be given of their measure of goods and services (*v.* Measures of Value, p. 197). Elsewhere these conditions do not hold good, but there are nevertheless certain forms of exchange by periodic exchange of goods and/or services. One role of exchange is in facilitating the productive process by allowing the factors of production to receive their reward in goods other than the object produced. Another role of exchange is in supplementing the resources of the community. It is usually of advantage to make a distinction between "internal" and "external" trade. The former includes not only the mechanisms of gift-exchange and barter between members of small, closely organized groups, notably kinship groups and village communities, but also exchanges with a money medium in local markets. The latter deals with more general phenomena of markets, trade with outsiders, and organized commercial expeditions. The investigator should make a careful description of all these processes. The existence of markets, their scale (turnover), supply area, periodicity, and organization should be noted. Consider the extent to which the existence of a market gives a stimulus to production. List the type and quantities of objects exchanged. Note whether they are then consumed, or exchanged further, or used for the production of further goods. It should be emphasized that what is sometimes termed the exchange of " non-utilitarian objects" may be as vitally important as exchanges of food or clothing. In the south-east of New Guinea the ceremonial exchange of highly valued shell ornaments between people of different islands is a most important part of the native economic life, though in fact these objects are rarely worn but are kept as wealth from which prestige is obtained.

The investigator should note whether there is a special class of traders. How far do they serve to bring the different bodies of producers into relationship? What return do they themselves obtain for this function? Note what the other parties who deal with the traders think of them and their role, since these ideas may react upon the exchange situation.

Great care should be exercised in the use of such terms as "money", "currency", "value". In our own economy most goods are regarded as mutually interchangeable in terms of a single exchange medium—money. In this system the value of any commodity is its price. In primitive communities it is frequently found that while there may indeed be some form of circulating exchange medium, it is not used to measure the values of most kinds of goods. There may be in fact different spheres or circuits of exchange, the items in each being capable of being measured against each other, but not of being exchanged against items in the other circuits. For instance, pots and hoes on the one hand, and goats and cattle on the other may be exchanged, but there may never be an exchange of cattle for hoes.

The value of goods is conditioned by many social factors in addition to their scarcity, or the labour required in making them, and these factors should be examined. For example, to what degree are traditional and religious associations important?

Even where European currency circulates in a native community, it may be used for only certain kinds of transactions. Moreover, its value may not be that of the region from which it comes. For example in the Nicobar Islands, where coconuts used to serve as a measure of value, a 2-anna bit (one-eighth rupee) was worth 16 coconuts, but a single rupee was worth only 100 coconuts; the difference was due to the fact that the coins were also used as ornaments, and eight small pieces thus had an additional local value to that of a single large piece of equivalent cash amount.

Therefore wherever money or exchange media of any kind are being discussed, a full description should be given of how such media are actually used. Strictly speaking where there is no true money there can be no "price" and no "value"; but the anthropologist nevertheless can usefully study relative "values". By means of a quantitative study of actual cases of exchange, including bride-wealth transactions, he can draw up an approximate scale of equivalents that will show the native standards of appreciation. He should then, if possible, attempt to analyse the social basis for these evaluations. How far are these capable of fluctuation in changing conditions of scarcity and demand? In situations of culture contact such as are now found in all parts of the world, native values are in constant process of change and modification. How far has such modification of native ideas already taken place? Can the present trend of change be estimated?

Consumption

Under this head we analyse the final uses to which the resources of the community are put. One of these resources is human labour, and it is important to study how labour time is divided between different types of work—in particular as between the production of productive equipment and goods that are to be consumed directly. In analysis of consumption an attempt should be made to estimate the *real* incomes, i.e. total receipt of goods and service less any outgoings in earning this income, of individuals over a period. This should be related to the expenditure of those individuals for the same period. Sampling will probably be necessary here.

This leads on to the question of standards of living. It must be realized that a standard of living must be discussed not only in terms of the needs of the people for food and shelter but also in terms of their wants defined by the values of the community in which they live. The problem is to ascertain how far these wants can be realized through the processes of production, exchange, and distribution which the people follow. The term "standard of living" should be used with caution. Terms such as "high standard of living" or "low standard of living" mean little unless the basis of comparison is also given. African or Asiatic standards, for instance, cannot be easily correlated with those of Europeans, apart from purely physical efficiency. A quantitative analysis of resources over a period may help to determine whether the standard of living is "rising" or "falling". But even this must be subject to qualification. A rapid increase in material resources may for a time disrupt the social and economic structure rather than improve general conditions.

An analysis should be made of the different ways in which wealth is used by different members of the community. It may be found that social distinctions can be correlated with economic distinctions, as expressed by a greater or less command over the available goods, and the different method of using them. In most societies studied examples of "conspicuous consumption" can be seen, i.e. of the use of wealth by individuals or groups in such a way as to call attention to their possession or disposal of it, and so gain social repute or prestige. Lavishness at a feast is a common example of such behaviour. An outstanding example is the *potlatch* of the Indians of British Columbia, a competitive display and transfer of valuables, culminating even in the destruction of highly valued property in order to achieve the maximum social effect. Instances of consumption calculated to raise the social position of the consumer by drawing attention to his immediate actions should be analysed and their relation to the form of the society examined.

Consideration should be given to whether a surplus of goods produced over those consumed is possible. This surplus allows of "saving". How far do individuals or groups deliberately set aside a portion of their income against future contingencies or possible investment?

In brief, the aim of the investigator should be to see the economy of the people he is studying as a dynamic system, as a "going concern", and to explain how present work and savings make for future rewards and available goods and services in terms of the values of the society. He should attempt to estimate how far there are strains and stresses in the economic system, manifest or latent, especially in regard to the uneven distribution of wealth and rewards for services. He should also estimate if possible what are the prospects of the economy at the existing level of population, technology and resources, and in the existing conditions of the social structure; what elements of change there are, and what their effect is likely to be, both quantitative and qualitative, upon the economic system.

CHAPTER VII

RITUAL AND BELIEF

INTRODUCTION

No people so far studied have been found to be without belief in supernatural powers of some kind. However effectively man deals with the problems of life by practical measures, there is always a margin of uncertainty, and often of anxiety, when propitiation of supernatural powers is resorted to in public or private worship or rite, and the accompanying emotions of awe and reverence can often be observed.

Beliefs that presume the existence of spiritual beings are commonly described as religious, while those referring to powers that do not presuppose the necessary existence of such beings are called magical. Often there is no clear separation in the ideas and practices of the simpler people between the two classes of beliefs. Although there is no agreement among anthropologists on the use of the terms "magic" and "religion", so that these words cannot be relied upon as technical terms, a body of behaviour which may be called magico-religious is generally recognized. As a rule every culture has a definite body of beliefs correlated with distinct observances and ritual practices. There may be organized cults with officials such as priests, priest-kings, chiefs with sacred functions, rain-makers, shamans, or sacred servants. These cults may be associated with sacred places or buildings and prescribed ritual, as well as rules of moral conduct. Certain rites in such systems could be regarded as magical if dissociated from the system in which they occur, or they may be given religious significance by some participants and magical value by others. People who accept an orthodox magico-religious system may also practise rites unconnected with the system which may be regarded as purely magical. Such magic may be the special province of experts or magicians, or it may be practised privately by individuals. It may be tolerated by the adherents of the orthodox system or be practised in open or secret opposition to it.

The connection between religion and ethics varies in different cultures; sometimes, judged by European standards, it may appear to be non-existent, but supernatural sanctions for moral values are nearly always to be found. In general, there is a very close connection between the religious system of a people and their social organization.

Though it is impossible to lay down strict definitions it is convenient to distinguish three main aspects of ritual and belief, viz.:

1. Religious beliefs and practices.
2. Magical beliefs and practices.
3. Witchcraft and sorcery.

It must be noted that though these aspects may be mutually exclusive, they are not necessarily so. The relative emphasis distinguishes the type. All three types may be found in the same culture. In order to understand the beliefs that motivate the behaviour of pre-literate peoples it is necessary to study not only the verbal formulations and explanations they give but also their ritual practices and their context. Ceremonies and rites represent the traditional mode of behaviour in which are reflected both implicit and explicit beliefs, ideas, attitudes, and sentiments.

Ritual may be considered in relation to four main categories of activities:

(a) Ritual concerned with the life of man and with extreme emotion.
(b) Ritual concerned with physical phenomena.
(c) Ritual concerned with economic activities.
(d) Ritual concerned with social structure.

These four categories overlap, thus a relationship between the fertility of the earth, animals, and man may be recognized; physical phenomena and economic activities may be interwoven; psychological elements are involved in any occasion that gives rise to ritual and any of these elements may affect social structure. All categories may involve any or all of the three aspects of ritual and belief.

Religious Beliefs and Practices

Ritual, like etiquette, is a formal mode of behaviour recognized as correct, but unlike the latter it implies the belief in the operation of supernatural agencies or forces. Religion is characterized by a belief in, and an emotional attitude towards, the supernatural being or beings, and a formal mode of approach—ritual—towards them. There are usually myths connected with the body of beliefs, and these are reflected in both the form and content of the ritual (*v.* p. 205).

It must be noted that various kinds of beliefs may be held in one society at the same time. Belief in a High God may coexist with belief in numerous other spiritual agencies, as well as with an ancestor cult or totemism.

Beliefs concerning Man

There is usually a belief in an *immaterial aspect of human personality, which may be called the soul*. This belief is often extended to the animal and vegetable worlds, and often to inanimate objects, or to notable examples of these categories (*v.* Totemism, p. 192). The soul may be thought of as an insubstantial replica or wraith, or as a formless emanation, it may be likened to the shadow, breath, or reflection, various parts of the body may be conceived of as the seat of the soul—head, heart, liver, etc. Injury to such parts may be believed to cause sickness (*v.* Medicine, p. 201). Danger or illness may be ascribed to temporary and death to permanent departure of the soul from the body. Dreams may be regarded as an activity of the soul. In some cultures multiple souls with separate functions are believed to exist. (For external souls, *v.* Shrines and Sacred Places, p. 183.) Everyone may be believed to have a soul, or there may be some stage, usually accompanied by ritual, in the life of the individual when the emergence of the soul is recognized. This may be naming, beginning to walk, or initiation. Sometimes the soul is the prerogative of one sex only—usually male—or of members of certain social classes.

It may be believed that some people have the power of transforming themselves into animals, or that the soul may leave the body in an unconscious condition and enter into an animal. In animal forms, such as wolves, leopards, or tigers, they may be extremely dangerous, more so than ordinary wild animals. Some such transformed animals give rise to cults (*v.* below, p. 179), in which case they are not necessarily wholly evil. Sometimes the capacity to take animal shape may be attributed to inborn witchcraft (*v.* below, p. 189).

Soul Substance. In Indonesia and Melanesia a conception translated as "soul substance" has been described. Human beings, animals and plants, and sometimes inanimate bodies, are said to be imbued with this substance; it permeates the human body, thus parts detached from the body, such as hair, nails, sputum and excreta, are supposed to be saturated with it. Portions of the substance can be taken from the body, and must be brought back or the body suffers. The soul substance of animals and plants can be acquired by eating them. Hence it forms the basis of a considerable body of magic. The conception is somewhat similar to the Polynesian *mana*, which seems to be a powerful mystical quality possessed by spirits, ghosts, and certain persons, and may be imparted to inanimate objects.

The practice of head-hunting is widespread in one form or another and the underlying beliefs that find expression in it are closely associated with those leading to ritual cannibalism (*v.* below) and to human sacrifice. These three cultural traits seem to be linked and to some extent interchangeable in several major areas—Indonesia, West Africa, and South America are instances, where each frequently appears in an area adjacent to one characterized by another of these three cults. All three appear to be based on a fertility cult into which the conception of soul substance enters. This view of life as a quasi-tangible, finite substance transferable between human, animal, and, to some extent, vegetable organisms associates it particularly with the head, where its presence is indicated by the pulsation observable in an infant's fontanelle, or in the minute human image to be seen in the pupil of the living eye. It is often closely associated with the tongue, no doubt because that organ acts as the vehicle of speech. Though it may be also located in the heart, liver, or blood, it is regarded as most easily transferable by the taking of the head, which may then be so disposed of as to make the life substance acquired in this way available to the individual who has possessed himself of it, or to the community to which the head-taker belongs. It is possibly some such doctrine as this which makes it so necessary in most communities which practise head-hunting, for a young man to take part in an act of head-taking before he is regarded as eligible for marriage. If he has no surplus life stuff, he is unlikely to be a successful begetter of children. Similarly any misfortune, such as an epidemic disease or a failure of crops, which indicates a shortage of life stuff becomes an occasion for a head-hunt, while in some tribes, such as the Wa of the Shan States, head-hunting is seasonally associated with the sowing of the crops, and is necessary to successful cultivation. Heads taken for magico-religious or prestige purposes are often preserved either *in toto* or in part. The skull, jawbone, or entire head from an enemy captured in war might also be preserved either for magical or for display purposes.

Conceptions similar to those just mentioned may be the motive for the custom found in some tribes of ritually eating certain organs or parts of the body of members who die, i.e. ritual cannibalism. Those who partake in this rite may be any close kin of the dead, or a limited group of kin or affines, or senior people in general.

This custom may be but is not necessarily associated with that of ritual preservation of parts of the body of a dead relative. Thus the skull, jawbone, or other parts of the skeleton are sometimes preserved and form part of the cult objects in ancestor cult (*q.v.*).

The ritual of death, burial, and mourning may be important for the fate of the soul after death (*v.* Death, p. 129). The term "*ghost*" *may be used for the soul of the dead if it is believed that it can be perceived by human beings.*

The After-life. There are very few peoples who do not believe in some kind of life after death. In some cultures individuals must prepare themselves for after-life by learning special formulae and other esoteric knowledge. Usually their welfare after life depends on the ritual carried out during funeral and mourning rites, and sometimes on the possession of objects buried with them or ritually burned (*v.* Death, p. 124). It must be noted that the belief in an after-life does not imply personal immortality, nor is the fate of the dead usually correlated with moral values. As a rule it depends on the ritual actions of living relatives and descendants, which, if not correctly performed, may anger the dead who may send disease, disaster, or death to the living. The intimate link between living and dead may be expressed as a cult of spirits of the dead (*v.* Wives to the Dead, p. 119; Adoption, p. 75), and the future need of the living for descendants to care for them after death may affect social practices. As a rule the dead send benefits as well as misfortune. The dead may visit their descendants and relatives in dreams or in some specified manner and at certain places (*v.* Shrines, p. 130). The reverence, propitiation, and gratitude shown to the dead form an important part of ancestor cult.

Abode of the Dead. This may be located geographically and/or mythologically, and may be correlated with the traditional home of the people. It may be underground, in the sea, or in the sky.

There may be a ruler or lord of the dead, who may be a culture hero, or god. Souls may sometimes be able to visit the abode of the dead during life (*v.* below, Sacred Places, p. 183).

Fate of the Dead. The dead may die again, or merely cease to exist after a span of years, or when they are forgotten or no longer have descendants to perform their ritual. In many cultures the dead are believed to have the same feelings, needs and occupations as they had during life. They may be born again, or reincarnated as animals. This idea is sometimes, but not always, connected with totemism (*v.* below, p. 192). They may "possess" the living (*v.* Shamanism, p. 181). Ritual relating to the soul of man may be observed at the crises of birth, illness, and death.

Reincarnation in human or animal form is not uncommon. A physical or psychological idiosyncrasy may be taken to indicate that a person is a reincarnation of someone deceased, usually a member of his family or lineage. Ceremonies may be performed, e.g., at the time of giving a name, in order to diagnose whose spirit is reincarnated. It may be believed that spirits are restless until they are reincarnated. If reincarnation is in animal form it should be noted whether the whole species is believed to be related to the dead person (*v.* Totemism, p. 192) or only some individual animal. If the latter it should be noted how diagnosis is made, e.g. some dangerous animal, such as a lion, may be believed to be friendly, and recognized as a reincarnation of an ancestral spirit. All behaviour of the descendants to such an animal should be noted (*v.* below, p. 192). Besides ritual concerned with the soul it is important in studying ritual associated with the life of man to observe also the *ritual of everyday life*, as well as the *ritual* of socially distinguished phases in *the life history of the individual* (*rites de passage*). Habitual as well as occasional actions, actions entailing danger or requiring special skill, may be protected by religious or magical rites.

Beliefs concerning Supernatural Beings and Agencies

Spirits. This word should be reserved for personified supernatural agencies not believed to be directly perceived by human senses. Thus, there may be unnamed spirits residing in trees, water, air, etc.; or natural forces such as wind, thunder, etc., may be spoken of as being or having spirits. All-pervading super-human forces, such as Heaven, Earth, the Sun, may be spoken of as spirits. We may speak of the spirits of gods, culture heroes, animals, or ancestors, performing such and such actions.

It is as difficult to define the term "god" as to avoid its use, so it should be used with caution. The term "High God" is often applied to a superior spirit, all-pervading and universal, with or without an organized cult, though it does not necessarily imply monotheism. High Gods are often considered to be too remote for direct worship, and the worship of lesser spiritual agencies existing side by side with the High God may be more prominent in the life of the people than that of the High God.

There may be malevolent spirits, such as the spirit of smallpox and other diseases, or of disease generally.

Besides the more important spirits whose protection is desired and whose anger is feared, there may be various kinds of minor spirits analogous to fairies, sprites, gnomes, imps, and goblins.

A cult is the sum total of organized beliefs and ritual concerned with a specific spirit or spirits, generally associated with particular objects and places, together with the ritual worship and officiants. Sacred persons (priests, divine kings) and sacred animals may themselves be objects of cults.

A cult may or may not be associated with a sacred place and/or objects. It may be spread over a wide area, even beyond tribal boundaries, it may be restricted to a locality, a clan, a lineage, a family, an individual, or a secret society. It may have a hierarchy of priests or officiants, or any member may be able to carry out its rites.

Persons, animals, plants, and objects associated with a cult are sacred. Their sanctity may be inherent, so that it is always necessary to treat them in a ritual manner, or it may be incidental to the performance of ceremonies, the presence of spirits, or during specific seasons or periods.

Individual Cults. Individuals may have spirit helpers or guardian spirits, there may be a recognized method by which the individual gets in touch with such spirits and learns their wishes. In some North American Indian tribes each adult should have a dream in which he gains spiritual experience (*v.* below, p. 187). There may be a ritual association between individuals and certain species of animals. This is often referred to as Personal Totemism because the sentiment is similar to that of totemism, but without the social grouping characteristic of totemism (*v.* p. 192).

Officiants and Ritual Experts. Officiants of a specific cult may be selected in special ways. There may be a special caste of priests, there may be hereditary officials or experts. Individuals may be dedicated from infancy to the priesthood, or some incident may occur which indicates that a certain individual is chosen by the spirits. The choice is often made manifest by some illness —generally of a type that might be described as psychological. There may be training requiring particular aptitude, so that not all who are hereditarily qualified become religious experts.

Divine kings, rain-makers and sacred chiefs may be the objects of a cult, and also the principal officiants in it; their lives, food, sleep, and sexual activities are regulated by ritual, and frequently they are not allowed to die a natural death. The virility of the divine personage and the fertility of the land are intimately associated. The traditional mode of death may be by the hand of one of the possible successors, by a member of a hereditary king-killing family, or by immolation, i.e. being walled up or buried alive sometimes at their own request (*v.* Death, p. 125).

A psychological characteristic often found in a religious expert, and sometimes developed to a very marked degree, is the capacity for auto-hypnosis and dissociation. In many cults, however suitable in other ways the religious expert may be, he is powerless unless he can evoke and become the vehicle of certain spirits. Dances and rhythmic movements accompanied by music usually lead up to the situation when the spirit or spirits take possession of the expert and direct all his actions. The expert when "possessed" may be in a genuine state of dissociation, or the pattern may be so thoroughly institutionalized that he can adopt it almost automatically. In this state he performs actions associated with the spirit or spirits which will be recognized as such by the onlookers, and he may also introduce new elements into the ritual. He may pronounce prophecies, give directions such as ordering sacrifices, etc. He may cure sickness by pronouncements or actions, utter blessings or curses or answer questions put to him by members of the community. *The condition of possession is typical of the medicine-men of the Siberian arctic regions, who are called shamans. The term has been applied generally for spirit possession of priests, and the manifestations have come to be called shamanism.*

In some societies it is not only the experts who become inspired or possessed, but any or all individuals may go through rigorous training, fasting or mutilations, etc., in order to obtain communication with spirits, receive mystic experience, obtain visions, or have sacred dreams. Such experience may be sought alone at special sacred places or may occur apparently spontaneously during some common ritual.

Besides the officiants of specific cults, which may or may not be attached to sacred places, there are often a number of experts who carry out their work by means of spirit assistance. Though medicine-men may often practise magic, it is common for them to appeal to spirits.

Experts for Supernatural Control of Special Activities. Any activity of life—hunting, fishing, agriculture, gardening, building, warfare, etc.—may have experts who are responsible for the ritual deemed necessary for success in that activity. Such experts may appeal to spirits, become possessed by spirits, or use magical ritual.

The social status of experts, whether the position is hereditary, and the manner in which they are paid for their services, should be investigated. It should be noted whether either sex may be experts, or whether certain departments are strictly limited to one sex. It may sometimes be found that while either sex may theoretically become experts, in some departments actually male or female experts predominate, being considered temperamentally more suitable. The technical and social contexts of their activities should be carefully considered at the same time (*v.* Knowledge and Tradition).

All taboos and personal ritual observed by experts should be noted: whether they wear any special dress or insignia of office when officiating or in private life; whether they observe special customs in regard to their person, i.e. wear the hair long, grow beards, etc., and whether they take special precautions with regard to their excreta, nail parings, etc.; whether apart from their official duties they carry on the normal activities of life.

Prophets. Individuals of outstanding personality, either officiants of a particular cult or initiators of new cults, may be called prophets. They may instigate new cults or revivals of old ones. Some cults have apparently swept over wide areas, their followers presenting the characteristics of mass hysteria. It should be noted whether their doctrine is in part influenced by Christianity or other foreign sources, whether they arise or their cult spreads in areas where disintegration of native culture has taken place as a reaction to foreign contact, and whether the prophets preach a return to a golden age and the annihilation of the European.

Where Christianity has been accepted, native schisms should be studied, especially with regard to all regulations that affect social organization, such as marriage and initiation (*v.* p. 108). The incorporation of native beliefs and practices into Christianity or other organized religions should be investigated.

Sacred Places. Natural features may be regarded as sacred, they may be associated with definite cults, and they may be the abode or temporary resting-place, during ritual, of spirits. These may be of diverse character, from High Gods to the spirits of the dead. Such places may be hills or mountain peaks, rocks, groves, single trees, rivers, or special pools in rivers. All myths concerning such places should be collected, and the manner in which their sacredness is observed noted. They may be feared and avoided, visits on ceremonial occasions may be made, accompanied by ritual and offerings; precautions may be taken by people who have to pass such places on non-ritual occasions. Such places or objects may be sacred in themselves or only because of their association with some deity or spirit, i.e. all trees of a certain kind may be sacred or only those individual specimens associated with a spirit.

Shrines. Any structure sacred to a cult where ritual is performed may be called a shrine. It may be a temple, cenotaph, household shrine, a tree, or in some cases merely a pot or potsherd in which some sacred object is kept. Graves are often treated as shrines, and when cattle are sacred the cowshed may become a shrine. Shrines may be empty, or they may form the shelter for objects of cult or the receptacle for ritual objects (*v.* below). They may be the shelters where spirits reside or rest temporarily, or the shelter for an "external soul". Animals, especially cattle, may be the abiding place of spirits and so become shrines. (For Household Shrines, *v.* p. 130.)

There may be private, household, kitchen or garden shrines to the ancestors or other spirits, and these may be elaborate or quite inconspicuous. It should be noted who is responsible for their upkeep, and whether food, drink, or other offerings are put there habitually or on specific occasions.

All ritual with regard to fire (*v.* p. 69) and food should be noted. Does eating together form a ritual or social bond? Are libations made to spirits when food or drink is taken regularly?

Sacred Objects. These may be (*a*) the objects of cult, i.e. objects that are worshipped or to which reverence is paid; (*b*) ritual objects, i.e. objects used in ritual. There is overlapping between the two classes; it may happen that a ritual object, such as a spear, may become so famous as to be in itself an object of cult.

(*a*) Images may be the representation of a spirit and be sacred at all times. Such images may be anthropomorphic (human form), zoomorphic (animal form), or represent some character, i.e. a phallus, or object of phantasy (*v.* Art, p. 310). In many cults and notably in West Africa, carved images and other objects of cult are the temporary abodes of spirits; when the spirits are immanent the objects receive ceremonial treatment and are worshipped, but when not present they receive no ceremonial treatment. Such images and objects are called *fetishes*. This word, which has been misused, should only be used in this sense. The word "idol" should be discarded altogether, because of its derogatory association.

(*b*) Ritual objects used in different cults, vessels, weapons, etc., may be similar to those in ordinary use, or be made with special care and special decoration (*v.* Art). Some may be deliberately constructed so that they would be useless for ordinary purposes, while others are objects designed only for ritual. Any such objects—a spear, an arrow, a bunch of leaves, or some elaborate structure—may become a temporary vehicle or resting-place for the spirit during ritual use. The spirit when evoked by a ceremony may be unable to manifest itself without the presence of such ritual objects.

Masks, effigies, sacred stools, drums, and other objects used in the ceremonies of the installation of kings and chiefs or in initiation ceremonies may be both sacred and of intrinsic value, because of the materials and the art and skill with which they are made.

Investigation should be made as to the custody and care of all ritual objects when not in use.

Forms of Ritual

Every cult has its forms of organized ritual. The three main forms of ritual are:

1. Prayer.
2. Offerings, sacrifice, and feasts.
3. Bodily gestures, and other activities.

Any religious ceremony may combine all three forms.

1. *Prayer implies the existence of some supernatural being.* It is oral invocation, which may be repeated by the officiants and/or the adherents of the cult at a set time and place or on any occasion. Prayer may be accompanied by prostrations, by special postures, by prescribed movements, and by music.

2. *Offerings* to supernatural beings may take the form of first-fruits of crops, or part of catch or hunt, or of valuable objects. Animals or human beings kept specially for a spirit or killed, part or all being offered to the spirit, are *sacrifices*. The killing is a ritual act, and its manner should be investigated. Special implements may be used. It is often believed that the spirit to whom the sacrifice is offered can use or enjoy the offerings, or some immaterial essence of them. Sacrifices may be burnt, and the sacrificial meat may be cooked and prepared, and part, or all, may be put in some special place for the spirit. This may be left to decay, or may be eaten privately by the officiants or given away. Often, however, the whole or some special part of the sacrifice is eaten by the community and is the occasion for a feast. The bones and skin and entrails are often treated ceremonially. If the animal dedicated to a spirit is kept alive, it may be so treated as to become a cult animal. A human being dedicated to a spirit may become a servant or officiant at a shrine. The occasion for all sacrifices should be investigated.

Feasts are very commonly an integral part of ritual, but at the same time have important non-religious aspects (*v.* Economics, p. 169). Feasts which at first sight appear to be entirely secular frequently follow a ritual pattern or have some ritual feature. The partaking of food in some ritual form may be a symbol of union, on the part of those who partake of it, with the supernatural being, or an imbibing of supernatural qualities.

3. *The actions* performed by experts, special groups, or the whole community may form an important part of ritual. These may be a series of rhythmic movements, processions, dances, mimes, or dramatic presentations, sometimes accompanied by music. Such ceremonies though performed on ritual occasions may be of a recreational nature. Fasting, bodily castigations, or complete immobility may be practised ritually (*v.* Shamanism).

In some cults all members may take an active part in its ritual, in others only the officiants. But even when a cult is general to a community, attendance and participation in ritual may be dependent on age and sex. This should be noted; the effect on women and/or children of being admitted to ceremonies, or excluded from them, is likely to be considerable.

The term taboo should be limited to describing a prohibition resting on some magico-religious sanction, infringement of which brings punishment automatically. Taboos are of many kinds, and often have great significance in the social structure. Recognized taboo signs may be used to protect private property from theft. All such signs should be described, and the ritual connected with their manufacture and use investigated. There may be taboos on eating certain food, as in totemism (*v.* p. 172), taboos on the behaviour of certain people in certain situations, i.e. menstruant women (*v.* Sexual Development, p. 107), people in mourning, engaged in special occupations, etc. A ceremony of purification may be necessary to remove a taboo.

In most cultures there are standard methods of obtaining guidance from supernatural agencies. Such guidance may be sought for personal reasons, such as to find an auspicious date for any undertaking, or the method may be part of the normal process of justice or medicine. The ritual employed may be conveniently described as magico-religious, for in some cases a spirit is invoked, in others the ritual is so performed as to give its own answer. Often the answer given refers the matter to spiritual agencies to whom propitiation should be made.

Oracles may be attached to definite cults, consulted by the officiant of the cult only on major issues; or they may be consulted by an expert unattached to a cult. The working of minor oracles or the reading of *omens* may be general knowledge and resorted to by anyone, much in the manner of tossing a coin before making a decision. Sometimes such methods are used in order to decide which oracle or diviner or other expert should be consulted for some grave issue.

Divination is performed by an expert. The two principal methods are (*a*) the diviner may manipulate various objects, e.g. bones, pieces of leather, or (*b*) the diviner may become the vehicle for the spirit, who speaks through him (*v.* Shamanism, p. 181).

Ordeal is a method of invoking the aid of supernatural powers to settle disputes or to test the truth of an accusation (*v.* Law, p. 147). It usually consists of severe and possibly dangerous tests (poison, fire, etc.), to which the parties to the disputes are subjected. Ritual precautions are taken to prevent blame falling on those administering the ordeal. Innocence is proved by coming through the ordeal unscathed. Sometimes the ordeal is administered to an animal in place of the human subject.

An *oath* is an invocation to a supernatural being or agency; punishment for an infringement of an oath falls automatically.

Dreams may be treated as omens, and there may be a recognized method of interpreting them. They may be considered as communications from the dead or from other spirits. There may be cults in which special dreams are sought (*v.* above, Individual Cults). Individuals may be considered responsible for their behaviour in dreams; thus adulterous acts in dreams may be treated as adultery.

Ritual Language. The language used in ritual is frequently different from that in everyday use. It may be archaic and still understood by the officiant and/or community, or it may be so different as to be unintelligible. It should be discovered whether the officiants understand it or merely repeat what they have learnt. All legends or history of ritual language should be investigated, and the actual words, whether intelligible or not, recorded.

Magical Beliefs and Practices

In magic no appeal is made to spirits. The desired end is believed to be achieved directly by the ritual technique itself, i.e. by the use of the appropriate actions, objects, or words. The action, formula, or object is believed to have dynamic power *per se* or to be set in force by the volition of someone who has the necessary knowledge. It is clear that many religious experts practise magic, and much religious ritual contains magical elements. Nevertheless, systems of magic apart from religion may figure in the life of a people, especially in connection with economic activities (*v.* p. 165), Law and Justice (*v.* p. 146), and Medicine (*v.* p. 201).

Magic may be used to fortify the individual or community in any undertaking such as love, war, hunting, gardening, and other economic pursuits. The virtue of magic may be held to lie in the objects used, the oral formula (spell), the person of the magician or magical expert, or in all three.

The magical expert may gain his power by knowledge and training; he may buy it, or inherit it. It may reside in some part of his person. He may have to observe special taboos or a special regime of life on account of his magic.

There are usually magical beliefs with regard to sex. The act itself may be considered dangerous, or dangerous in certain circumstances, or specially beneficial. There may be magical means of protection, or the act may be taboo in certain circumstances, or performed ritually on some occasions (*v.* Ritual Union and Ritual Abstinence, p. 122). A ritual act of incest may be sometimes necessary to gain certain magical powers. There may be a great fear of impotence, which may be due to magic, and protection by magic may be sought.

It is common for the female physiological characteristics of sexuality to be considered dangerous. All taboos on women and girls during menstruation, pregnancy, and childbirth, their termination or continuation at the menopause should be noted, especially with regard to food, cooking, and association with male activities, etc. (*v.* also Sex and Age, p. 66).

Charms are objects invested with magical power; they may be of many different forms and for many purposes; they may act as protection against accidental harm or against evil magic, i.e. charm against the evil eye, etc.; they may give power, luck or health. They may be worn on the person as amulets.

Magical practices characteristically imply a homeopathic principle whereby it is assumed that like produces like. Magic may be believed to act: (*a*) by contagion, e.g. carrying part of an animal to acquire its characteristics, or as a protection against it; or an object similar in some way to the desired object is carried, kept, or used as a medicament; (*b*) by association, e.g. some object may be used, or an image may be made and an imitation of the action desired to happen to a third person be performed, spells recited, and charms prepared to bring about desired results.

There may be a close connection between the personality of an individual and parts of his body. Thus hair, nails, dirt from the skin, or excreta of a person who is sacred may also be regarded as sacred. Or magic may be worked on such parts of a personal enemy to do him injury. Such beliefs may influence habits of sanitation (*v.* Daily Life, p. 99).

It may be believed that esteemed qualities such as courage or power lie in certain organs, e.g. the liver, or in the whole body, and that these qualities may be acquired by eating the organ or body of someone endowed with them. This belief may become the motive for ceremonial cannibalism (*v.* also Soul Substance and Sacrifice).

Witchcraft and Sorcery

There is not always a hard and fast distinction between good and bad (black and white) magic, though there usually is a distinction between socially approved and anti-social magic. *Sorcery and witchcraft are ritual means of working harm against an enemy.* Though usually anti-social they are not necessarily so; sorcery in some cultures is used to detect and punish a criminal. It may be permissible to seek revenge on an evildoer by injuring or killing him or a member of his group by witchcraft. *A sorcerer is a person who wittingly directs injurious magic on others.* He may be able to transform himself into animal shape; he may be able to injure by the power of thought, or may have the "evil eye". Such persons may keep their power secret, or it may be known and they may derive power from it. They may be regarded as public enemies, or may be tolerated and employed to wreak evil on personal enemies.

Witchcraft is distinguished from sorcery in that it is generally believed to be a power, more for evil than for good, lodged in an individual himself or herself (the witch). It may be inborn, hereditary or acquired by undergoing special rites. It may be believed to act on others without the volition of the witch, bringing sickness, death or other misfortune on the victim. In many societies evil witchcraft is regarded as a crime and proved witches are punished by banishment or death. Elsewhere it may be regarded as a sickness which can be cured by magico-religious treatment. Where there is witchcraft, there are usually special cults or ritual experts for detecting and combating witches. These cults may be elaborately organized, with priests and other servants, shrines and peculiar ritual customs, affording protection to its members against witches—able by oracular means to expose a witch. Otherwise the detection of witches may be carried out by ordeals or oracles administered by an expert, or chief.

Ritual in Medicine and Therapy

The belief, so widely held, that ill-health, accident, and death do not occur from natural causes alone, brings much of medicine under the heading of ritual and belief (*v.* Medicine-men, p. 201). The investigator should, therefore, observe in addition to the practical methods of treatment by drugs, massage, manipulation, inoculation, etc., the ritual connected with these practices.

Diagnosis. This is often a ritual to get in touch with spirits causing disease; divination may be used, or the expert may become possessed by spirits. The illness may be regarded as due to (*a*)

G*

wrongdoing on the part of the patient, his relatives or clan; (*b*) evil wishes and magical practices directed towards the patient by some third person; or (*c*) the action of some spirit.

Treatment. All ritual connected with the collection or preparation of the *materia medica* should be observed. The ritual behaviour and any payment between patient, or his intermediary, and expert should be noted, and the psychological attitude of the patient to the expert.

The treatment may consist of reparation to and propitiation of the offended spirit, or some method of appeasement, reparation to or coercion of the third party to make him remove his evil wishes or magic; or such treatment may be combined with other practical therapy. Treatment may be private, but sometimes when spirits are invoked the ceremonial is public (*v.* Shamanism, p. 181) (For recognition of symptoms, and practice of medicine and surgery, *v.* p. 201.)

There may be various kinds of manipulative treatment by which material objects are believed to be extracted from the body of the patient.

Preventive medicine often includes magical measures to ward off evil influences, e.g. evil eye. For this *charms* may be used, or means may be taken not to arouse envy; for instance, a child of rank may be poorly clad, or left dirty; derogatory phrases may be used instead of praise, etc.

Ritual and Beliefs concerned with Physical Phenomena

Magico-religious beliefs concerning the earth, water, fire, the heavenly bodies, etc., with accompanying ritual to ensure normality in their manifestations, are very common. In studying them, it is valuable to make a ritual calendar noting all seasonal rites and festivals, as their most significant features are correlated with the geographical and climatic conditions and the economic activities of the people (*v.* Knowledge and Tradition).

The earth and heaven and any of the heavenly bodies may be personified, and may be of either sex; each "element" may have its particular sphere of influence and corresponding cult and officiants. It should be noted whether the earth itself is sacred. Such ritual acts as taking a sod of earth to a fresh locality when a person or community changes its abode or any other ritual indicating the sacredness of earth, apart from the fertility of the earth, should be recorded. All ceremonies seasonal or occasional to ensure the fertility of the earth should be noted. The fertility of man may be associated with that of the earth. It should be noted whether fire is considered sacred and whether there are any rituals connected with lighting and extinguishing fires both on ceremonial occasions, seasonal festivals, initiation ceremonies and deaths, or in daily life. If so, the method of fire-making on these occasions should be noted (*v.* Fire, Material Culture, p. 240).

All beliefs and ritual concerning eclipses and other astronomical phenomena and also thunder and other storms on land or at sea should be noted, as well as how deaths due to storms are treated.

Ritual and Belief concerned with Economic Activities

Besides technical activities directed towards the use of natural resources, ritual is usually considered necessary for success in agricultural, pastoral, hunting or fishing pursuits, especially with regard to the fertility of the earth and the succession of the seasons.

There may also be ritual for housebuilding, canoe-making, trading operations, both communal and private, raiding parties, etc., and for any of the special crafts. It may be necessary to ascertain auspicious days for beginning work, and all participants may have to carry out taboos or special regulations.

Ritual and Belief directly concerned with Social Structure

It is commonly found that the most important social and political relationships recognized in a society between individuals or groups have a magico-religious aspect. This applies also to leading positions and offices.

The ritual connected with birth, naming, puberty, marriage, and death commonly gives expression to the social structure.

Beliefs that certain modes of behaviour are ordained by spiritual agencies, or that infringement of certain rules brings automatic punishment on the wrongdoers or misfortune on the whole group, are commonly held. Thus the regulation of *incest* (*v.* Marriage, p. 113) is usually based on the belief in supernatural disapproval. The alleged results of incest and any ritual means taken to circumvent them should be noted.

Totemism (v. Clan, p. 90).

The term *totemism is used for a form of social organization and magico-religious practice, of which the central feature is the association of certain groups (usually clans or lineages,* p. 89) *within a tribe with certain classes of animate or inanimate things, the several groups being associated with distinct classes.* In the widest use of the term, we may speak of totemism if: (1) the tribe or group said to be totemic consists of groups (totem-groups) comprising the whole population, and each of these groups has a certain relationship to a class of object (totem), animate or inanimate; (2) the relations between the social groups and the objects are of the same general kind; and (3) a member of one of these totemic groups cannot (except under special circumstances, such as adoption) change his membership. It should be noted that totem relationship implies that every member of the species shares the totemic relationship with every member of the totem group. As a rule members of a totem group may not intermarry.

There are often obligatory rules of behaviour for the members of each totem group, sometimes the prohibition of eating the totem species, sometimes special terms of address, decoration or badges, and a prescribed behaviour to the totemic object. Thus, if the totem is a wild animal, not only will the members of the totem group not hunt it but they may sometimes put food for it in the bush, bury a dead specimen, and may believe that their life and property will not suffer from its depredations. Often there is a sentiment of kinship between the members of a totem group and their totem. Though animals are the most common totems, plants, inanimate objects, and natural forces may be totems. Sometimes a group has more than one totem, or one totem may be linked with several groups.

Beliefs concerning the relation between a group and its totem may be expressed in myths in which the ancestor of the group and the totem species were associated in some miraculous way, and the bond of gratitude has been kept up by the descendants of both lines. Or the totem species may have given birth to a human, or to a human and a totem as twins, or the human ancestor may have become transformed into an individual of the totem species. Such myths may be associated with beliefs in the reincarnation of humans as members of the totem species (*v.* above, p. 179), either of all members of the group or only of notable personages, and with the belief that twins of human and totem species may be born at any time. Occasionally a certain locality may be associated with the myths, and this place may be the centre for totemic ritual.

In some cultures the ceremonial side of totemism is strongly marked, in others it is not important. Members of a totem group may meet to perform ceremonies connected with their totem, often to increase its fertility; or a member of a particular totemic group may perform ceremonies for the whole community, e.g. if there is a totemic group whose totem is a pest—a bird or insect which may destroy the crops—a member of that group must perform the ceremony to protect the crops (*v.* Political Systems, p. 138).

Dual Organizations (*v.* Kinship Structure, p. 92).

The beliefs associated with this form of social organization may be of great importance. Seasonal activities may be regulated by ceremonies based on belief in a dual organization of man, nature, and cosmic processes. Certain rituals will be carried out by one moiety of the organization while others will correspondingly be the duty of the opposite moiety. In a similar way life may be regulated by ceremonies of a *calendrical* order. Myths concerning both systems may exist.

Culture Heroes

The cult of a *hero* or *heroes* may be a central feature of some social group, clan, or tribe. All information, historic or mythological, should be collected. The hero may have led the people to its present habitation, or have introduced new arts or crafts. His spirit may be reincarnated in successive rulers (*v.* Divine Kings, p. 181); he may be Lord of the Dead, or worshipped as a god. Often the alleged number of generations between the hero and his descendants is known. It is common for the birth and death of the hero to be miraculous, as well as some of his exploits.

The ceremony of *investiture of chiefs* may be important. A chief may not have authority until the correct rites are performed. There may be special sacred objects—spears, drums, stools, thrones, insignia, or regalia—whose possession is necessary for a reigning chief. The position and authority of a chief frequently rest on religious sanctions (*v.* Political Systems).

Other forms of ritual which may be essential to social structure are those of peace-making, and ceremonial exchange of feasts (potlatch) between different communities.

Secret Societies

These may be recognized parts of the social structure, or they may form organizations working in opposition to social order. They usually consist of one sex only, male societies being more common, but female societies existing in some localities. A *secret society* is an association, membership of which is usually selective and attained either by purchase or by a ceremony of initiation, or both. It is sometimes public as regards membership, ceremonies, etc., but generally knowledge of its purpose and main proceedings is withheld from non-members. Secret societies may have no clearly explicit purpose, but they may have certain administrative, educational, religious, military, or economic functions. Secret societies may have great influence, owing to the general belief in their magical powers. Members may be believed to take animal form, to practise cannibalism, etc.

Inquiries should be made as to how membership is acquired. Where the secret societies are powerful it may not be possible to get information from members as to the ritual and function, but information from non-members, though possibly unreliable as regards ritual, will be valuable in showing the effect of the institution on social structure. There may be a tendency for new secret societies to arise during periods of political tension; detailed histories of such societies are of great interest.

Ritual concerning Warfare

The conduct of war is almost always accompanied by numerous ritual activities and observances. The leaders often have to observe special taboos before and during hostilities. Protective and aggressive magical actions may be taken on the outbreak of war, medicines prepared by specialists, and offerings given to ancestor spirits or gods. The conclusion of peace generally involves ritual acts as well, which are designed to extinguish blood guilt, or bring together erstwhile enemies in peace, or restore the normal order of life (*v.* above, Head-hunting, and Warfare, p. 143).

Myth (*v.* Knowledge and Tradition, p. 205).

CHAPTER VIII

KNOWLEDGE AND TRADITION

AMONG pre-literate peoples the traditional knowledge which is an essential part of every culture is of necessity handed down orally from generation to generation. Much of this is practical knowledge based on observation and experience, but myth, magic, and supernatural beliefs are also preserved and may be mingled with the empirical knowledge. Specialized knowledge, kept in the hands of experts, must be distinguished from knowledge that is general. It is important to investigate the methods by which knowledge is passed on, e.g. by special instruction (*v.* Initiation); by apprenticeship; by parent to child in the family setting (*v.* Family).

Methods of Recording and Communication

All methods of recording and communication other than speech should be noted (*v.* Gesture, Sign language, Signals, p. 208). It should be discovered what methods are employed to send messages or summon people to a gathering. In sending messages are sticks notched, knots tied on string, marks made on wood, bark or stone? Symbolic objects may be used, the meaning of which may be traditional. Are records of events ever kept in a similar way, carved on stone or wood? Are pictures painted or carved on rocks as a record of events? Is any kind of picture-writing used either as a record or a form of communication? If so are the pictographs conventionalized? Is any form of syllabary or alphabet used? Is there any record of the invention of the system or the source from which it was borrowed, or the person who introduced it?

Is there any method of recording numerals? Is there any kind of printing by stamps, blocks, or type? Are there owner's marks on tools, weapons, or other objects, maker's marks, brands on cattle, or marks cut out on the bark of trees, set up on posts, or carved on rocks as land marks or to mark boundaries (*v.* Taboo Signs, p. 234).

Reckoning and Measurement

Methods of counting and reckoning should be recorded as they are actually practised.

Modes of Reckoning. Many peoples reckon by simply grouping the objects to be counted; and display little interest in accuracy in large numbers. When the question of number is complicated by that of value, as in all forms of exchange, the procedure may be difficult either for Europeans to follow, or for the people themselves to explain. Note therefore all words for numerals, including compound numerals, such as "five-two" for seven; for numerical order ("first", "second", "third", etc.); for arithmetical processes, and also all words and phrases for particular groups of objects, such as "handful", "dozen", "score"; all gestures and other unspoken aids to reckoning, and all written *figures* for numerals and values. Some of these are self-explanatory, such as the outspread fingers for five, the word "man" for twenty. Of large numbers, note whether they are used precisely, or merely to express multitude. Record the people's own explanation of all descriptive numerals and processes, and how such arose. Note in what order the fingers and toes, or other parts of the body, are counted, and whether these are true numerals or merely a tally. If any kind of counters, counting-boards, or tallies are in use, obtain specimens, and learn how to use them. Note what the dimensions are, and whether they are equal, and supplement your observations by setting simple arithmetical problems.

Measures and Weights are means for comparing the size and weight of an object with accepted and accessible standards. Such *measures* and *weights* are often natural objects of uniform size or gravity, or parts of the human body, or customary kinds of effort, such as lifting or pacing, or objects specially made for the purpose, i.e. Ashanti gold weights (*v.* Medium of Exchange, p. 169). Multiples and mutual relations result partly from the system of numeration, partly from experience of the number of times that one standard or unit is measured by another. Sometimes the same seeds, pebbles or other objects are used both as *counters* and as *weights*.

Measures of Distance commonly employed are the breadth or length of finger or thumb or fingernail; or the span either from thumb to little finger-tip or to forefinger-tip; from top of middle finger to elbow (cubit, ell), or over outstretched arms from finger-tip to finger-tip (fathom); or various kinds of pace, the length of a spear, or other implement, or customary lengths of cord or chain, spear-cast, bowshot, or a day's journey.

Measures of Surface are expressed in terms of such units as the area of an ox-hide, mat, or cloak; or of the day's ploughing of a yoke of oxen, or of the land which can be sown with a given measure of seed. Note the shape of the land unit; in particular what the dimensions are, and whether they are equal or different, and what account the people give of each of them. How is land measurement actually performed? How are the areas calculated and boundaries adjusted?

Measures of Capacity include the hollow of the hand, the handful or armful, the load of a man, beast, wagon or boat; the content of an egg, gourd, or other natural object; or of some manufactured object in common use, like a basket. Note all measures used specially for grain, beverages, or any other commodity.

Measures of Weight are often seeds of plants; and for large quantities the customary load of a man or beast or wagon. What measures, if any, are reserved for particular commodities? Obtain examples of apparatus for weighing. Sometimes a standard of weight is formed by a measure of capacity filled with a particular commodity, such as water or grain.

Measures of Time are commonly derived either from some kind of human endurance or (for longer periods) from the observation of sun, moon, or stars. Note what subdivisions of the day are in use, and how they are estimated; is there a standard subdivision irrespective of the seasons, or are the periods longer or shorter according to the seasonal daylight? How is the night-time measured? Are any instruments such as *sundials, sand-glasses*, used to measure time? Are any time-tables or calendars in use? If so, what account do the people give of their origin? Obtain specimens of all such devices, and learn to set and use them (*v.* Seasons, Calendar and Weather, p. 198).

Measures of Value. Familiar examples of currency are natural products, such as salt, fruit, grain, seeds, fish, shells, stones, drugs, timber, and even livestock, and wholly or partly manufactured products such as tea, sugar, spirits, dried fish, worked stones, hides, skins, feathers, domestic utensils, charms and spells, beads, personal ornaments. It should be noted how such articles are measured, whether roughly or precisely, and whether conventionally marked, stamped or moulded, either officially or privately, to show their value. If the currency consists of tokens or articles not domestically usable, they may be either natural products, such as cowries; or manufactured articles, such as mats, imitation spear-heads, hatchets, hoes, knives, lumps or bars of metal of special forms. Note whether there are higher and lower values for money, and how they are arrived at. It may be by size; by multiplication, as in strings of beads or teeth, belts, bundles of feathers, shells; by intrinsic value, due to skill, labour, or difficulty in manufacture or production; or to rarity or antiquity, as with some kinds of beads; or by convention and custom. Sometimes the articles used for money are obtained by trade and imported, e.g. wampum beads, Venetian and Aggry beads, or cowries (*v.* Exchange and Money, p. 169).

General Questions. Note, in all cases, what degree of accuracy is observed by the people themselves; whether artificial standards of length, capacity or weight are in use; whether such standards agree together? What is thought of persons who use inaccurate measures or weights? Whether any authority is responsible for testing measures and weights, or generally maintaining standards. Record all multiples, customary "tables" of comparing larger and smaller measures and weights.

Cosmology, Seasons, Weather and Calendar

All methods of estimating time should be recorded, whether there are named divisions of the day and night, months and seasons, whether the year is divided into lunar months and whether this reckoning is corrected with reference to the solar year. All knowledge and myths concerning the sun and moon should be recorded. The planets and constellations may be recognized and named, and the rising and setting noted and correlated with seasons and prevalent winds. It may be that only experts have such knowledge. The attitude of the people to the normal procession of the seasons, as well as eclipses, storms and catastrophic natural phenomena, should be noted and all ritual and myth connected with such events should be recorded. Weeks, or recurrent periods of definite numbers of days, may fix the occurrence of markets or other activities.

A clear and reasonably precise statement of the annual climatic cycle should be made, taking special note of native nomenclature and theories, especially in so far as it may affect economic and other activities. A weather diary recording the occurrences, intensity and duration of rain, frosts, strong sunshine, fogs, mists, etc., is of great value, and should be accompanied by statements of native views of the normality of particular conditions and the effects they are believed to have on plant growth, etc. An attempt should be made to draw up a comprehensive calendar of seasonal changes in the natural environment and of the sequence and duration of associated human activities. These should be set out together on a calendar diagram, the degree of correlation between them investigated, and inquiry directed to native views concerning relations between them. The native concepts of time and the criteria of time-lapse should be carefully considered in this connection.

It should be noted how personal age is reckoned, the birth of a child may be associated with some event or natural phenomenon and the years counted from then. The selection of particular natural and other phenomena, e.g. moon phases, star risings, horizon position of rising sun, etc., for calendar purposes should be related to the general physical and social conditions. It may be found that periods nominally defined in terms of particular physical conditions are not in practice closely restricted to their duration but correspond to phases of activity related somewhat loosely to those conditions. The mythology associated with such phenomena should be recorded as well as the role of rain-makers and other experts.

Geography and Topography

The investigator having gained knowledge of the geography of the area under observation (*v.* Technique, p. 40) should note all native names for the types of land and geographical features and the history and myths associated with any of them. Any methods used to record geographical positions, routes and landmarks should be noted. All knowledge of the sea, its tides and currents should be noted.

Vegetation

All knowledge regarding natural vegetation and cultivated plants should be observed as well as knowledge of types of soil. The uses made of wild and cultivated plants for food, medicine, industry (timber fibre, textiles, dyes, etc.), art and ritual should be recorded (*v.* Economics, p. 165). Knowledge and myths regarding growth, fertilization and germination with accompanying ritual should be observed (*v.* Ritual and Belief, p. 190).

The character, distribution and seasonal changes of natural and cultivated vegetation and their cultural significance should all be reviewed. This may be based on counts over sample areas.

Information about the density and height of trees and the frequency of useful species is often of great use. Pressed specimens including leaves, flowers and fruits of plants of economic and ritual importance will permit of their later identification and may have significance for comparative studies over wider areas. Notes should be made of the abundance of such plants and the conditions under which they flourish. Besides the existence of flora utilized by the native, the presence or absence of insects and other pests should be noted. All nature knowledge and belief on the subjects should be recorded. The methods adopted by the natives for protection against these and the mythology connected with them should be investigated (*v.* Totemism, p. 192). (For further consideration of some of the points referred to here, see the sections on Economic Life.)

Man and the Animal Kingdom

The origin of man or of some of the human characteristics is usually recorded in myth (*v.* p. 205). It should be noted whether such myths of origin apply to mankind or only to the tribe of the informants. It may be that people recognize physical differences (traditional or real) between themselves and neighbouring tribes; these should be recorded, as well as their observation of the differences between themselves and Europeans and other aliens, together with their explanation of such phenomena.

All traditional and practical knowledge of human anatomy and physiology should be recorded and its correlation to the treatment of disease. Any custom showing a knowledge of anatomy should be recorded, i.e. the removal of foetus from the corpse of a pregnant woman for separate burial; all knowledge concerning blood, semen and excreta, bone-setting, obstetric manipulation, etc., should be recorded.

All knowledge and legends concerning the animal kingdom should be recorded. This will vary with the occupation and economic life of the people. Wild animals may be regarded mainly as objects of chase or as dangerous carnivora or marauding herbivora. Domesticated animals may be a source of wealth, food or prestige, or other economic value; they may be sacred or kept as pets. It should be noted whether the young of wild species are ever brought home and domesticated.

The attitude to animals generally as well as to distinct species or varieties and to individual animals should be investigated. Is there any belief in the kinship of man and animals (*v.* Totem, p. 192)? Are any animals believed to have human characteristics, such as speech or the understanding of human speech, or to have benevolent or malevolent characters? Are any or all animals believed to have souls? What attitudes and beliefs are associated with the killing of different species of animals? (For further investigation of animals, see Livestock, Food, Economics, Ritual and Belief.) Is reproduction in animals believed to differ in any way from that of man? Are there any (common) customs or practices indicating a knowledge of animal physiology and anatomy?

Medicine and Surgery

Information obtained from the local medical officer as to the prevalence of specific diseases will help the lay investigator to make useful observations on native medicine. The common attitude to health and disease should be investigated, the presence or absence of anxiety concerning health, the common practices for relief of pain and treatment or accidents recorded and the accepted theories with regard to disease (*v.* Ritual and Belief, p. 176). A list of native names for all recognized diseases, and the main symptoms and the usual treatment should be made.

Practitioners. On what occasion is a professional healer or expert consulted? Are there specialists among "medicine-men" for specific troubles? The training, status and remuneration of practitioners should be investigated.

Diagnosis. Besides magico-religious diagnosis, all methods of recognizing symptoms should be recorded (*v.* Ritual and Belief—Experts, p. 189).

Treatment. It should be noted that besides the magico-religious practices of experts there may be a considerable body of practical treatment. All forms of bathing, fumigation, disinfection and fomentation, cupping, bleeding, leeching, counter-irritation by blisters, "firing", kneading, massage, etc., should be noted; also the use of a tourniquet, emetics, purgative or abortifacients, and inoculation. Are there any customs indicating the knowledge (even if only partially understood) of contagion, infection, or infestation due to insects or worms? Are endemic or epidemic conditions recognized? Beliefs regarding the origin and causes of diseases should be recorded as well as the history of the outbreak of any disease previously unknown. Some specially dangerous diseases may have cults of their own, with guardian deities who are both malevolent and protective, i.e. smallpox.

The attitude to and treatment of insanity, feeble-mindedness and idiocy must be investigated. Insane people are often believed to be possessed by evil spirits, and the patient may be beaten and ill-treated to drive away the spirit, treatment occasionally being continued until the patient is killed. On the other hand an insane person may be believed to be possessed or inspired by a deity or good spirit and be given complete licence. Cases of mild feeble-mindedness are often treated with indulgence and believed to be favoured by the deities. In all cases of mental derangement the theory concerning the derangement should be ascertained.

The attitude to deformities, albinism and other abnormalities should be investigated. Infants with certain types of congenital abnormality may be destroyed at birth.

The diagnosis and treatment of all disorders of the reproductive system, male and female, should be investigated. Special attention should be paid to sterility (*v.* Demography, p. 62) and impotence, and their alleged causes and prescribed remedies. Are venereal diseases diagnosed and the method of infection recognized? Are any special diseases or misfortunes believed to be due to sexual intercourse or to intercourse with prohibited persons and at prohibited times?

Surgery. What methods on the surgical side are employed to stop bleeding (mechanical, ligature pressure, chemical); to ensure rest of injured part, to maintain position of a fractured bone (splints), to reduce dislocation, to repair wounds; how and what dressings are applied; are ligatures and sutures employed, what are they made of?

Is removal of tissue by operation practised; amputation of digit or limb, or eye; removal of cataract; extraction of teeth; trepanning of the skull; puncture and incision of abscess? What treatment is adopted for mortification, burns, snake-bite, arrow and spear-wounds, ulcers, etc.?

Are the following practised: cauterization by heat, manipulation and massage? Are any means employed to produce general or local anaesthesia?

Midwifery. If possible detailed descriptions of confinements should be recorded. Are any persons specially skilled or trained in midwifery? If so what is their training, status and remuneration? All preparations for the delivery should be recorded, the habitual postures adopted during the several stages of labour, any forms of manual or mechanical assistance rendered during normal, abnormal or protracted labour. Other aids invoked before, during and after confinement, whether medical or magical, should be recorded (*v.* Life Cycle, Birth, p. 105). Records should be made of maternal and infant mortality and of miscarriage (*v.* Demography, p. 62).

Materia Medica. Drugs may be of animal, vegetable or mineral material. Records of native names of all drugs and the purpose for which they are used should be noted and specimens collected for examination, if possible in sufficient quantity to allow chemical analysis.

It is often believed that drugs only become efficient if the correct ritual is carried out in the collection, preparation and administration of them; this should be investigated. It may be that plants or other substances with specific medicinal or poisonous properties are to be found in natural conditions in the area; it should be ascertained whether these are known to the natives, and whether and how they make use of them.

Poisons. Obtain specimens of native drugs, vegetable, animal, or mineral, and information as to their uses. If of vegetable origin, the flowers or inflorescence, fruits, seeds, leaves, etc., should be obtained for identification. Stems bearing ripe seeds should, if possible, be taken from the plant. Other specimens in the original state should be packed in air-tight receptacles. Herbarium specimens should be brushed with a solution of corrosive sublimate (four grains to the ounce of methylated spirit); fleshy fruits may be preserved in formalin. Sufficient material should be collected if it is to be investigated chemically.

Some peoples as a whole, or certain sections, families or individuals, may have great reputation as dangerous poisoners. Investigation should be made to discover whether such reputations rest on facts or on the fear of their magical powers.

Sanitation.—Hygiene. Are there any general rules of health in regard to: (1) preparation of food; (2) collection and storage of water; (3) bodily cleanliness; (4) disposal of refuse and sewage; (5) insect prevention; (6) prevention of epidemic and communicable disease?

Under (1) include any custom or taboos in regard to the preparation of food by menstruating women and as well as practical and magical restrictions on mixing different kinds of food, e.g.

meat and milk. Under (2) remark the use of copper vessels, small amounts of copper in solution will kill the organism of cholera; (3) habits as regards washing, shaving the head and pubes to get rid of lice, etc.; (4) exact notes as to habits in regard to defæcation and urination are important, also in regard to disposal of house refuse. Under (5) mention any special measures adopted against flies, mosquitoes, midges, fleas, bugs, lice, itch, jiggers, ticks, the floor maggot, etc., all associated with disease. Are there any customs which obviously have for their object the sanitary welfare of the community without being so recognized? Concerning (6) are there any methods adopted for combating epidemics and other communicable diseases, segregation of the sick, excommunication of lepers, inoculation for smallpox, etc.? Is disinfection by washing or by fumigation practised? Describe all rites and practices in connection with smallpox, leprosy, etc.

History and Myths

It is important to note what people think of their own past, both when their accounts are mythological and when they are historical or in part historical. Among peoples with a central organization there is often an official recorder, whose duty it is to recite traditional history on state occasions, such records sometimes going back for several hundred years. People with a social organization based on unilateral kinship may remember the names of their ancestors for many generations. Some members of the lineage may be forgotten, but if a number of genealogies are taken and events are traced back to an ancestral hero or person noted in the traditional history some idea of the lapse of time since its happening may be formed. It is usual to allow 25 years as an average for a generation, but with the classificatory system this may prove misleading and careful checking is necessary. Reported events may be purely mythological or they may refer to migrations, conquests, or to the invention or introduction of arts or customs.

Sometimes such events may occur in the histories of neighbouring tribes and migrations and conquests can actually be traced. Sometimes reference may be made to the arrival of some historically known foreigner, or to an eclipse, or some other datable natural phenomenon. Monuments or other memorial structures, cairns, earthwork paintings or sculptures on rocks, designs cut on turf, etc., may be made in honour of some person or some event.

It may be possible to discover the relative date of such a monument by reference to the number of generations of certain living persons to that of those who made the monument or in whose honour it was made. Legends are frequently told of persons who may be historical. The fact that such legends may relate to miraculous events does not necessarily indicate that the heroes of them are not historical.

Myth. Sacred tales or *myths* play a very important part in the social and magico-religious life of peoples, and they come into play when rite, ceremony, or a social or moral rule demands justification, warrant of antiquity, reality, sanctity. *Myths of origin* cannot be dispassionate history since they are made to fulfil a certain sociological function, to glorify a particular group, or to justify an anomalous status. Myth though retrospective is an ever-present, live actuality; it is neither a fictitious story, nor an account of a dead past, but is alive in that its precedent, its law, its moral, still rule the social life of the people. In many types of magic there is a story, myth of magic, which tells, not of magical origins, but when and where that particular magical formula entered the possession of man, how it became the property of a local group, how it passed from one to another. It justifies the sociological claims of the wielder, shapes the ritual, and vouches for the truth of the belief. Myths of love, of death, stories of the loss of immortality, of the passing of the Golden Age, of the banishment from Paradise, myths of incest and sorcery, play with the very elements which enter into the artistic forms of tragedy, of lyric, and of romantic narrative. Myth as a statement of primeval reality which still lives in present-day life, and as a justification by precedent, supplies a retrospective pattern of moral values, sociological order, and magical belief. It is therefore neither a mere narrative, nor a form of science, nor a branch of art or history, nor an explanatory tale. It has the function of strengthening tradition, which it endows with a greater value and prestige by tracing it back to a higher, better, and more supernatural reality of initial events. Myth is therefore an indispensable ingredient of all cultures. The heavenly bodies and fabulous animals may occur in myths. Myth and ritual are closely related: it should be noted if any objects related to heroes, masks, or effigies of heroes, legendary or fabulous animals, are represented in ritual or dramatic performances. Mythological or legendary persons and animals may be represented in art in the designs on ceremonial objects and those for ordinary use (*v*. Art, p. 310).

Stories, Sayings and Songs

The repetition of stories, proverbs and traditional sayings may be an integral element of culture, corresponding among illiterate peoples to literature among the literate. It should be noted how much the art is cultivated, whether stories are told to children as much for their education as amusement, whether children are taught to recite stories or sayings with the traditional intonation.

In some cultures there are professional story-tellers. It should be noted when and where they practise their art, what is their status, and the remuneration they receive. It should be noted whether stories with special themes are considered appropriate for definite occasions, seasonal festivities, weddings, etc. Stories may be related at any informal gathering by anyone gifted to do so. It should be noted whether the general content and themes vary when such gatherings are of persons of both sexes and of men and of women only. Are there special stories told to children? Are such stories mainly amusing, frightening, or admonitory? Do the people themselves classify stories as histories, i.e. stories of heroes, famous deeds and wonders, legendary, fabulous (stories of animals, etc.), topical, moral stories and stories for amusement? Which types are considered most popular? Is there private property in songs?

Proverbs, traditional sayings and riddles may be educational or a form of intellectual recreation.

Songs are best recorded on a recording machine, but texts should be recorded in writing (*v*. Music, p. 331). The place of song in the life of the people should be studied. Song may be an accompaniment or an integral part of ritual of all or specific kinds; of communal or solitary work, of war, of dancing, of recreation and of courting. It should be noted whether each class of activity has its appropriate themes, metres and tunes. Some poetry may be intoned rather than sung. Are there professional singers? What is their status, on what occasions do they perform? What types of songs do they sing and what is their remuneration? Do they compose songs? Are there solo songs, burdens, refrains, and choruses, set rules of rhythm and metre? Are there songs specially sung by men, women or children?

General Directions. All narratives should be written, if possible, in the native language, and with the native idioms exactly rendered, and should be read over to the narrator for correction. The name, age, sex, residence and occupation of the narrator should be recorded, and where and from whom the story was learned. Variants and fragments of stories should also be recorded; but they should be kept separate, not pieced together, or used to "correct" other versions. There is sometimes a rhythm in tales, and there are often long "runs" which are repeated; in transcriptions, these are tedious and apt to be omitted, but they should nevertheless be indicated as they occur, otherwise the literary structure of the tale is destroyed. Tales should be listened for when people are talking at leisure among themselves. The importance of collecting the unwritten literature of a people is being increasingly recognized. An attempt should be made not merely to collect odd songs, stories or histories, but to get some general picture of the whole literary field. What literary genres or types occur, and in what proportion? Is there epic poetry or lyric? Is poetry always sung? Is there a body of history with some kind of literary form? Are songs and stories usually traditional or impromptu? Or are they a mixture of the two? All stories should be studied in their social context and for their literary form as well as the content.

CHAPTER IX

LANGUAGE

Gesture, Sign-language and Signals

Gesture is largely used as a means of expression, especially where verbal communication is imperfect, as between persons of different speech. Note all significant gestures of the head, face, hands, arms or body, narrating the actual incidents in which they were observed; for example, gestures expressing assent, denial, approval, disapproval, invitation, repulsion, anger, grief, shame, entreaty, prayer, command, blessing, cursing. How do people point to objects, near and distant? How do they beckon? How do they count, or indicate numbers? How do they express the comparative and the superlative? Do the gestures of men and women differ? What gestures are resented as insulting? Gesture is also a means of artistic and emotional expression, closely connected with dancing. Note any descriptive, symbolical or aesthetic use of gesture or posture; any performances in dumb show (*v.* Dancing and Drama, pp. 331–3). Note what signs are depicted in pictorial art.

Sign-language. The use of gesture is sometimes developed into a sign-language, more or less systematic, in which objects and ideas are represented by postures and movements of the hands, arms, head, and body, imitating the most conspicuous outlines of an object or the most striking features of an action. These signs may be abbreviated or conventionalized in use, to make them more intelligible at a distance.

Careful inquiry should be made to discover the existence of such a system; where it is not in general use it is sometimes used by one sex only, preserved in the memory of old people, used in ritual, or guarded as a secret art. In each system note what sort of ideas can be expressed by signs. Give a full vocabulary, if possible; if not, typical examples.

Can connected narratives or speeches be expressed? Are there signs for "beginning" and "ending" a message, to indicate a question, or otherwise to qualify or explain any sign or group of signs? Are the signs used as an accompaniment to spoken language? Are they used in hunting, war, bargaining? Is the sign-language deliberately taught? Is it used between people who speak the same language, or chiefly for communication with foreigners? In what tribes, and over what area, would these particular signs be understood? Are new signs invented and adopted? Are mimes performed for amusement? Is there any artificial system, like the finger-alphabet taught to European deaf-mutes, or the Morse code?

Record, if possible, connected speeches or narratives and phrases which will show the order in which the signs are performed, and reveal the syntax of the gesture-language; also all variations of manner or speed, with the people's own account of their significance. Native legends and explanations of the signs should be recorded.

Significant gestures and sign-language may be recorded in simple sketches. Time and trouble will be saved by preparing a number of front- and side-view outlines of a man on cards or slips. Dotted lines may be used to indicate movements to place the hand and arm in position to begin the sign, and not forming part of it. Short dashes to indicate the course of a rapid movement; longer dashes to indicate a less rapid movement; broken lines to indicate slow movement. > indicates the beginning of a movement. × represents the end of a movement. ⊙ indicates the point in the gesture line at which the position of the hand or finger is changed. The use of the kinematograph affords a far more reliable record than the above.

Signalling. Native methods of conveying information at a distance should be described, noting whether the interpretation depends (*a*) on a code of signals generally known, (*b*) on individual judgment, or (*c*) on a code prearranged for the occasion. The methods named below are well known, but are summarized here to suggest actions and occasions of which the significance should be ascertained. No doubt others will be discovered.

Signals and messages are sent (1) by gestures of the arms or body; (2) by waving something held in the hand, e.g. a flag, blanket, green branch, torch, or by throwing dust in the air; (3) by running or riding to and fro, or in circles, at various paces; (4) by signal-fires or smoke—the number and position of the fires and the amount of smoke may be significant; separate puffs of smoke are produced by raising and lowering a hide or a wet blanket over the fire. Fires are generally "attention-signals", meant to invite a visit, announce the return of friends, or give warning of a marauding party; they may be distinguished from casual fires by their sudden disappearance; (5) by flashing a mirror, shooting fire-arrows, or striking sparks from flint and steel; (6) by "blazing" trees, tying leaves and grass, arranging stones, sticking branches in the ground. Such devices are used both to convey information and to mark boundaries or warn off trespassers; ignorance of their meaning has sometimes led to disaster; (7) by marking pictures or conventional signs on the ground, on rocks, on the bark of trees, or on pieces of hide. These marks may show where a party has gone, what it has done, and whether intentions are friendly or warlike (*v.* Writing, p. 195); (8) by shouting in a particular manner, whistling, blowing horns or trumpets; (9) by sounding drums or gongs, or beating a tree, a canoe, or a shield. In some places a drum- (or gong-) language is known to be highly developed, and all details should be carefully noted. The best method of recording sound signals is afforded by the recording machine (*v.* p. 329).

Spoken Language

It goes without saying that some knowledge of the language spoken by the people he is investigating is valuable for an anthropologist. Fortunately for him linguistic research in the lesser-known languages of the world is steadily if slowly progressing. Whereas, therefore, the anthropologist of fifty years ago had frequently to tackle an unknown or undocumented language, his successor to-day may be more fortunate. Two different kinds of linguistic situation may be envisaged and different methods of dealing with them are suggested. He may, in the first place, find that a good deal is already known about the language of his special area, that it possesses adequate grammars and dictionaries, and that expert teaching of it can be obtained. On the other hand, he may be confronted with a language about which little or nothing is known, and which he must therefore investigate for himself to the extent which will enable him to use it as a tool in his anthropological inquiry. In both cases it is necessary to expose the fallacy, still too widely prevalent, that the best way to deal with a language is simply to "pick it up" on the spot from the native speakers. There are rare individuals with exceptional linguistic gifts for whom this method, or lack of method, may be adequate. But the vast majority of people need a technique whereby the "picking up" process may be rendered effective. Working with a language informant, for instance, is not a simple matter. It needs to be tackled methodically.

(*a*) In the first case where a good deal is known about the language, the prospective field worker should seek such expert teaching as is available both in pronunciation and the structure of the language. In rare cases it may be possible to find a trained native speaker, and such cases will, it is hoped, become more frequent. But at present it will usually be necessary to obtain expert non-native teaching in conjunction with a speaker of the language in question. Reliance on books alone is hopelessly inadequate for the learning of a spoken language. Gramophone records, however, may be helpful if used in conjunction with a teacher. It is sometimes maintained that a language should be learned from the beginning on the spot, where it can be used in its social context. If expert teaching is available on the spot, well and good, but it rarely is, and even if it were it would still be necessary for the student to spend many hours in the classroom as well as in the market-place. It is, therefore, possible to exaggerate the drawbacks of preliminary language study in academic conditions, particularly as a trained teacher will be fully alive to the contextual aspect of language.

(*b*) *Undocumented or inadequately documented languages.* The number of languages about which little or nothing is known is still considerable, and there are very many more for which documentation is inadequate so that the anthropologist must conduct a certain amount of research into the language himself. The fact that there is little help in the way of grammars or dictionaries makes it all the more important that the student should have a really good course of linguistic training. Moreover, study of the construction and phonology of a better-documented cognate language will often be helpful. Since language is for the anthropologist only a tool, he needs to economize time and energy in learning it. The phonetic, grammatical and semantic aspects of language will here be considered separately. The student must in the first place get a good grounding in practical phonetics, and in the principles of phonology.

Phonology

(1) Linguistic information, if it is to be of any scientific value, must rest upon a foundation of accurate phonetic observation,

Our familiarity with the sounds of our mother tongue, and the ease with which we perform the necessary physical processes, are apt to persuade us that we are naturally fitted to observe and record the sounds and processes of other languages. There is probably no branch of science that more urgently requires a total suppression of the personal factor than this branch of linguistic observation, and it should be the investigator's first task to become acquainted in detail with the nature of the processes that bring about those acoustic elements in his mother tongue upon which intelligibility rests. He will have to rely, in the field, upon his ear, which hitherto has been trained to pick up the significant acoustic features of one language only, viz., his mother tongue. If the investigator is to provide reliable linguistic information, he should endeavour to obtain adequate ear training.

For anthropological purposes, theoretical phonetics is of little direct use: the investigator should make his knowledge of the subject as practical as possible, and his practical training should be carried out under the guidance of an experienced phonetician.

(2) From the acoustic point of view, a language may be said to be made up of

    (*a*) sounds,
    (*b*) attributes of sounds:
        (i) relative length of sounds;
        (ii) relative pitch of sounds;
        (iii) relative loudness or prominence of sounds.

(3) Accurate and reliable information must be given on all these points; the isolated sounds must be described in modern scientific terminology. Loosely descriptive adjectives such as "hard" and "soft", "thick" and "thin", "clear" and "dark" should not be used to describe speech sounds; all sounds should, as far as possible, be described by reference to the physical movements or articulations by which they are produced. The number of technical terms required for this purpose is not large and can be learned from any good modern text-book of phonetics.

Observations must also be made of the modifications that speech sounds undergo in connected speech, and some care is needed to determine whether any new word is a significant variation or an accidental assimilation of a word already known. For example: if English were being recorded for the first time, it would be unnecessary to record *thish* as a significant variant of *this*, even though it is almost invariably used in ordinary speech in phrases like *thish shop, thish shoe,* etc. (and often in *thish year*), because by testing both pronunciations the investigator would discover that *this* in such positions would invariably be equally intelligible.

Any noticeable variations in the length of sounds, vowel and consonant, must be marked, and the most careful observation made upon the intonation of the speaker. Our familiarity with the intonation scheme of our mother tongue is apt to persuade us that there is no such thing, or that our habitual tonal modifications are "natural". But our English tonal system is in reality a complicated affair, deeply involved with our syntax, our sentence structure, and our unique stress accent.

The tones of a language are as essential to intelligibility as its sounds, and in some cases, intelligibility may be more dependent upon tone, i.e. pitch, than upon the sounds.

The investigator should make it his business during the period of his training to get some practice in taking down a tone language, i.e. a language in which the meaning of words and the grammatical structure varies with variations of tone, such as Chinese, Yoruba, etc. His aim in the field should be to present a written version of the language that shall adequately represent all those features in the pronunciation of the language that are essential to intelligibility, and to present the version in the way that will be of the most practical assistance to those whose business it is to observe, to classify or to learn to speak the language. Such a version is an accurate *phonetic transcription* of the language, which is something entirely different from an orthographic version of the language.

The production of a phonetic transcription requires of the observer two qualities:

(*a*) the ability to hear, distinguish and classify a very much larger variety of sounds and tones than he is accustomed to;

(*b*) sufficient familiarity with an adequate and practical system of notation to represent on paper what he hears.

(4) *Ear Training.* Regular training may be obtained at most modern Departments of Phonetics. Such training should comprise some practice in taking down the actual pronunciation of native speakers of African and Asiatic languages. He should also be able to write down accurately the sounds of his mother tongue. He should be familiar with the principles upon which the modern system of classifying vowel and consonant sounds rests, and he

should be able to recognize and describe in detail the principal speech sounds of the most important languages in the world.

He should further be able to make these sounds, so that when the occasion arises, he can verify his conclusions by actual pronunciation. It is only by repeatedly pronouncing a language to a native speaker, testing the actual spoken word in every detail of sound, length and pitch, that the observer can arrive at any reliable conclusion.

(5) *Notation.* It will be realized that some sort of alphabet is necessary, and the wise observer will resist the temptation to invent an alphabet of his own, for such an alphabet will be of little use to anyone but himself. There are alphabets in abundance in existence, and the student need not add to their number. Recent research into the various desirable qualities to be postulated of a phonetic alphabet will be found summed up in the *Memorandum on the Practical Orthography of African Languages.*

Too much must not be expected of a phonetic alphabet: it must be remembered that no system of visual symbols will ever adequately or accurately represent a system of sounds, for the very elementary reason that sound and sight are irreconcilable. A phonetic alphabet is at best but an approximation, but it has the merit of being shorn of the arbitrary conventions that exist, in most historical alphabets, between sound and symbol. Relationship between sound and symbol in all languages rests upon conventions that vary from language to language, and, indeed, from word to word, in any given language. A phonetic alphabet, if it is scientifically designed, must reduce these conventional relationships to a minimum, and be capable of representing as adequately as is necessary for practical and scientific purposes, the pronunciation of any and every language. The alphabet of the International Phonetic Association, which is the joint production of a number of linguistic scholars, and is based upon many years of practical experience in taking down unwritten languages, is quite adequate for any scientific purpose.

(6) In actual practice, the observer should repeat to the native speaker every word, phrase or sentence taken down, imitating as closely as possible the native pronunciation. If the observer's pronunciation is rejected, then he should make an effort to see precisely what feature in his pronunciation is at fault. If his pronunciation is accepted, it must not be assumed that it is correct, for most native speakers are over-tolerant of a foreigner's

attempt to pronounce their language. Deliberate mispronunciations must then be tried, to see whether the native speaker is willing to accept any approximate version, and these deliberate mispronunciations must be carefully designed to test for:
 (i) sounds,
 (ii) length of sounds or syllables,
 (iii) pitch of sounds or syllables.

Not until the observer is satisfied that he is capable of producing at will a pronunciation that the native speaker will accept every time, should he attempt to draw any conclusions as to the essential acoustic features of the language under observation.

It is recognized that any worker other than a linguistic specialist is unlikely to use or need the whole of the alphabet of the International Phonetic Association. A somewhat simplified form, *Practical Orthography of African Languages and Cultures*, 1929, 2nd edition, price 1s. should be obtained and studied.

The use of a large number of letters of special type, though necessary in linguistic work, is inconvenient in anthropological work and can often be avoided. It may be mentioned that only a certain number of the sounds that require special symbols will be necessary in each language. If the worker prefers to use diacritical marks he should be careful always to make a note equating it with the correct phonetic alphabet, but the psychologist, the pedagogue, and the typefounder all condemn diacritical marks. In any case he should make it clear what system of transliteration he has adopted.

General Phonetics by Noel Armfield (Heffer and Sons, Cambridge) will be found useful, and the International Phonetic Association has numerous publications. (Address: Secretary, University College, London, W.C.1). Westermann and Ward's *Practical Phonetics for Students of African Languages* (Oxford University Press) is the most useful for workers in the African field; Ward's *Practical Suggestions for Learning an African Language in the Field* (International African Institute) and Nida's *Learning a Foreign Language*, are also very helpful.

Grammar

An investigator equipped with a practical training in phonetics can at once begin noting down words and short sentences. Simple questions and answers, such as "What is this?" "It is an egg", and so on, greetings, commands, are good material to begin on.

Wherever possible, they should be noted in the context in which they occur, and it is, of course, only thus that they can be obtained unless a speaker can be found who has at least a modicum of the languages both of the investigator and of the people he is studying. Nowadays it is unlikely that some kind of an interpreter will not be forthcoming. If found, he must be used with great caution and as much counter-checking as is practicable. If he is not obtainable, the anthropologist must proceed by pointing to objects and getting their names, taking care to point in the manner that is socially acceptable in the particular community. Shooting out the lips, for instance, rather than pointing with a forefinger, may be the local method. The learner can also perform and watch actions and get their verbal equivalents, and listen to greetings and commands. Longer conversational sentences and finally continuous texts can then be taken down. In each case information about the particular speaker should be recorded, name, age, sex, place of origin, dialect if any, degree of sophistication or education, knowledge of English, etc. The linguistic data of each informant should be checked against that of others. But the data of each should be kept separate so that each individual collection forms a consistent whole. This is necessary for any good linguistic work and particularly so for any language which has different dialects. The investigator will soon have material from which he can make at least that minimum study of the structure of the language which is essential for his purpose. For this he should have had some training in the principles of grammatical analysis.

Much work on language structure in the past has been vitiated by the fact that an untrained individual tends to impart into the new language the grammatical categories of his mother tongue. If, for instance, he is used to adverbs as in English, he will be tempted to find something in the new language which he will call an adverb whether in fact a particular grammatical form of the kind exists or not. If his mother tongue makes great use of inflexion, he may neglect the importance of word order in determining form in the language he is learning. He also tends to incline towards a notional rather than a grammatical basis for his forms with resulting confusion in his analysis. The attempt to equate natural sex and grammatical genders is an obvious example. It is tempting to list the word "man" as masculine, and the word for "woman" as feminine. In French this is correct as their grammatical behaviour is different. They each take, for instance, a different article: but in a language where they are in no way distinguished grammatically, there is no justification for considering them as being of different genders. The main principle of grammatical analysis is that grammatical forms should be identified and defined by their grammatical behaviour, the word "behaviour" being used to include both morphology and syntax. Grammatical behaviour and not "meaning" or notion is the criterion of grammatical form. One form may have many "meanings" or possibilities of usage, and from the semantic angle this is all-important. But at the grammatical level this is not the relevant consideration and will only hinder a true appreciation of the structure of the language.

To grasp the principle that grammatical form is identified by grammatical behaviour is one thing, but to apply it is another. It is here that the student needs a course of training with an expert, and he cannot be too strongly urged to obtain it.

Any sound system of orthography must be based not only on the phonology but on the grammatical structure of a language. As the investigator proceeds with his grammatical analysis he will therefore gradually modify the predominantly phonetic spelling (notation) with which he started.

Semantics

Here again the student is strongly advised to get training. It is not possible to do more than indicate the complexity and richness of this aspect of language study. But an elementary appreciation of what it implies should at least put the investigator on his guard. A language must be studied in its social context if it is to be understood, but again certain broad principles can be grasped beforehand. To imagine, for instance, that a word in one language which is given an equivalent in another is therefore "understood" is to be deceived. This happens all too often. Translate the word "father" by "nna" in a West African language and the next step is to imagine that one knows what "nna" means. Its "meaning", however, can only be discovered by a study of the sociology of the people who use the word (*v.* Genealogical Method). Still more is it misleading to find a verbal equivalent for an abstract word. "He is a good man." The equivalent in another language means nothing apart from an understanding of the scale of ethical values in the culture of which the new language is part. Words are signs or labels. Their significance is covered rather than revealed by the label. To know the "meaning" of a word or phrase is to know the particular context—the context of situation—in which it is used. Malinowski has stressed the fact that speech is basically a form of action rather than a conveyor of ideas, and that it enters dynamically into the situation in which it occurs.[1] In this connection it is important to distinguish between the emotion and the referential functions of language and to take due account of both.[2] Words may be used to convey information as in the statement "Two and two make four", but they are more often used to evoke a state of mind or feeling even though they may take the form of a statement. If we say "He is as brave as a lion" we are often appealing for sympathetic admiration rather than conveying information. It is, therefore, desirable to discover whether in the language into which we are translating a lion arouses the idea of courage or some quite different and perhaps contrary notion.

It is unnecessary to labour the extent of the subtlety and delicacy of which a language is capable and which only prolonged and deep study of it will reveal. The use of metaphor, for instance, may perhaps be suggested as of great interest and significance.

Careful attention should be paid to the gestures, facial expression and signs of emotion which accompany or at times replace speech. It is important to note, for instance, not only that a beckoning gesture is made, but the exact action whereof it consists. The European action of beckoning with an upturned finger is an insult to many Africans, who in the same circumstances use down-turned hand.

---

[1] See Malinowski, *Coral Gardens and Their Magic* (Allen and Unwin, London, 1935). Also Ogden and Richards, *The Meaning of Meaning*, supplement I. (by Malinowski): "The Problem of Meaning in Primitive Languages" (Kegan Paul, Trench, Trubner, 1923).
[2] See Stebbing, *Introduction to Modern Logic*, ch. 2 (Methuen and Co., London, 1930).

PART III

MATERIAL CULTURE

Introduction

The study of all aspects of the material side of a people's life is of great interest and importance not only from the intrinsic interests of the artefacts themselves, but for sources of invention, and questions of diffusion. Further, artefacts and techniques have great importance by virtue of their relation to the whole social organization and to religious and other ceremonial practices. The anthropologist will find that an intelligent interest in artefacts and technological processes is an excellent way of gaining the confidence of a people. The ritual aspect of material culture is of great importance. Many ritual practices have become so interwoven with technical procedure that they may be regarded as an integral part of a given technique. As such they may inhibit the development of new methods, or at least guide them along predetermined channels. Any account, therefore, of a particular occupation of a people must give full weight to both ritual and ceremonial acts on one hand, and to technical processes on the other. The attitudes of the people towards their own techniques should be noted and recorded, and their effect on the practical operations investigated.

It sometimes happens that particular plants are cultivated with much more ritual than others and some without any at all; an endeavour should be made to discover which are so treated and why. Whereas there may be relatively little or no ritual in the building or first occupation of ordinary dwelling-houses, that of the "men's houses" and other ceremonial structures may have a rich and profoundly significant ritual. The same distinction applies to the construction of ordinary small canoes and the large kinds used for fishing and war, and so with other industries. Further the ritual may be confined to an implement or to its manufacture, but not to the things made by the implement.

For the sake of convenience the various arts and crafts and appliances are considered in separate sections of this book, where the problems peculiar to each technique are considered separately, but there are certain fundamentals which apply to the study of techniques and their relation to economics which should be borne in mind when studying any of the subjects set forth in the following pages. All techniques of manufacture require expenditure of time and labour and it is important, especially from the point of view of primitive economics, to estimate this quantitatively in relation to the output of the finished article, and to the demand for it. The sources of motive power open to primitive man are his own muscular energy, animal power, and in some of the more advanced communities the forces of wind and water. The anthropologist should therefore apply himself to the following questions: How much can be produced by the technique in question? What is the degree of complication and skill introduced in the process? Are there any particular motions or attendant conditions which make it either psychologically or physically more suitable to men or women? To what extent is it a specialist's activity, i.e. what proportion of the population practise it, and what proportion of their time is devoted to it?

Implements and appliances should be considered in relation to the above questions. For example, hoes vary considerably in size, shape, angle of blade and so on; the shape of the hoe and the nature of the stroke employed should be correlated. Does a man use a heavy hoe requiring a long swing while a woman uses a light one requiring short quick strokes? If so, do men and women do different work? Questions of this kind can be applied *mutatis mutandis* to other appliances. The full value of a study will be missed if attention is concentrated on a single instrument, or a single phase of manufacture. The best results will be obtained by a study of all phases of a particular industry or occupation. For example, a finished textile represents not only the product of a loom but of a chain of integrated processes starting with the harvesting of the cotton or the shearing of the sheep or obtaining the necessary fibre, and continuing through the various stages of cleaning, carding, spinning, weaving and dyeing. Each phase should be studied in relation to the demands of the succeeding phase, or where traded goods take the place of the earlier processes, the means of exchange should be considered.

STATUS OF THE CRAFTSMAN

In the simplest societies every individual can and does perform all the secular activities characteristic of his community, except so far as there may be artificial restraints on his so doing, and as a rule he can make any of the implements he requires; but, even so, there are usually to be found men who are more expert than others. Thus, in some societies this expertness gives rise to special craftsmen; or certain men may in their spare time specialize in certain crafts. Where special craftsmen are found it is necessary to make inquiries as to their status, and how they are paid. Iron-workers in Africa may form a despised or pariah class, or they may have special privileges and position. Canoe-builders, or other carpenters, may have a high social status. These craftsmen should be considered from economic, social, magical, religious, legendary and mythological points of view.

Personal Care and Decoration

It is characteristic of man habitually to reinforce and enhance his natural qualities by artificial means. He takes pains to maintain his person in proper condition; he enhances, disguises, or alters peculiarities of personal appearance, such as features, complexion, and growth of hair; he decorates himself with personal ornaments, and covers himself with clothing, to protect him or to distinguish him from his fellows.

Cleanliness

Personal cleanliness, the standard of which varies greatly, seems to be a tribal characteristic, and is often quite independent of the accessibility or scarcity of water. Note should be taken both of performance and omissions. Do the people wash themselves habitually, at regular times, before or after eating, before or after formal acts, such as visits or on ritual occasions? How do they clean their teeth? Have they regular bathing or washing places, a distinct apparatus for washing? Do they use *soap*, or any substitute for soap? How is this prepared? How do they dispose of the water in which they have washed? Record all characteristic acts and gestures. Do they rub their body with oil or fat, or smear it with fine ashes? Is the purpose of such customs to keep the skin in good condition, as a protection from insects, or from external parasites?

Most, if not all, peoples have a recognizable odour, more or less well marked, and some peoples are more sensitive than others to such odours. It is well to remember that to some peoples the natural odour of Europeans is quite as offensive as that of their odour to European travellers. It is impossible to describe odours. Are the people aware of their own odour or that of others? Is the odour of clothing or implements used as an indication of ownership?

Perfumes. The natural odour is often disguised by the use of perfume, either intentionally or as a consequence of the odour of unguents, paint, or other cosmetic. Perfumes and the wearing of scented substances are frequently employed to enhance sexual attraction. Note should be taken of the substance which is used; its name, preparation, purpose, and mode of application. Samples should be obtained if possible, and should be preserved, like all other native products, in air-tight bottles or cases, if there is any fear of evaporation or decay, or if of an organic nature, in formalin or spirit.

Sanitation. Other sanitary observances should be noted with the same care as personal ablutions. Are the houses and their surroundings, streets and other public places kept clean? If so, by whom? What is done with dirt and rubbish? What provision, if any, is made for the disposal of excretions? Are there public or private areas set apart for daily use, or are there public or private latrines? Are the children taught cleanly habits, or left to themselves?

Personal Appearance

Quite apart from cleanliness, are the people careful of their personal appearance? In particular, do they modify in any way their natural appearance? If so, what reason, if any, do they themselves give for what they do?

Hair-dressing. This is one of the commonest modes of enhancing or altering the personal appearance, either for convenience or for ornament, or to distinguish individuals, sexes, or social ranks, such as married or unmarried chiefs, officials, and private persons. In every case the practice should be carefully observed and described before questions are asked as to the reason for it. In what way is the hair dressed? Hair is sometimes felted into more or less permanent shapes of traditional style. For this purpose hair shaved off the head of another individual is sometimes worked in to increase the bulk. Is the additional hair bought, or given by relatives, or inherited? Is the hair allowed to grow to its full length? If it is cut, what becomes of the cut ends? Is the hair partially shaved off, such as above the forehead, or so as to leave patterned tufts of hair, or a tonsure, etc.? Is it totally shaved, or cut very short? Are these practices associated with age, status (such as marriage), mourning, or religion? Who cuts or shaves the hair, and what implements are used? Is the hair plaited, twisted, or curled into ringlets? If so, by what means is it kept in place? What ornaments are worn in the hair, such as combs, pins, beads, feathers, or flowers? Are *cosmetics* used, such as oil, grease, or clay? Is the hair intentionally or unintentionally coloured? If the former, what methods are employed, and why? More or less bleaching of the hair may be the result of substances used to kill lice; this should be inquired into. To what extent does swimming in the sea affect the hair? Are *wigs* in use? Are beards or moustaches worn? If the face-hair is not worn, is its absence due to natural causes, or to depilation, or shaving? What reasons are assigned for removing face-hair? How is depilation effected? When a beard or moustache is worn, is the hair allowed to grow naturally, or is it cut into shape? Does depilation or shaving of any of the body-hair occur, and, if so, why?

Nails. Most peoples trim their finger-nails regularly, and some the toe-nails also. Others perforate or stain them. If stain is used, note the colour and the material used for staining and obtain a sample of it. What becomes of the nail-parings? If the nails are allowed to grow, what reason is given?

Teeth (*v.* p. 227).

For the magico-religious attitude towards parts of the person separated from it, such as hair, nails, teeth, excreta, and even clothes (*v.* Ritual and Belief, p. 188).

Deformations

It is a widespread practice to mould the body in accordance with some preconceived ideal of beauty, or for ritual purposes or, by some surgical operation, to provide for the attachment of some ornament which may alter the normal shape of the organ to which it is attached. It should be ascertained whether deformations are performed on both sexes, or on one sex only; whether they are common to the whole population, or peculiar to a certain rank or professional class.

Particular pains should be taken to discover the native reason for such practices; whether the alteration of form so produced is accidental or designed; whether an alteration which is now regarded as the end to be attained was once only an accidental consequence of wearing a particular article of clothing or ornament; whether a given form of deformation exaggerates a natural tendency characteristic of the tribe or people under observation.

For questions as to the mode of operation, the status of the operator, and the significance of the practices, see General Questions at the end of this section, p. 231.

Cranial Deformation. The head of the new-born infant, being largely membranous, may easily be moulded into a form often markedly different from the normal. Head deformation may be divided under two main headings:

(1) *Accidental.* For example, a decided flattening of the back of the head may be caused if the child lies in a cradle with a hard base, or is fastened by swaddling bands to a board; or the head may become asymmetrical from being laid to rest continually on the same side or nursed on the same arm; or, again, the shape may be modified unintentionally by the use of a tight form of head-dress, producing a flattening of the forehead, or a conical form of occiput, or both.

(2) *Intentional.* Deformation is effected (*a*) by simply moulding the head of the child with the hand; (*b*) by the use of bandages; (*c*) by the use of one or more boards or pads applied to the head by means of bandages; (*d*) by the use of a particular form of cradle, or of some appliance fixed to the cradle.

The first of these methods, which is often practised, has, probably, but small effect upon the ultimate shape of the skull; but the other three give surprising results. Note the nature of the appliance, the method of affixing it, the length of time for which it is affixed, the ages at which it is applied, and abandoned, and the form which is ultimately given to the skull. Photographs, drawings, and measurements should be secured.

Well-known types of cranial deformation are (1) flattening of the forehead, with or without increased projection of the occiput; (2) flattening of the occiput with an increase in height or width of the head; (3) flattening both of forehead and occiput, combined with marked increase in breadth; (4) conical or cylindrical lengthening of the occiput or of the crown.

Note the effects, if any, produced by cranial deformation on the health or on the mental or moral qualities of the subject. What is the native opinion of this practice?

Facial Deformations affect chiefly the nose, ears, cheeks, lips and tongue.

The *nose* is most commonly deformed by (1) simple moulding of the infant's nose by the mother or nurse, either with a view to depress it or to render it more prominent; (2) piercing of the nasal *septum*; (3) piercing of one or both *alae*; (4) making a hole in the *tip*.

If ornaments are worn in or on the nose, note their material, form, dimensions, and weight, together with the mode of inserting and securing them by plugs, hooks, rings, wires, or thread, and any possible effects they may have in modifying the shape of the organ. When are ornaments first inserted? Once inserted, are they fixed or easily removable? Are they of graduated sizes as the wearer grows older, or are they inserted at once of the full size? Is there any method of mending the nose if it should be torn?

The *ears* are often pierced with one or more holes, either in the lower part or *lobe*, or in the shell-shaped *helix* or upper part. Ornaments may be worn in the holes or attached to the ear by means of them. Sometimes the size or weight of the ornaments is gradually increased until the aperture or apertures are remarkably distended. The following points should be noted: (1) The number and position of the punctures; (2) the material, form, dimensions, and weight of ornaments worn in, or attached to, the ear, the method by which they are attached, and their effect, if any, on the ear; (3) methods of increasing the size of the apertures; (4) methods of repairing the ear, if by chance it becomes torn; (5) whether the ear is ever torn intentionally; and, if so, for what motive.

The *cheeks, lips and tongue* may be pierced, and ornaments are sometimes worn in the holes. State distinctly whether observations refer to the upper or lower lip, or both.

Dental Deformations. There are four methods of treating the teeth:

(1) *Colouring.* Note the colour used and, if possible, obtain a sample. How is it prepared and applied, and to which of the teeth? Distinguish accidental colouring or incrustation due to betel-chewing or diet.

(2) *Inlay* and *Incrustation.* Note the materials used, the method of preparing them, the manner of drilling the tooth, and the method of affixing the inlay, incrustation, or plating.

(3) *Chipping* and *Filing.* Note the implements and methods. Secure photographs and describe the result in diagrammatic form, showing which of the teeth are treated.

(4) *Extraction.* Note the number and position of the teeth extracted, the method of extraction, the instruments employed, and the subsequent treatment of the patient.

The operation of chipping, filing, or extraction of teeth may be performed on individuals or upon groups of young people of either sex. Inquiries should be made as to whether the persons operated upon at one time form any kind of special group. All ceremonies associated with the operations should be recorded.

Deformations of the Limbs and Trunk. The limbs are sometimes modified by one or more of the principal methods: (1) by constricting the limb by means of tight bands or heavy ornaments; (2) by compressing the extremities by means of bandages (as the Chinese deformed the feet of women); note also deformation due to rings or toe-rings, ill-fitting shoes, and sandals which are gripped by or between the toes; (3) by amputating one or more fingers or toes; (4) by allowing the nails to grow untrimmed or by piercing them for the attachment of ornaments (*v.* Nails, p. 225). Careful note should be taken of *callosities, wrinkles,* or other unintentional deformities due to habitual postures or occupations, either of a whole people or of the members of particular classes of industries.

The *breasts* of women are sometimes modified so as to produce an elongation of the nipple or of the whole breast, usually by simple manipulation or by bandages, or both. Note the purpose of such practices.

The *waist* of men or women is sometimes compressed by a simple band or by some more elaborate apparatus corresponding to a corset.

The *hands.* Sometimes objects are continually held in the hands or rolled between the fingers with a view to giving a definite shape to the hands.

For other modifications of the limbs and trunk, *v.* Tattoo (p. 229), and Cicatrization (p. 230).

Deformation of the Genital Organs. The penis may be deformed by *incision* of the upper layer of the foreskin (prepuce), in which case the latter may form an irregular projecting mass beneath the glans; or the whole foreskin may be removed, *circumcision.* The glans may be perforated for the insertion of a foreign body, or foreign bodies (stones or gems) may be inserted under the skin of the penis. Part of the urethra may be slit open, *sub-incision,* the so-called *mica* operation. There may be total or unilateral *cas-*

tration. In the female, the labia may be lengthened by manipulation, or by other means. Some contrivance may be attached in such a manner as to prevent copulation, *infibulation*, or the desired result may be caused by *excision* of almost the whole vulva including the mons, scar tissue being formed so that only a small posterior orifice is left, or the labia may be scarified or sewn up and caused to adhere together; in both cases an incision is necessary before the consummation of marriage. The commonest operation in the female is the excision of the clitoris. Describe any varieties which occur in any of the above, the mode of performing the operation, the implements employed and the social status of the operator.

It is important to ascertain ideas concerning the origin and purpose of every deformation of this kind, and to record all ceremonies preceding, attending and following such operations, as well as the age at which they are performed. If possible, observations should be made on the effect of such practices on the frequency of conception.

Decoration of the Skin

The modes of decorating the surface of the person may be classified under: Painting, Staining, Tattoo, and Cicatrization. The older writers often confused cicatrization with tattooing, and some modern writers are equally remiss.

Painting includes all modes of decoration by means of coloured substances (powder, mud, lime, etc.), or definite pigments laid upon the surface of the body which do not permanently discolour the skin itself.

Staining affects the colour of the skin itself more or less permanently. Local differences of pigmentation, due to sunburn or individual complexion, are sometimes so marked as to be mistaken for artificial stain. The nails of the fingers and toes may be stained.

Tattoo, properly so called, consists of pricking pigment into the skin, leaving a smooth even surface. *Moko* consists of grooves in the skin produced by a small chisel or adze-like instrument, pigment being rubbed into the grooves. *Kakina* is produced by a needle and thread covered with soot being drawn through under the skin; by the Eskimo, the point of the needle is rubbed with a mixture of the juice of *Fucus* and soot, or gunpowder. Note the nature, method of preparation and mode of application of the pigment or pigments. Is any stamp employed to print the design on the skin?

If a certain colour prevails, is this due to actual preference for this colour, or to the fact that this particular pigment is more easily obtained? Are the pigments imported and, if so, whence? The technique and instruments employed should be described.

Cicatrization is effected by scratching, cutting, piercing or burning the skin, or otherwise causing the formation of scars. The wounds may be allowed to heal naturally, forming plain scars, which are usually slightly depressed; or they may be aggravated so as to form deep gashes; or raised scars called *keloids* may be produced by continued irritation of the incisions, by the insertion of foreign matter, and the resultant proliferation of regenerative tissue. Is any form of ornamental scar (*moxa*) produced by the application of some caustic material; and, if so, are the resultant scars permanent or temporary? Describe the method of producing keloids and the irritants employed. Are foreign bodies inserted in the wounds or introduced under the skin? Is pigment introduced into the wounds? What is the colour of the scars as compared with the surrounding skin? Care must be taken to distinguish between intentional scarification, and the scars which are incidental to cutting or burning as a sign of mourning, and those resulting from cuts, punctures, burns, etc., which are made to relieve pain; in all cases the implements employed should be described.

There is so much in common between these four groups of skin decoration that the following remarks may apply to any or all of them. A person may be both tattooed and scarified, and on occasions may be painted as well. As a general rule, scarification is confined to those with very dark skins, and tattooing is employed on skins which show up the designs. Designs may be borrowed from other peoples, and those who formerly only scarified have been known to adopt tattooing as well; where these are both employed inquiries should be made on the subject, and the original source of all designs noted.

Is the operation completed in one stage or in several? In the latter case what portions of the body are ornamented at the several stages, and what are the periods of life at which the various stages are performed? Is the ornament confined to any particular part or parts of the body? Is it the same in both sexes? Is the design intended to emphasize and enhance the natural features and contours of the body, or give it an independent scheme of composition, pictorial or conventional? Are any designs or details of designs peculiar to any portions of the body? Is the design sketched out on the body beforehand, or is any pattern used as a guide to the operator? Does the design seem to be in imitation of ordinary personal ornaments?

Collect native names, with their meanings, of all designs or details of designs (*v.* Decorative Art, p. 311). Note whether any marks on an individual signify his rank, caste, trade, clan, tribe, or religion, also whether certain classes, such as criminals or prostitutes, are habitually tattooed, etc. Very little is known about tribal marks in the true sense of the term. In the case of clan marks, do they represent a totem, and is this indicated realistically, or conventionally, also what reasons are assigned? Is the design, or any portion of it, hereditary or peculiar to the individual? May designs be exchanged or temporarily adopted by those not entitled to them permanently? May a design indicating some special renown be also made on the man's wife or child? Note the significance in all cases where group or individual designs occur on objects and artefacts (*v.* Recognition Marks, p. 233). Endeavour to discover the meaning of all forms of personal decoration, and whether their significance is social (to stimulate physical development, to indicate puberty, completion of marriage contract, number of children, death, mourning, membership of a secret society, as a sign of prowess or homicide, etc.); or magico-religious (before undergoing a dangerous enterprise, as a love charm, for amuletic, prophylactic or other "magical" reasons, to indicate kinship with a totem, dedication to, or assimilation with, a deity, to benefit the individual in the after-life), or whether the decoration is simply ornamental (to enhance or preserve bodily charm, or to prolong youth by repeating the operation). The observer will very frequently be informed that decoration is purely for this last purpose when there is every reason to believe that it has a deeper meaning. Where colour is employed, note whether it is symbolic (*cf.* Art, p. 314).

It is most important that sketches of designs and patterns should be obtained, since these are of far greater value and less troublesome than a description. It is usually very difficult to photograph tattooing even when isochromatic plates and a screen are employed; it is usually necessary to paint the pattern accurately with a black pigment.

General Questions

The following queries may be taken to refer to any of the foregoing methods of deformation, mutilation, or decoration of the skin:

Function, Origin and History of the Practice in Native Belief. What is the alleged reason for the practice, and what is its alleged origin? Give any legends connected with it. Is it mainly ornamental, or social, or religious in character? Does the operation form any part of an initiation ceremony? Is there any trace of the operation having been customary in former times and having now become obsolete? Is any reason given for its disuse? Is the operation still performed in full, partially, ineffectively, or in pantomime? Is any such practice spoken of, but not found in general use, or supposed to be practised secretly? What is thought of those who practise it? Can any of these practices be traced to foreign influence?

The Operation. Describe the process as minutely as possible, giving native names for all processes, instruments, and other details (*v.* Surgery, p. 202). If some portion of the body is cut off, state what is done with the severed part; and if the operation involves the shedding of blood, whether there is any special custom with regard to the blood shed.

The Patient. Is the operation peculiar to either sex, or to any state, rank, caste, trade, clan, tribe, or cult? Is it performed at any particular age or within any limits of age? Is the operation ever performed on the dead? Does the patient undergo any preliminary treatment (seclusion, fasting, purification, etc.)? Does the patient undergo the operation alone or in common with others? If the latter, note on what system the patients are grouped, whether any particular bond is supposed to exist between them subsequently, and its nature. Is any special dress or ornament worn by the patient? Is any name given at the time or subsequently? What is the subsequent treatment of the patient (exemption from labour, special food, etc., independent of any regulations or taboos connected with a puberty ceremony)? Is anything done to heal the wound? If the idea of ceremonial is attached to the patient at any period, state when, its nature, and the purificatory ceremonies. Do those who have submitted to the operation have any privileges which are denied to others? Does an individual who has not been so treated suffer any disabilities, real or imaginary, as regards marriage, hunting or fishing, particular foods, burial, or in the future life?

The Operator. Must the individual operator belong to any

particular age, sex, caste, trade, clan, tribe, or religion? Must he stand in any definite relation to the patient? Does he enter into any relationship with persons on whom he performs the operation? Must he undergo any preliminary ceremonies or observe certain taboos? Has he any special place or status, either at the time of the ceremony or generally? Is he regarded at any time as ceremonially impure, and, if so, what purificatory ceremony is undergone? Does he receive payment for his work and from whom? Describe all ritual connected with any type of deformation.

Personal Ornaments

Many peoples, irrespective of the amount or absence of clothing, wear ornaments which may be attached to any portion of the person. With equal opportunities for procuring ornaments there may be considerable variation in the wearing of them. The distinctive ornaments of neighbouring peoples should be observed. Are the people ready or unwilling to adopt new types of ornaments? Ornaments may be stated to be solely for decorative purposes, but even so very many have, or have had, a "magical" and protective significance, and thus are *amulets*; *talismans* are worn for good luck. Some represent the reserve of wealth and are readily traded. Others are signs of social status, and when worn only by officials in virtue of their office they may be termed *insignia*, but this term may also be employed for definite ornaments which can be worn only by those who have performed some noteworthy feat, such as homicide, or by those who have proved themselves successful warriors or hunters. Ornaments are so frequently associated with social or religious events that a description of them is appropriately made when describing those events. Special note should be taken of the *wearing of flowers* or *feathers* and other objects of natural beauty, in which case the kind and colour should be recorded. *Beads* of all kinds deserve particular attention. In every case note any reasons which may be given for wearing ornaments. Have any ornaments a known individual history?

Recognition Marks—Tribal and Personal

Various forms of deformation, more especially marks on the skin, such as cicatrices, tattooing, or painting, may be made in order to identify the individual, and thus they have a social meaning. In some cases totem representations, conventions, or symbols indicate the clan of the individual, and may thus serve as a warning against incest. In other cases the marks may have a tribal significance. In many cases the natives affirm that the marks are solely for decorative purposes, but this statement should be critically examined. Distinguish the marks indicating different kinds of social grouping: the clan, caste, occupation, tribe, and so forth. Give the part of the body on which the designs are placed, and make photographs, careful drawings or tracings.

Many people mark or deform their bodies in various ways, and wear peculiar and distinctive clothes, ornaments, or badges, and it is important to discriminate between those decorations which are purely individual and those which have tribal or social significance. Warriors on the war-path are usually distinctively coloured, or have weapons, head-dresses, and other ornaments, which differ from those in use on other occasions; investigate the reasons for such distinctions. Sometimes the chief intention is to distinguish the opposite sides. If so, are there individual variations which nevertheless keep to one common type? Collect all the variations you can, and endeavour, with native aid, to trace out the sequence of them.

If animals, trees or other natural objects are marked, do such marks indicate personal or collective property? Endeavour to trace the signification of all marks.

Marks or patterns are sometimes made on weapons or domestic utensils to denote the owner or maker. Thus a distinctive mark on an arrow will indicate its owner, and establish a claim for the killing of game, etc. Pottery-makers may imprint their individual signs or patterns on their ware. Care must be taken not to confuse such marks with others that have a different significance as, for example, tally-marks on weapons which denote the number of persons killed with that weapon; in one sense these are owner's marks, but they do not serve to distinguish one owner from another.

CLOTHING

Do one or both sexes go entirely nude? The term clothing covers the range from a mere band to a multiplicity of articles of dress; thus *clothing* may consist of one or more *garments*, or of particular kinds of *dress* or *costume*. The quality and shape of each garment or article may be determined by secondary uses, as for instance to mark the social, political, or religious rank of the wearer, or the particular act or occupation in which the person is presumed to be engaged while wearing it.

There is considerable variation in the idea of what part of the person it is indecent to expose. Some people when given clothes do not think of wearing them to cover the genital region, but they soon adopt the custom if derided by a neighbouring clothed people, or when they are likely to meet strangers, though they may revert to their own habits when by themselves. Women in some societies, for instance, consider it more necessary to cover the mouth in public than the genital regions.

Every article of dress, wherever it may be worn, should be considered from the point of view of whether it is worn from a sense of shame, or for decency, for "magical" protection, for relogous motives, as a protection against the weather, or to attract attention and enhance sexual charm.

Some peoples merely cover the genitals; this may be done by a shell or gourd, or a perineal band, a small piece of fabric, a fringe, or merely a leaf or leaves. In the case of the men, does the article cover only the free end or the whole of the penis, or all the organs? Does the removal of the covering of either sex, however small or restricted it may be, give rise to a sense of shame?

The significance of wearing *belts* should be inquired into, and also the tails or other objects which may be inserted into or suspended from belts. Are belts, armlets, etc., worn for themselves or to secure other articles of clothing, or to sustain objects, or to act as receptacles for small objects?

Forehead-bands, head-coverings, and *veils* (face and mouth-coverings). It should be ascertained whether they have ritual significance or are worn for convenience either by men or by women. What special precautions are taken for protection from the weather? Are grass or fibre coats worn for rain? Are sunshades used? Is anything worn to protect the eyes? What head-dresses are worn in social events, war, and in rites? Is the use of a head-covering obligatory, or forbidden to either sex, to any class, age, or social status, and what reasons are given?

Are *gloves* or *shoes* or *sandals* worn? Of what materials are they made? How are they fastened? Are they worn habitually or only on occasions?

So varied are the details of perineal bands, fringes, belts, loincloths, petticoats, pants, and other garments and costumes that no enumeration of them can be made here. Examples of all characteristic garments and complete costumes should be secured, with precise details of their use and the rank and occupation of the wearers. Photographs, drawings, paper-patterns and native-made models, and descriptions are necessary when the garments themselves cannot be obtained. Particular attention should be paid to all decorations of garments, and their significance noted, i.e. whether purely aesthetic, "magical", religious, recording exploits, illustrating traditions, etc.

It should be noted at what age either sex adopts certain articles of clothing, and whether the first wearing of clothes is accompanied by any ritual. Do any garments incidentally or intentionally deform any portion of the body?

Manufacture. Information should be obtained of materials used for making clothes, whether native or imported, what parts of animals or plants are used (*cf.* Skin-dressing, Bark-cloth, Weaving, pp. 284–94), how they are worn or draped, the method of fastening a garment or the parts of one, whether a perforated needle is used or the thread passed through a hole, the kind of thread, and the process of sewing. Are the clothes made by each individual or family, or are there recognized makers of clothes?

Use and Significance. Is there anything which corresponds to "fashion" in clothing? All variations in clothing should be noticed, according to the season of the year, for festivals, for indoor or outdoor wear, for everyday occupations, for protection against weather. What clothing is worn at night? Are any costumes or articles of dress peculiar to social groups, trades, localities, to men who have killed an enemy or some formidable animal? (*cf.* Personal Ornaments, p. 233), or for specific occupations? The garments worn by warriors, medicine men, priests, and all persons in authority should be described, and it should be noted how these distinctions are enforced. Does any clothing serve the purpose of armour? What importance is attached to the *regalia* or other *insignia*? Are any portions of clothing removed on saluting or visiting a superior or a sacred place? Are clothes destroyed after sickness? What becomes of the clothes after the death of the owner?

HABITATIONS

This section includes all forms of shelter and dwelling-place, whether temporary or permanent, natural (such as caves, rock-shelters and trees), or more or less completely adapted or constructed by man. Besides the purpose of each structure, the planning and arrangement of its parts, the materials used, and the mode of construction, note should be made of the situations chosen for habitations, and the customary grouping of the separate huts or houses in a composite group. If possible, a number of such groupings should be recorded and planned, so as to secure ample material for comparative study.

Caves, Trees, and other Natural Shelters are used either habitually or as places of refuge. If no longer is use, are there traditions of their former use or place-names indicating such use? Have they been improved by excavations or by adding artificial walls or screens? Are the walls or roofs of caves ornamented in any way by painting, engraving, etc.?

Marsh-dwellings. Are attempts made to solidify and raise above water-level the floors of hut-structures? Describe the methods employed and the habitations erected over the raised floors. Is there evidence of the floor-level having been successively raised from time to time by fresh additions of material?

Lake-dwellings and other Structures on Piles surrounded by Water. If these are in use, ascertain the reason. Describe the mode of construction, especially the manner of driving the piles; the modes of access; the allotment of spaces or huts to individuals or social groups; the provision for maintenance and repair, for defence, for protection against fire, for disposal of rubbish. Are pile-dwellings found inland? In river channels? Or at a distance from water? Are there traditions of place-names indicating the former use of pile-dwellings or lake-dwellings? Lake-dwellings of early date have often left instructive traces. All small islands near the shores of lakes or rivers should be examined to see whether they have been inhabited; whether piles of wood have been driven in round the margin; and whether there has been communication with the shore by means of a causeway. Preserve all relics found on or beneath the surface, and make a plan of the locality. Native fishermen, especially if they use drag-nets, often bring up manufactured objects from the bottom in the neighbourhood of ancient lake-dwellings.

Portable Shelters and Tents. Describe the structure and materials. Is any provision made for hearths, smoke holes, and other ventilation, windows and doors, floor covering, subdivision into rooms? Are special places assigned to the owners or other members of the household, and to guests? Who erect the tents and to whom do they belong?

Permanent Houses and Huts. Are the houses, though ordinarily permanent, so constructed that they can be easily taken down and re-erected? What is their form and plan, and is this arrangement uniformly observed? How are they constructed? Is the work of construction individual or communal, or assigned to a special class or guild? Are they specially orientated? Are there few or many separate rooms? If so, for what purpose? Or for what class of persons? Are they structurally separate? Grouped round a central space? Or in series along one or more passages or corridors? Or are there passage-rooms communicating with one another directly, without passages? Where the ground forms the floor is it prepared in any way by beating, or cementing or tessellating with fragments of pottery? Are there floors or raised platforms in some parts of the house? If so, for what purposes? Is any part of the house below ground? Is there an upper story or more than one? How is it approached? By inclined planes, ladders, or staircases? If so, are they external or internal, portable or fixed? Describe their construction in detail. Note the material and the design of the roof; the way in which timberwork is fastened, whether by lashing, by pegs, or nails, or by mortice and tenon; and all devices for hinges, latches, locks, and bolts. What provision is made in the houses for light, warming, cooking, eating, sleeping? Is there a special fireplace and how does the smoke escape? Is part of the house set apart for the men or for the women in general, or for particular persons, or for guests, or for domestic animals, or for stores? What is the average life of houses, and are they destroyed on being abandoned? Who erect the houses and to whom do they belong?

Furniture. Are there fixture-fittings forming part of the house, such as clothes-pegs, spear-racks, shelves and cupboards, hammock-slings, sleeping-berths? In addition, what movable furniture is customary? What purpose does it serve? Is it regarded as part of the house or as the private property of members of the household?

Ceremonies in Construction. Are any ceremonies observed in selecting a site for a house? Or in preparing material, felling trees, digging foundations, laying the first stone or post, or during the building, or on completing the structure, or on entering into occupation? Is any part of the house sacred, or specially reserved?

Or used to contain sacred objects, heirlooms, or trophies? Or believed to be inhabited by any kind of spirits? If so, what is believed about their nature and habits?

Appropriation and Use of Houses. Does each family have a house? Or how is the population distributed among the habitations of a community? For example, are there communal dwellings or club-houses for the unmarried men, or for other sections of the population? How is space assigned in them? Are there guest-houses? If so, to whom do they belong? To whom does a house belong? (*v.* Property, p. 149). What happens to a house when its owner is dead? Is the owner buried in or under his house? Are different houses, or different kinds of house, or different parts of the house occupied at different seasons of the year? Are there special houses for unmarried youths and girls? If so, describe them in detail and note any secondary purpose which they may serve. Are there separate houses for the storage of food, or for the protection of cattle? Describe all decorative work in house construction. (*v.* Sanitation, p. 224.)

Arrangement of Camps, Villages, and Towns. Besides the descriptions and plans of individual houses and other buildings, it is important to record the general outlay of the whole settlement; the situations chosen for villages and towns, e.g. valleys, plateaux, slopes, forest, grass, etc.; the reason for the choice, water supply, drainage, flies, access and defence; the customary grouping of the buildings; the position, construction, and purpose of public or ceremonial buildings, such as temples, official residences, court-houses, assembly-rooms, and club-houses; the provision for streets, open spaces, markets, wells, and other water-supply, surface-drainage, refuse-heaps, sanitation. Describe all defensive structures, such as stockades, walls, trenches, gates, and outlying forts (*v.* Warfare, p. 142). A sketch-map should be made, if possible, of every village and town which is visited, since much may be learned by comparative study of a large enough number of examples. The name of the owner or headman of the house should be ascertained, and also the clan or group to which he belongs. The social grouping of houses is of importance, and it should be noted if in a hamlet, village, or town they are arranged or grouped in such a manner as to indicate social distinctions. How long does a village remain in one place? When is it moved, and how far, and why?

Fire

The use of fire is general in human societies, and statements as to tribes which either use no fire, or do not know how to make it, should be viewed with suspicion. The important aspects of the subject are fire-making, fire-keeping, the uses of fire, and its place in social and religious ceremony.

Fire-making. Fire is usually obtained by wood-friction or by percussion. The friction methods comprise: sawing with one piece of wood across the grain of another (*fire-saw*), the "saw" being either a stiff piece of wood or bamboo, or a flexible strip, e.g. of rattan, as in the *sawing-thong* method; ploughing along the grain of one piece of wood with another piece (*fire-plough*, or stick-and-groove); or drilling with a cylinder of wood into a pit in the passive piece or "hearth" (*fire-drill*). Bow-drills and *pump-drills* may also be used for making fire (*v.* p. 258). In addition to observations of the type and construction of the apparatus, notes should be made of the kinds of wood and their relative hardness. Note the method of working, whether two persons co-operate, the usual length of time that elapses before a spark is obtained, and any precautions or devices that are determined by environmental conditions.

In fire-making by percussion, e.g. with *flint and steel*, note the materials and the form in which they are used, the details of any kind of holder or container, and also the use of any mechanical method of producing the concussion.

The *fire-piston* depends for its working on the development of heat by the sudden compression of air in a confined space, and it has a restricted distribution; note the details of construction and the constituent materials (wood, horn, metal).

In all fire-making observations record should be made of the kind of *tinder* that is used (if any) and its source or method of preparation. Cases of partial or complete displacement of native by foreign methods should be recorded. Ascertain whether any methods, obsolete for ordinary use, are employed for ritual purposes.

Fire-keeping. How is fire maintained? Can it be carried about? What device is used? What is the customary *fuel*? What precautions are taken against damage by fire? Do neighbours co-operate to extinguish fires?

Light, Warming, and other Uses of Fire. How is artificial light obtained? From the house-fire only, or from torches, candles, or lamps?

If possible, obtain specimens of all such devices, even the commonest, and discover how they are made and by whom. How are huts and houses warmed? Describe all fireplaces, ovens and stoves, portable braziers, fire-irons, and other implements, smoke holes and ventilation (*v.* Habitation, p. 236). Fire may be used in wood-working, in boat-building, in fishing and hunting, in agriculture, in time-measuring, and in other ways.

Traditions and Observances. Are there legends of the discovery of fire, or invention of fire-making appliances? Is fire itself, or the means for making it, personified? Is there any legend of the introduction of the present mode of fire-making? Is there a god of Fire?

Are any beliefs connected with the household fire? Is it placed in charge of any particular person? Are any ceremonies connected with it? and by whom are they performed? Who has charge of making and keeping up the fire? Is it sacred, and is its extinction unlucky? Are the household fires ever put out and kindled from newly produced fire? On what occasions and by whom? and is the new fire made by any special methods or with any ceremony? What account do the people give of the practice? Are fires used as beacons? Are there any occasions on which fire is made publicly, or when all fires are extinguished and new fire made ritually? Investigate all beliefs concerning fire and its use in ritual. Is fire used to purify from uncleanness, blood, death, moral guilt, etc.? and how is it applied? Is fire a means of driving away evil demons? Is ordinary fire used for any of these purposes, or is fire specially made?

Is fire-making in any way connected with moral (especially sexual) purity?

Is there any custom against wounding or polluting fire? Are offerings given to it or consumed by it? For what reasons?

Food

Foodstuffs and their Preparation. For the preservation and collection for subsequent identification of foodstuffs, see General Note on the Collection of Specimens, p. 361. Give in detail the staple and the accessory articles of diet, and identify the principal varieties of animals and plants employed in either category: the former may include anything from worms or grubs to man, and the range of plants is equally varied; also note whether the foodstuffs are derived from wild, domesticated, or cultivated species. When are the principal foodstuffs in season? Does any seasonal migration or trade depend upon this? Describe any ceremony that marks the opening of a seasonal change of food, the first kill, fishing-haul, harvest, etc., and how this differs from subsequent customary rites of like nature. Collect all myths dealing with the origins of certain forms of food, and those referring to their procuring, cultivation, preservation, and treatment.

When poisonous plants or animals killed by poison are used as food, how are the noxious qualities extracted, and by what appliances? Does any wild or cultivated cereal, pith, roots, or other substance provide material for making bread, cakes, puddings, etc.? What methods are employed for pounding, grinding, and otherwise preparing grain and other foodstuffs as flour, or to break up any other kind of food? How is the flour prepared for eating?

What animals, or parts of animals, or plants are thought unfit for food, and why? Is marrow much sought after, and how is it extracted? Is any particular kind of fat prized, and why? Are any parts of animals reserved for medicinal use? The diet of sucklings and children should be investigated.

Milk. What use is made of milk? From what animals is it obtained, in what vessels is it collected? Who does the milking? Are any ruses employed in order to milk the animals? Are there any preparations of milk made, such as curds, butter, cheese, etc.? Is milk ever boiled? Describe all customs and beliefs connected with milking and the use of milk. Is it believed that the use to which milk is put will affect the animal milked?

Preservation and Storage of Food. Are there storehouses for food, how are they constructed and protected from the ravages of animals? Are they the property of individuals, families, clans, or villages? Has the storing of food ceremonial importance? Does it enhance social prestige? Are there any devices or appliances for keeping food out of the reach of vermin? Is any food preserved by smoking, salting, or drying in the sun or over a fire, and is such food consumed without further preparation? Does prepared food feature as a medium of exchange?

Cooking. What articles of food are eaten raw? Is food preferred fresh or "high"? Is meat preferred slightly or well cooked? How are the various foodstuffs prepared before cooking, and what methods of cooking are employed and by whom? In roasting, broiling, grilling, etc., are spits or other appliances used? In frying, what grease or oil is preferred? In baking or steaming, what ovens are used, of what form, and what are they made of? There are various kinds of earth-ovens, or pits, and the whole procedure should be described; are they temporary or permanent? Are hollow trees, termite-hills, or other natural objects employed? In stewing and boiling, what cooking vessels are used, are they cleaned after use? What other appliances are used? How are cooking vessels supported or suspended over the fire? Are hot stones used for boiling, or natural hot springs or "fumaroles"? Are special cooking vessels preserved for individuals, ritual occasions, etc.?

Condiments. Are any vegetable or animal oils or fats used in cooking, how are they made and by whom? Make inquiries about the use of *salt*, the method of obtaining, preparing and storing, and concerning all customs and taboos connected with it. Is sugar, honey, or any other sweetening substance used? Are there any stimulants to the appetite in use? Are any vegetables cooked with the meat? Are any broths or stews made with the vegetables? Is any leaven, yeast, or similar substance employed? Is there any mode of preserving fruit or vegetables by cooking with sugar, fermenting, pickling, etc.?

Observances and Traditions. Is the cooking carried on in the dwelling-house or in a separate building? Are there any spots where it is definitely forbidden to cook? Is cooking performed exclusively by one sex, and are there any rites or beliefs connected with it? Is the food for men and women cooked together or separately? When cooks form a separate class, do they rank high or low in the community? Are there traditions as to the origin of the art of cooking? All ritual in cooking should be described, both in the household and on those occasions when food is cooked for public ceremonies and feasts. (*v.* Field Antiquities, p. 340.)

Is the preparation of any particular substance used as food or in cooking, such as suet, in the hands of particular persons?

Meals and Eating Customs. Are meals at set times, or dependent on the accidental supply of food or on individual inclination? Are they common to a household or village, or does each person eat separately? Does all the household eat together or is there a distinction of sex, ranks, or ages? How are foods served and in what receptacles? Is there any particular sequence in the order of the dishes? Can any estimate be made as to the quantity of food eaten at each meal? If possible work out an average daily dietary for both sexes. Is it the custom to eat to satiation? Are any attentions paid to invited guests or strangers? Is the food ready cut up or does each help himself? Is there any order observed in helping the persons present or in giving drink? Give names of all implements used at meals. Note any peculiarities in the mode of eating or drinking; if food is taken with the fingers are both hands used or only one, if so, why? Does the method of eating differ according to sex? In times of temporary scarcity or on a journey, are any means employed to deaden the pangs of hunger? Are there any ceremonies used at the beginning of meals, such as washing of hands, offerings to divinities, etc., or any rites connected with them? Is there any ritual connected with the disposal of food left over from a meal? On what occasions are great *feasts* held? and are they given by individuals, or by kinship or local groups? Are any special foods associated with particular feasts? Are ancient habits observed at feasts, such as old dishes or ways of cookery? Are there special forms of address, healths, etc., at feasts? Is there any special licence as to language, sexual intercourse, etc., allowed at feasts? Is the ceremonial exchange of feasts, or of food, a definite part of the customary or legal obligations of kinship, chieftainship, priesthood, etc.?

Prescribed and Forbidden Foods. Is any special article of food prescribed, restricted, or prohibited by custom, taboo, or special enactment of a chief, priest, or medicine-man? Do these restrictions apply to individuals either permanently or in some special crisis such as pregnancy, during menstruation, childhood, puberty, illness, etc.; to families or clans, either permanently or seasonally, or on special occasions; to special societies, age-grades, or ranks? Are the prohibited animals, plants, etc., objects of cult? What reasons are given to account for such prohibitions? What penalties are believed to follow the breaking of the restrictions? Are there any occasions when cooked foods or warm foods and drinks are taboo? Note particularly any restrictions on the use of milk or of other foods in combination with milk.

Exceptional Foods. In seasons of scarcity or famine, are any unusual substances used as food, such as bark, clay, etc.? Is the eating of earth known apart from this? What is the nature of the earth? what effect has it on those who eat it? and what is the reason for eating it? When going long journeys or undergoing hard labour, is any kind of substance of a peculiarly invigorating nature eaten? Is there any marked difference in the food of the chiefs or rich men, and that of poorer persons?

Cannibalism. Is it frequent or exceptional? Is human flesh looked upon in the same light as other animal food, or partaken of as a matter of ritual? When human flesh is naturally eaten, do the partakers believe they obtain the qualities of the deceased? What is the native attitude to the practice? Are any reasons assigned for it? Are the victims generally men, women, or children? Are they enemies slain in war, captives taken in war or by deceit, or slaves, or other persons selected for the purpose? Has the cooked human flesh any name of its own, euphemistic or otherwise? Is it prepared in the usual cooking-places, or are there special cooking-places set apart or constructed for the purpose? Are any special vessels or implements used for cannibal feasts? What parts of the body are eaten, and why? Are any parts considered delicacies? What is done with the bones? Are any of them used for implements or ornaments? Record all connected beliefs and ideas. Are the victims considered sacrifices to the gods? Is the use of human flesh confined to any class or sex? Is an individual considered unclean after joining in a cannibal feast (i.e. is there a distinction between a dead body in the ordinary sense and one intended to be eaten?). Is the sacramental eating of parts of the corpse a mortuary rite, and with what beliefs is it associated? If cannibalism no longer prevails, are there any traditions as to its once having been known?

Water and other Natural Drinks. Is water commonly drunk? Is it drunk in sickness? What devices or utensils are used for collecting, and what vessels used for transporting and storing water; who do this, how are water vessels carried, and what methods are employed to prevent spilling? (*v.* Waterworks, p. 249.) Are any means used for purifying it? What method is employed when drinking from a stream or pool? What substitutes are used when water is scarce? What natural drinks other than water are used?

Manufactured Drinks. What drinks are prepared and how? By *solution*—dissolving sugar or honey, kava, chicha, etc., in water; are these solutions allowed to ferment? By *suspension*—mixing meal and water. By *extraction*—treating with cold water. By *infusion*—treating with hot water, as tea, coffee, etc.—are they of native growth or imported? By *decoction*—boiling. *Fermentations*: are fermented drinks made, and, if so, are they known to be indigenous, or introduced? Are they made from naturally occurring substances, and, if so, how collected? Are special crops grown for this purpose? Are fermented drinks of the nature of (*a*) *wine*, by the fermentation of fruit juice, tree saps, etc. How is it made and stored, and how long will it keep? (*b*) *beer*. Is beer known? From what grain is it made? Is the grain used raw or malted? If the latter, how is it turned into malt? Are any ingredients employed, such as hops, to flavour the beer or make it keep? Describe the mode of brewing and fermenting. Is any substance added, like yeast, to cause fermentation? How is the fermentation checked? How is the beer stored? and how long will it keep? What are the utensils used in making, storing, and serving out the beer? Is beer-making a distinct vocation or does each family brew for itself? (*c*) *spirits*, by distillation, when a fluid, usually fermented, is heated in a closed vessel to separate the more volatile part, which is collected by condensing the vapour. Is any fermented drink made from milk? Are any ardent spirits known? Are they of native manufacture or imported? If native, from what substance are they made? and how? Describe the still and other appliances. Is any flavouring employed? Is there any tradition as to the source from which the art of distilling was learned? What is the approximate strength of the alcohol?

Uses and Observances. Are the prepared drinks used alone or with food? Are they prepared by men or women? at the time of drinking or beforehand? Are they used on ceremonial occasions? Are any ceremonies connected with drinking? Are special vessels reserved for particular drinks? Are they said to be nourishing, medicinal, stimulating, narcotic? Note their effects. Does blame or loss of reputation attach to the use of any particular drink, or to drinking in excess? Are there any myths or traditions connected with drinks or drinking customs?

Stimulants and Narcotics

What stimulants are used other than fermented and distilled drinks? What narcotics are used? Are they indigenous, or imported? In either case, give any traditions as to their discovery or introduction. Give the name of each and its meaning, if this is known to the people. Are they now an article of trade? What is the method of preparing them for use? Are they used in a pure state, or mixed with other substances? Describe and photograph the process of preparation and the utensils employed. On what occasions are they used? Are any beliefs connected with their use? Is the use of any of these confined to one sex or class? Is its use carried to excess, and what is thought of those who exceed? What effects on the physical and moral condition of the people can be observed from the use of such substance? Is narcosis induced for surgical or other purposes? (*v.* Medicine, p. 201).

Tobacco and its Substitutes. Is tobacco grown? How is it prepared? Are any plants such as hemp or other material smoked, chewed, or used as snuff? How are they prepared for use? Obtain specimens of raw materials and apparatus for smoking or inhaling, and photographs of smokers. If pipes are used, of what material are the bowls and stems made? Is the smoke drawn through water, fibre, or other filtering substances? Is each pipe smoked by a single individual, or are pipes passed from one person to another? Are any ceremonies or myths connected with smoking?

The Food Quest

The collection of an adequate food supply is the first charge upon the time and energies of all peoples, and the character of the food quest goes far to determine the nature and amount of their supplementary occupations. In the sections which follow, evidence is sought as to the way in which a people actually maintains itself under the given geographical conditions: soil, climate, vegetation, and the like. Historical information may often be obtained in the course of inquiry into present conditions, but it should be recorded separately and authenticated with the name of the informant, and also, if possible, with a note as to his means of information. The subject may be divided into the principal headings of:

1. Food-gathering.
2. Plant cultivation.
3. Domestication of animals.
4. Hunting.
5. Fishing.

General Questions are put together at the end (p. 257).

Food-gathering

The most primitive method of obtaining food is the simple collection of such wild animal food as insects, shellfish, etc., or wild plants and seaweeds. A list of such wild produce as merely requires finding and gathering should be obtained; describe and collect appliances used for reaching, or detaching, or raking together, and for carrying. The method and seasons of collecting, and the share undertaken by each sex should be noted.

Plant Cultivation

This section considers only the arts of horticulture and agriculture, by which a people obtains food and other commodities from cultivated plants. The social activities which are characteristic of societies which depend mainly on cultivated plants for their food supply and livelihood are considered separately (*v.* Economics, p. 160).

Note cases in which wild trees or other plants are protected or fostered in their growth. How, to what extent, and by whom are clearings made in forested or bush land? Are grass and bush burnt, and for what purpose? What implements are employed to prepare the ground? digging-stick, spade, pick, hoe, mattock, plough, roller, etc.? Describe the forms of the tools, materials, and construction. Are they of home manufacture, or imported? Make drawings and, if possible, photographs of them in use, and obtain specimens or small scale-models of them. Is any domestic animal employed in ploughing or harvesting, or otherwise? Are *irrigation, drainage, terrace-cultivation* practised, and by what methods? (*v.* Waterworks, p. 249).

What plants are cultivated, whether for food, food for animals, stimulants, narcotics, or ornaments? What parts of the plant are used for food? List all introduced plants in cultivation and inquire into native traditions or beliefs as to their origin. Are there any legends respecting the introduction of any of the food-plants, or their creation by culture-heroes or deities? At what seasons and in what manner are the plants sown? How are the proper seasons recognized? Is any instrument or appliance used to indicate or ascertain times and seasons? Is any attention paid to the growing crops? Are hedgerows, or plantations, or other devices used to protect crops in exposed situations? How are the crops protected from depredation, and against what animals? Describe scarecrows or other means of frightening away birds or animals. Are any charms placed in the field, or rites or dances performed to make the crop good?

How and with what implements is the harvest gathered and carried home? How is it stored? Describe all stacks, caves, pits, or specially constructed granaries. What methods and appliances are used for cleaning, threshing, or winnowing grain, or for preserving fruits or leaves, and for the preparation of food from the raw material? (*v.* Food, p. 241). Are the poisonous or other peculiar qualities of plants well known? Is the same land tilled again and again till it is exhausted, or is fresh ground cleared and tilled yearly, or after a few years only? Is fallowing understood, or is there any idea of rotation of crops, or of the use of manure of any kind? What are the relative values of the different cereal or other grains, fruits, leaves, or roots? What produce is imported and exported, and whence, and whither?

Are any wild or uncultivated plants commonly used as food, or resorted to in time of scarcity? If so, what are they? Are any diseases attributable to their use? Are any of the cultivated plants apparently derived from indigenous wild ones?

Waterworks. Describe the source of the people's water supply. Is it from rain-water, springs, streams, standing pools, or wells? If from wells, how is the water raised? (*v.* Mechanisms, p. 258). Is water stored, or must it be sought as required? Obtain, if possible, examples of the ordinary buckets, ladles, and water jars. How is water transported, for domestic use or for agriculture? Describe all cisterns, channels, and conduits, dams, sluices, and floodgates. Note any cuttings made to drain off surface water; these should be distinguished from irrigation canals. Is water-power employed for mechanical purposes? If so, give working drawings, and full descriptions of the water-wheels, or other mechanism. Is water used to measure time, in a *water-clock*?

Are there any legends or ceremonies connected with water? Is water sacred? Is it capable of ceremonial pollution? Is it personified or regarded as the abode of spirits?

Livestock

This section considers only the arts by which animals are kept under human control and made serviceable for food or other purposes. The social activities characteristic of societies which depend mainly on domesticated animals for their food supply and livelihood are considered separately (*v.* Economics, p. 166).

Give a list of the domestic animals kept by the people, with their native names; add photographs or drawings, and full notes of all peculiarities of breed, colour, shape, humps, length of horns, etc. Are any marks or peculiarities in the animals considered lucky or unlucky? If so, what reasons do the people give? Are the same or similar animals found wild in the neighbourhood, or elsewhere? Is there any knowledge or tradition as to the origin of domestic animals?

Species and Uses. If *cows, sheep,* and *goats* are kept, are they horned or polled? Are any malformations practised on their horns? If so, is this done to all or to special beasts, favourites, leaders of the herd, cows, entire or gelded beasts? Are they used for draught purposes? Is their flesh eaten or blood drunk for food (as among the Masai), or are they kept only for dairy or religious purposes? Describe fully the dairy establishment, and, above all, dairy processes (*v.* Food, p. 241). What use is made of the hair or wool? (*v.* Weaving, p. 287). If it is made into *felt*, describe the process. If *camels, reindeer,* or other animals are used, state for what purposes. Are *horses* used for food? Is the mare's milk used?

What breed of *dogs* is kept, and what are they used for? How are they treated? How do they behave to their owners; to their owners' families; to other members of the same social group; to strangers? Are *cats* kept; of what kind; are they fed, or are they expected to provide for themselves? What other vermin-killers are kept (mongoose, ferret, etc.)? Is *poultry* kept? Describe the breed. Is cock-fighting customary, or any analogous practice? Are *bees* kept? Describe the hives. Is wild honey collected? Describe method of collection, utensils, etc. How is the honey taken? Is any kind of *silkworm* kept? On what are they fed? Are any animals (including insects) kept for their music, for fighting, or other purposes? Are any animals kept specially for use in hunting or fishing or as pets?

Rearing and Taming. Are domestic mammals and birds kept or bred for sale, or only for the owner's use? If sold, to whom, and at what markets, and what is the relative value of each? Note any beliefs and customs in regard to *breeding*. Are certain breeds considered better than others, if so, why? What care is taken to preserve purity of breed, or to improve the breed? Are the females mated with wild males of the species? Are any hybrids or *mule* animals bred?

How are animals broken in and trained? If they are caught and tamed, describe the mode of catching (*v.* Hunting, p. 253).

Are they shod? If so, how? Describe the mode of recovering half-wild animals or animals at pasture, and of catching and herding the flocks. What cries or instruments are used to call animals, to encourage draught animals, etc.? Is the *sling*, or the *lasso*, or the *bolas*, or the *pellet-bow* used in herding? Are domestic animals well treated?

Is gelding practised? If so, what is the process, and what is the alleged purpose of the operation? Who performs it? Is it accompanied by any ceremonial? What is done with the removed organs? What diseases prevail among the domestic animals, and what remedies are employed to cure them? What becomes of animals which die naturally? How are the animals killed, butchered, and cut up? Describe how each class of domestic animal is housed, fed, and tended. Is the care of any class of animals regarded as the privilege of a particular class, caste, or sex? What precautions are taken to protect the animals from wild beasts? Is any hay or other fodder stored for winter use? If so, how is it prepared, how and where stored? Describe all brands and other marks of ownership which are put on the animals, noting the process, and adding drawings of the marks. Also whether brands are personal, belonging to the clan or the group. Are cattle the absolute property of the individual, or are individuals allowed only the use of certain beasts which may be recalled at the will of a superior? (*v.* Property, p. 150.) To what extent do cattle-lifting and other concomitants of pastoral life lead to war? Does the necessity for finding pasture lead to definite seasonal migrations of the herdsmen? Describe the conditions. Describe all beliefs, observances, and customs involving domesticated animals, or dairies, or implements used in connection with animals. Whips, yokes, saddles, bridles, etc., may be considered here or under Travel and Transport (p. 300). Note any charms that are used either on the harness or elsewhere, for the protection of animals. Are women under any circumstances forbidden to have anything to do with any class of domestic animal? What is the reason alleged?

Hunting

(1) *Hunting, or Active Pursuit practised by Individuals* (sometimes in pairs) usually consists of (i) tracking; (ii) stalking; (iii) lying in ambush by water, salt-licks, game-runs, etc., with or without a prepared bait or decoy; (iv) overtaking by greater speed or

endurance. In each case give a list of animals and birds so obtained, and describe methods and *weapons*. State whether portable screens or disguises are used to conceal the hunter when stalking. In some cases the hunter makes use of the active services of some trained animal, such as a dog, cheetah, or hawk. The method of training and use should be described.

(2) *Hunting by a Number of Persons acting in concert*. In this case the object of the party is usually to drive the game, by fire or by beaters (often assisted by dogs), into nets, pits, enclosures, or up to a party lying in ambush. Fences are often constructed beforehand to force the game to take a certain line. The tactics employed during a hunt should be described, and also the weapons carried by the various parties into which the hunters may be divided. Are the weapons poisoned? Describe preparation of the *poison* (*q.v.*) and secure samples. The formation of the groups of persons who hunt together should be investigated. Are they bound by any social or local tie?

Traps, Baits, Nets, and other Hunting Gear. The sections which follow, on the appliances used in hunting and fishing, are given in greater detail than some others, as an example of the kind of classification which may be attempted in every department of technology, when circumstances permit.

Traps may be divided into the following classes: (1) Traps set in motion by a concealed operator; either (i) unbaited traps (cage traps, net-traps, etc.), or (ii) traps used with a *bait* or decoy (*v.* below).

(2) Traps caused to act by the game itself. (i) Unbaited traps (such as pits, with or without stakes, snares, dead-falls, springes, bow-traps, spear-traps, etc.), or (ii) traps used with a *bait* or *decoy* (*q.v.*), e.g. cages, nets, springes, etc.; (iii) *bird-lime*: the preparation should be described and the method of laying the trap.

(3) *Poison* is sometimes laid for game. The method of preparation and laying should be noted, and a sufficient quantity of the poison obtained for analysis (p. 203).

The localities in which traps are set, the method of setting, and the animals for which they are intended should be carefully noted. Native names for traps and parts of traps should be recorded. Diagrams should be prepared showing manner of operation and the position of the parts when the traps are set. In such a matter as this a simple diagram involves less labour, and is far more comprehensible than a detailed description.

THE FOOD QUEST

Baits are attractive objects, used with land traps of all kinds, and also in fishing. They may be divided into (i) baits proper; (ii) decoys; (iii) lures; (iv) flares.

(i) *Baits proper* are the actual food of the animal or fish it is desired to capture. *Live-baits* may be any living animal; state the baits used for particular game and the method of securing them, etc. *Inanimate baits* may be animal or vegetable; state whether used fresh or putrid, whether poisoned or not; give the baits for particular game.

(ii) *Decoys* are animals (or imitations of them), the presence of which attracts other animals of the same or allied species. Animals may also be attracted by means of a call or cry, produced with or without an instrument. The living decoys may be either untrained or trained.

(iii) *Lures* are occasionally employed to attract land-animals as well as fish; they usually appeal to the sense of colour and curiosity rather than to the appetite.

(iv) *Flares* are mainly used over water to attract fish, which are speared or netted, but they are sometimes used to attract land-animals.

Fishing

Fishing is carried on either (*a*) without appliances or with (*b*) nets, or (*c*) traps, or (*d*) dams or weirs, or (*e*) lines, or (*f*) arrows, spears, harpoons, and gaffs, or (*g*) poison, or (*h*) trained or wild animals (birds, beasts, fish).

(*a*) *Without appliances*, fish are often caught with the bare hand; the various methods of approaching and seizing the fish should be described in detail.

(*b*) *Nets* are of many fabrics and designs. (For the mode of manufacture, *v*. String, p. 286). Nets are designed either to be manipulated by one or more persons, or to be self-acting. The latter class merges with the self-acting or basket-traps.

Nets manipulated by one or more persons may be classed as follows: (i) *Hand-nets*, with or without a frame of cane or of wood; (ii) *Cast-nets*, weighted at the edges; when thrown upon the water they sink and enclose any fish which may be beneath; describe the weights and the method of casting and recovery, with photographs if possible; (iii) *Seines* are long nets, furnished along the upper edges with floats, and along the lower edges with sinkers; these are "shot" into the water so as to enclose as large an area as possible, and then drawn to land; describe the floats, sinkers, and method of "shooting" the net; (iv) *Trawl-nets* are bag-shaped like hand-nets, but weighted so as to sink to the bottom, and stayed open to admit the fish when the net is towed along.

Self-acting nets are stationary, and are usually supported by stakes, which may be either (i) set in one straight line across the set of the current or tide, so that the fish become entangled in the meshes; or (ii) arranged so that the net forms a chamber or series of chambers, into which the fish find their way, and in which they are left by the receding tide or falling river. In the latter case a plan should be given, the position of the stakes being indicated by dots. Often the place of the net is taken by a wicker fence; in this case a plan should be given and the construction of the fence described in terms of Basket-work (p. 274).

(c) *Traps* are either manipulated by the fisherman or are self-acting.

Traps manipulated by the fisherman include (i) *basket-traps*, simple conical structures, placed rapidly over fish seen in shallow water (to be described in terms of Basket-work, p. 273); (ii) *Nooses*; (iii) *Cage-traps*, of which the door is lowered or closed when the fish are seen to be within.

Self-acting traps include (i) *basket-traps*, of the "lobster-pot" and the *thorn-trap* patterns, which simply prevent the escape of the fish once they have entered; (ii) *automatic traps*, such as cage-traps, net-traps, and spear-traps, used with a bait, disturbance of which sets the trap in action. Give plans and diagrams.

(d) *Dams and Weirs*. Streams can be dammed and the water scooped out of the pools, the fish being captured with the hand, baskets, spears, etc. Describe the construction of the dams (v. Waterworks, p. 249), and all implements used in connection with them. Weirs and dams may be built across streams or pools, or on the seashore below high-water mark, which are filled by the rising river or tide. They are often designed with openings in which basket-traps (*q.v.*) are set.

(e) *Lines and their Tackle*. Describe the material of the line; the method of holding or securing (by a reel or winder). Is a rod used? If so, give full description.

1. *Lines* directly operated by the fisherman include (i) lines furnished with bait, simply tied to the line, or placed on a hook, or on a gorge (see below for explanation of these and other fishing terms); (ii) lines furnished with a lure, either simply tied to a line,

or attached to, or forming part of, a hook, or of a gorge; (iii) lines furnished with a gig. Lines unconnected with the fisherman may be (*a*) stationary lines, such as pegged lines, long-lines, night-lines; (*b*) trimmers. In either case describe how these are secured, and how the bait is attached.

2. *Tackle* may be described briefly and accurately with the help of its special vocabulary, which is copious and precise. The most important terms are as follows: The function of a *hook* is to form a support for a *bait* or *lure*, and to capture the fish by penetrating some portion of the mouth or gullet. The parts of a hook are the *shank*, the *bend*, and the *point*; the last may be furnished with one or more *barbs*. Hooks may be divided into (*a*) solid, and (*b*) composite. Describe carefully the materials and the method of manufacture. A hook proper should be carefully distinguished from a hook-shaped *gorge*.

A *gorge* is simply a support for a bait, which the fish swallows and is unable to eject. (*a*) A straight gorge consists of a straight, or nearly straight, more rarely angular, piece of wood, turtleshell, metal, etc., to the centre of which the line is made fast so that it forms a toggle; this turns athwart the gullet of the fish when the line is pulled. (*b*) A hook-shaped gorge is furnished with a recurved point or barb under which the bait is fastened; the fish is captured not because the point has penetrated any part of the mouth or gullet, but because it is unable to eject the bait and its support.

Bait may be either "live bait" or inanimate: note what bait is used, and how it is attached to the tackle. A *lure*, as opposed to a bait, is an artificial representation of some object upon which fish feed, or it is an object which attracts fish by its bright colour, or rapid motion through or on the water. *Lures* may be (*a*) a brightly coloured piece of textile, a piece of sponge, or spider's web, simply attached to the line in place of a hook, to entangle the teeth of the fish or to act as a gorge; or (*b*) fragments of shell, beads, imitation fish, etc., attached to hooks; or (*c*) the lure may be itself a hook made of shell or some other gleaming material; these are often so cut that they acquire a spin when they are drawn through the water. A *flare* is a fishing-torch sometimes used to attract fish at night.

Floats are used to sustain the bait at a certain depth in the water; they may serve further as a tell-tale to inform the fisherman when a fish has taken the bait. The nature and construction of all floats should be noted, and any method of weighting described. Ordinary floats are in direct connection with the fisherman; in this category may be included *kites*, which are sometimes employed to keep the bait tripping along the surface of the water. *Trimmers* are floats which are not connected with the fisherman, but are allowed to wander as the wind or stream may carry them. The buoyancy of the float and the resistance which it offers to the water assist in tiring the fish.

Methods of using Tackle. In *bottom-fishing*, the bait is on or close to the bottom. State how the bait is weighted. Floats are not essential here, but are often used as tell-tales. The following terms may be useful in description. A *ledger* is a tackle in which the sinker is attached to the line above the bait, so that part of the line and bait lie on the bottom. In a *paternoster* the weight lies on the bottom, and the bait is attached to the line a short distance above it. In *mid-water fishing* the whole is suspended; there is usually, but not always, a float. In *surface-fishing* there is no sinker, and the bait is simply cast upon the water, as in fly-fishing, or suspended from a kite.

(*f*) *Appliances for transfixing Fish.* Of these the commonest are *Spears*, *Arrows* and *Harpoons*. Describe each variety and state whether it is thrust with the hand or thrown, with or without the aid of some appliance; whether it is attached to the fisherman with a cord; whether used from the bank or from a specially prepared staging, or from a boat; whether the spear is thrown or thrust at random, or whether the fish is stalked; whether flares, cressets, or other artificial lights are used in spearing fish at night; how are they made and used? Do they attract the fish like a lure, or simply reveal them to the fisherman? *Arrows* are sometimes used for fishing: describe and state whether they are of simple form, barbed or unbarbed, or of a harpoon type; and whether a cord is attached to them by which they can be recovered. (*v.* also p. 265, Weapons.) *Gaffs* are really fish-hooks with the shank prolonged into a handle, or lashed, or otherwise fastened to a handle; describe in detail. A double gaff has a Y-shaped handle with a barb fixed on the inner face of each of the arms. A *trident* has three arms, each furnished with one or more barbs; a *leister* has four or more. A *gig* is a hook which is lowered among a shoal of fish with the design of "snatching" one of them. *Rakes* are sometimes used for transfixing small fish which congregate in shallow water.

THE FOOD QUEST

(g) *Poison* is sometimes placed in the water of pools and rivers so as to kill or stupefy the fish. Describe its nature, preparation, and use; and secure samples in sufficient quantity for analysis.

(h) *Use of Animals.* Fishing with the aid of an animal is rare, but of exceptional interest. Note whether the animal is trained like a dog, otter, cormorant (*v.* Livestock, p. 200), or an untrained animal like the remora or sucker-fish.

Social Observances arising from the Collection of Food

How are hunting and fishing parties organized, and under whose leadership? Do both sexes participate? If so, are particular functions or methods reserved for each sex? How are the spoils divided, with special reference to (i) the chief; (ii) the person who inflicted the first wound; (iii) the person who killed the animal; (iv) the owner of the land, water, or foreshore; (v) the owner of the net (in hunting or fishing); (vi) the owner of the canoe? Are women forbidden to touch or approach hunting or fishing appliances, either altogether or when in a particular state of health? Do hunters or fishers form a particular class or caste? If so, is the status inherited, or by what method can entrance to the class or caste be obtained? Are any records kept of prowess in hunting, or are distinctions assumed by successful hunters? Do any ceremonies take place before setting out on, or on returning from, a hunting or fishing expedition? Must hunters or fishers observe any particular prohibitions (from certain foods, sexual intercourse, etc.) before setting out? If so, for how long? Must women observe any prohibitions during the absence of men on hunting expeditions? Give any details as to hunting or fishing rights. Are they hereditary? How is trespass incurred and punished? Are there any beliefs or legends relating to luck or omens in hunting or fishing, or to "mighty hunters" in the past? Are any reasons given for any ceremony, prohibition, or other practice connected with hunting or fishing?

Is hunting or fishing the principal mode of obtaining food, or are these associated with agricultural or pastoral occupations? In the latter case, has there been a *recent* change from a nomad, hunting condition to a more sedentary habit?

TOOLS AND MECHANISMS

The simpler tools of backward peoples may consist of natural objects or materials used with little or no preparation, as in the case of shells and boars' tusks for cutting and scraping; or attachment to a haft or holder may be usual, as when a knife is made by attaching a row of sharks' teeth to a strip of wood. The casual use of natural objects should be distinguished from the regular use, and in all cases the origin, constituents, construction, and methods of use of the tools should be recorded. Methods of hafting should receive special attention, since these cannot always be precisely ascertained by inspection of the tools themselves. Except for general utility tools which may serve indifferently for hacking, cutting and scraping, the English names of tools may generally be used for classification, but it may be necessary to employ such terms as scraping-tool, graving-tool, planing-tool, adzing-tool, where the form and construction do not justify the use of the unqualified term. Care should be taken to distinguish between the axe and the adze, since they differ both in construction and in function. It must be noted, however, that many weapons are used as tools (*v.* Weapons, p. 261). In all cases an attempt should be made to ascertain if, and how far, the materials or the construction of a tool have been modified by foreign influences. The chief types of simple tools that may be looked for are: *For striking and driving*—hammer-stones, mallets, sledges, hammers; *for hacking, cutting and dividing*—stone choppers, knives, saws, axes, adzes, wedges; *for surface-working*—gravers, chisels, scrapers, planes, adzes, rasps, files; *for boring*—awls and drills (*v.* Stone Implements, p. 279); *for holding*—tongs.

Mechanisms. These may be broadly defined as tools in which, during use, one part moves on another, and rotary motion is frequently an essential factor. Thus, boring or perforating may be done by means of a simple drill twirled between the hands, or the rotation may be effected by means of a bow (bow-drill), or a thong (thong-drill), or a cord attached to the top of the drill and to the two ends of a crossbar (pump-drill). Rotary motion is also involved in such mechanisms as pivots or hinges, pulleys, windlasses, potters' wheels, rotary querns and grindstones, and lathes, though none of these appliances is of common occurrence amongst the more backward peoples. The principle of the lever is involved in forceps, tongs, pliers, and scissors, but it may also be used in other ways, as in the lifting of heavy weights. The use of rollers, wheels, presses, and tackle of any kind should be recorded, and any devices in which the elasticity of wood, the weight of the atmosphere (as in pumps and bellows), are made use of; and also all other appliances in which the force exercised undergoes a change of direction or amount by the utilization of mechanical principles or natural laws. In all cases full details of materials, construction, and working should be recorded, as well as instances of the employment of wind, water, and animal power. Any data on the amount of useful work done with any tool in a given time are important and have frequently been neglected in the past. It will be noted that many of the mechanisms referred to above, and also many tools of simpler make, are associated, wholly or in the main, with activities treated elsewhere in this volume, but they do not thereby lose their interest as tools and mechanisms.

Weapons

In dealing with weapons, three points should be kept in mind. (1) All descriptions should be constructive, i.e. the description should follow as far as possible the process of manufacture. (2) Outline sketches with diagrams of cross-sections are valuable when dealing with objects of such infinite variety of form as weapons. (3) Particular care should be taken that the material or materials used in the manufacture of a given weapon are expressly stated. This may seem too obvious to mention, but experience shows that an observer very often omits this most necessary piece of information, no doubt owing to the very fact that it is so evident to his eye. A weapon may be simply "cut from the solid", but if it is composite, the material of each part should be stated.

General Notes. Is poison applied to any weapon? If so, describe the nature and method of preparation and application, and secure samples for analysis (*v.* Poisons, p. 203). Is any "medicine" or any charm applied or fastened to weapons? If so, describe with details as to the supposed effect. Is any incantation or analogous ceremony performed over weapons (*a*) before or (*b*) after use? If so, state the alleged purpose. Are any weapons peculiar to either sex? to any rank or class? to any individual? (as a mark of distinction for bravery or otherwise). Are women forbidden to touch or see weapons, either altogether or when in any particular state of health? Is any reason given for this? Are there any restrictions as to the carrying of weapons by certain individuals, in certain places, at certain seasons, on certain occasions? Are any reasons given? What penalty, if any, for infringement? Who may own weapons: the individual, the chief, the community? Where are they kept? Is there any penalty for the loss of any weapon in fight? Describe any decoration of, or ornamentation applied to, weapons, and give the alleged meaning or reason for it. Are conventional copies of any weapons made in the same or other material for ceremonial use? If so, when, and how, and by whom are they used? Are weapons or imitations of them ever used as currency? (*v.* Exchange, p. 169.)

History of Weapons. Such history may be either explicit or implicit. By the first is meant any tales or legends referring to the origin, invention, adoption, use, obsolescence, of any particular weapon. The second, which is equally important, depends not on tradition, but on observation; valuable indications of the early history and migrations of a tribe may often be gleaned from the use among them, for ceremonial purposes or as currency, of weapons and forms of weapons which are no longer employed in war or hunting. Such weapons are often highly conventionalized and difficult to identify, but sometimes the native name will give a clue as to their real nature. The presence of a weapon, in actual use, which is obviously ill adapted to the present environment, e.g. weapons suited to open country, in a forest tribe, may give a clue to an earlier home.

Weapons may be classified according as they are used for *offence* or *defence*.

I. *Weapons of Offence*

These may be classified according to their mode of use, as follows:

(A) WEAPONS HELD IN THE HAND

(i) *Natural Objects* should be noted, which are either (*a*) simply held in the hand so as to give additional force to the blow, or (*b*) used to strike a blow.

(ii) *Manufactured Objects.*

(1) *Ornaments,* such as rings and wristlets, are sometimes furnished with spikes or a cutting edge so that they can be used for offensive purposes; state whether any form of protection, such as a sheath, is affixed to the spike or edge when not in use.

(2) *Clubs* and analogous weapons are designed to strike a crushing blow, and for this reason the centre of balance will almost invariably be found to be situated nearer the head than the handle.

Some clubs made of hard wood are furnished with an edge which will inflict a cut. The following parts may be distinguished in a club: (1) the grip; (2) the shaft; (3) the head.

(*a*) Solid clubs and maces. Describe material and shape, giving outline and sections at various points. Is the grip roughened or otherwise treated as as to afford a firmer hold? Is there a "stop" at the end of the grip, or a wrist-loop, to prevent the weapon slipping from the hand? How is the weapon carried when not in use? How is it used—with one hand or two?

(*b*) Composite clubs, maces, and hammers call for records of constituent materials and of details of construction.

(*c*) A *flail* or *ball-and-thong club* differs from a club proper in that the head is connected by a strip of pliable substance, a chain, or a hinge-joint so that it swings more or less loosely on the end of the shaft. See queries above and note in addition how the head is connected with the shaft.

(3) *Axes* and analogous weapons have primarily the same functions as clubs, to strike a crushing blow, but the axe has a secondary action of cutting or piercing; it is further distinguished by the fact that this cutting or piercing portion is set at an angle, usually not greater than a right angle, with the handle. When this weapon is furnished with a cutting edge it is termed (1) an *axe*, if the edge lies in the plane of the stroke; (2) an *adze*, if the edge is at right angles to the plane of the stroke; (3) a *pick* if intended for piercing. The following parts may be distinguished in an axe or adze: the *grip*, the *haft*, the *head* of the haft (usually in composite weapons only), the *blade*, and the *edge* or *point*. The adze is very rarely used as a weapon.

Axes, adzes, and picks may be either *solid*, made all in one piece, and so passing over into beaked or sharp-edged clubs or *composite*, in which case note especially besides queries above, how the blade is fitted to the haft, whether (i) by bending a pliable stick round the blade so as to grip it; in such a case, is the blade grooved or notched to receive the stick, or is an adhesive used to strengthen the fixing? (ii) by a tang or spike entering or passing through the head of the haft; (iii) by a socket into which the bent head of the haft fits; (iv) by a hole through which the haft passes, or a socket into which the head of the haft (which in this case is not bent) is fitted; or (v) by any other method. Have any of the axes an elastic or flexible haft?

(4) *Spears* include all long-hafted weapons used to pierce with

a thrusting or stabbing action (*v.* Throwing-spears, p. 264). Like clubs and axes, they may be divided into solid or composite. A solid spear is either quite plain, or, if it has parts, they can be best described in terms of a composite spear after noting the fact that the weapon is "cut from the solid". The parts of a composite spear may be (*a*) the head; (*b*) the fore-shaft; (*c*) the shaft; (*d*) the butt; (*e*) the counterpoise; and (*f*) the sheath.

(*a*) The *head* may be composed of (i) a blade or point; (ii) a shank; (iii) a socket or a tang. In the *blade*, the following points should be noted: the number of blades or points, and how arranged; the outline (triangular, leaf-shaped, lozenge-shaped, or spatulate, when the greatest width is nearer to the point than to the shank, etc.); the presence or absence of a *mid-rib* between the wings or flat margins of the blade, of grooves or openings in the thickness of the wings, and of barbs; the transverse section of the blade, and of the cutting edges, whether convex and wedge-shaped, or concave (sometimes called "hollow ground"). Note whether barbs are cut from the solid, or affixed; give their total number; their arrangement (alternate or irregular, unilateral, bilateral, trilateral, etc.). In the *shank*, note (as above) the section, the presence or absence of barbs. The spear-head may be secured to the shaft either by a *tang* or spike which penetrates the shaft, or is lashed to one side of it, or by a *socket*, which is itself hollow and encloses the end of the shaft. Note how the head is secured to the shaft or foreshaft, whether by accurate fitting or by gum, binding, nails, or rivets.

(*b*) The *fore-shaft*, where present, forms a connecting link between the socket or tang and the main shaft; note the section and the method of securing to the shaft, and whether (if the head has a tang) the fore-shaft has any binding or collar to prevent splitting. Is the fore-shaft of heavier material than the shaft?

(*c*) The *shaft*. Note the section; the presence or absence of binding (as above), or a "grip" for the hand; the decoration.

(*d*) The *butt* is a separate protection or fitting on the proximal end of the shaft. Note the outline (spiked, spatulate, bifurcated, etc.); the section; whether it has a socket or a tang; how it is secured to the shaft.

(*e*) The *counterpoise* is usually alternative to a butt, and may be a knob cut from the solid on the shaft itself, or some heavy object attached to it. Note the method of attachment.

(*f*) The *sheath*, when present, protects the point or cutting edge.

Describe how the spear is used, and how it is carried when not in use; also whether it is used in conjunction with other types of weapons.

(5) *Swords* are designed either for cutting or for piercing, or both; the centre of balance of the weapon is almost always nearer the hilt than the point; they may be divided into (*a*) solid and (*b*) composite, each class being subdivided into (i) cutting swords (*broadswords, cutlasses*), and (ii) thrusting swords (*rapiers*).

The parts of a sword are (*a*) the blade; (*b*) the guard; (*c*) the hilt, with its pommel or counterpoise; (*d*) the sheath.

(*a*) The *blade* may be described in the same terms as a spear-head, but note whether it is designed for cutting or thrusting, or for both; if for the former, note the extent and form of the cutting edge, whether there is more than one cutting edge. Describe how it is fastened to the hilt (by spike, or tang reinforced by gum, binding, nails, or rivets).

(*b*) The *guard*, if present, may be a spur or cross-piece, or sometimes it forms part of the blade; if it is a separate piece, state how it is fastened to the blade or hilt; or a more or less concave shield for the hand, sometimes very elaborate, as in the "basket-hilt" and "cup-hilt". The "counter-guard" unites the guard and the pommel.

(*c*) The *hilt* may be subdivided into (i) the grip and (ii) the pommel. The *grip* has its own outline and section; note whether it is roughened or furnished with a binding to prevent the hand slipping, or with a wrist-loop, or other contrivance to prevent it flying from the grasp. The *pommel*, or knob, serves the double purpose of a counterpoise and a stop to prevent the hilt from slipping through the hand; note the material and shape, and if it is a separate piece, state how it is fixed to the grip.

(*d*) The *sheath*. Describe fully the material, manufacture, outline, section, ornaments, and fittings, especially the rim and the *chape* which guards the point, and the method of attachment, if any, to a belt or baldric.

Describe how the sword is carried and on which side of the body; how drawn; how used (with one or both hands). Do the first, or first and second, fingers grip the cross-guard? Is it used in conjunction with other weapons?

(6) *Knives* may be divided into (*a*) *Knives* proper, for cutting; (*b*) *Knife-daggers* for a cutting-thrust; (*c*) *Stilettos*, for thrusting only. State whether they are used with or without poison; if with

poison, describe the materials and method of preparation (*v.* Poisons, p. 203. For the parts of a knife, *v.* Swords, p. 263).

(B) MISSILE WEAPONS

Practical tests should be arranged and statistics collected to show both the extreme and the effective range of each kind of missile, and also the accuracy with which they are used. Missiles may be classified as follows: (i) *Natural Objects*, such as pebbles, are sometimes selected for size, shape and weight. (ii) *Worked or Manufactured Projectiles*, sling-stones, stones thrown from the hand, and native-made pellets and bullets, and other projectiles for pellet-bows or firearms should be noted. (iii) *Throwing-clubs* are either clubs proper or boomerangs. The clubs proper should be described in the terms of (A), p. 260. *Boomerangs* are curved and flattened throwing-clubs, some of which, when thrown, describe certain evolutions in the air and return to the neighbourhood of the thrower; they are invariably flattish (flat, planoconvex, biconvex) in section, and curved in outline, and in returning boomerangs the surfaces do not always lie in the same plane, but have a slight spiral twist. Describe, with diagrams where possible, the method of preparation, of holding, and of throwing; the course taken by the boomerang in the air, noting especially whether it returns to or towards the thrower; the range; whether they are used for practical purposes or only for amusement. Collect specimens of which you have observed the flight, and learn to use them yourself.

(iv) *Throwing-spears* form a large class, including all javelins, harpoons, darts, or arrows, differing mainly in size and mode of propulsion.

(*a*) *Javelins* (or throwing-spears proper) may be described like other spears (p. 261), but note in addition: (1) Whether there are "wings" at the butt end to assist the flight of the weapon; their material, shape, position, and method of attachment; (2) the presence of any loop, notch, or other appliance attached to, or forming part of, the shaft, to assist the act of throwing; (3) the method of holding and throwing, and the range and accuracy; whether the points are notched so as to break off in the wound; or poisoned (p. 259).

(*b*) *Harpoons* include *retrieving-javelins* and *arrows* in which a cord is attached to the shaft, so that the spear or arrow may be recovered after being thrown; and *harpoons* proper, which are spears of which the head is removable, and usually connected with the thrower or with a float by a long line of cord, or with the shaft by a shorter one. The parts of a harpoon are (i) head, with blade, shank, and socket or tang; (ii) fore-shaft; (iii) shaft; (iv) butt; (v) wings; (vi) line with buoy or float. For the head, fore-shaft, shaft, and butt, see Spears, p. 261, but note also how the lines which communicate with the thrower and with the shaft are attached, and how the head is fitted to the shaft and released from it. Describe also the wings, if present; the material, make, and length of the lines; the buoy, if present; its nature and method of attachment. Does the shaft itself when detached act as a buoy or drag?

(*c*) *Darts* are made small enough to be projected from a blow-tube (p. 266), and consist of a shaft, which may be simply sharpened, or furnished with a head or point of some other material. They are furnished with a butt of pith, leaf, etc., which fits the tube, or with a packing of cotton, feathers, or other soft substance, fastened round the shaft.

(*d*) *Arrows* include all small missile spears designed to be propelled from bows (p. 266). These may be described in terms of Spears or Harpoons. Note further (1) *wings* or *feathers* added to steady their flight; their material and number; their position, straight or spiral; and the method of preparing and affixing them; (2) the *nock* or *notch* for the bowstring; its shape, depth, and binding; also whether the notch is cut in the end of the shaft-butt itself, or in a plug of harder material inserted in the butt. Describe how the arrows are carried: in the hand, hair, belt, quiver; how the *quiver* is constructed, and how it is carried and attached to the body; whether it forms part of the bow-case, and whether it is furnished with a point to stick in the ground.

(v) *Throwing-knives* are in some cases simply knives which are thrown, and should be described in terms of *Knives* (p. 263). Others are of extremely complicated outline, with many blades and edges, and need a sketch to explain their form.

(C) APPLIANCES FOR HURLING OR DISCHARGING
any of the foregoing missiles, so as to attain a longer range than with the hand and arm; some of these appliances are attached to the missile; most forms are not themselves thrown, but remain in the hand of the thrower.

The principal appliances are flexible spear-throwers, rigid

spear-throwers, slings, blow-tubes, bows, and firearms. (*a*) *Flexible spear-throwers* or *throwing-cords* are attached to some portion of a javelin either temporarily, so as to be retained by the user, or permanently, in the form of a loop fixed to the shaft and accompanying the javelin in its flight. (*b*) *Slings* may be used for sticks, stones, or artificial bullets. In all cases note the material, dimensions, all knots and loops, and the mode of use. How are the cords affixed to the javelin and how released? In slings, what ammunition is used, and how is it released?

(*c*) *Rigid spear-throwers*, sometimes called "throwing-sticks", are used to add length to the arm and effectiveness to the grip. There is usually a *peg* at the end of the spear-thrower to fit into a cup-shaped hole at the butt end of the javelin; note the material of the peg and whether it is cut from the solid or affixed (describe the fixing); note outline, section, grip, method of holding. If, as in some cases, there is a socket to receive the butt end of the javelin, note its shape, depth, position. Does the spear-thrower serve any other purpose than spear-propulsion, e.g. as a tool, or in fire-making?

(*d*) *Blow-tubes* through which darts are propelled by the breath of the operator may have two parts, *tube* and *mouthpiece*, each of which should be described and classed according to material. A length of hollow reed or bamboo may form the tube, but in some cases the tube is bored in a solid wooden rod; note the manner in which the bore is produced, the length and calibre, and the binding, if any. Some blow-tubes have a delicate inner tube protected by a stronger sheath, or the tube may be made up of two grooved halves. Any other method of construction should be recorded. If *sights* are used, describe their material, method of attachment, and position on the tube. Is a spear-head attached to the muzzle end of the tube?

(*e*) *Bows* have the following parts: the stave comprising the grip, the limbs and the horns, the string. The *belly* of a bow is the surface facing the archer when drawing the bow; the *back* is then remote from the archer. Note dimensions and form of the curve; the form of the stave of the bow; the section at various points; the method of construction. Bows may be classified as: (1) Plain or "self" bows; (2) compound bows; (3) composite bows; (4) pellet-bows; and (5) cross-bows which may be of any of these four kinds.

Bowstrings, methods of release, bracers, and gloves, need separate treatment.

Plain bows are made of a single stave, of which the natural elasticity is not reinforced by any foreign substance. It is important to note the section of these bows.

Compound bows are built of more than one piece of the same class of material, e.g. a bow of hickory "backed" with yew.

Composite bows are built of two or more different elements which combine to make a resilient shaft, e.g. reinforced with sinew, or layers of horn, wood, whalebone, and the like, usually combined with sinews. Where a "backing" of sinew is used, is this in the form of cords or is the sinew moulded in layers upon the surface of the stave? Such bows should be described constructively, following the method of manufacture. Special note should be taken of the form they assume when unstrung (reflex curve). Thin transverse slices cut from composite bows are essential for detailed examination.

Pellet-bows are adapted for bullets or pellets instead of arrows. Note specially besides the form of the stave, whether single or double, plain, compound, or composite, whether the string is double or single, how the pellet is held, and the position of the pellet-holder in relation to the grip of the bow. What is the material of the pellets?

Cross-bows have the stave fixed transversely to a *stock*. The stave may be plain, compound, or composite, as above. Note the material and form of stave and stock, and also particularly the devices whereby the bow is bent and the string held and released and the kind of missile.

Bowstrings may usually be described in terms of String (p. 286). Note the materials and method of preparation; how the string is fastened to the bow; by a hole or eyelet; or resting in notches cut in the horns; or on stops cut from the solid or affixed, etc. Is the end of the string looped? Is it attached directly to the bow, or indirectly, e.g. by being spliced to some other material? Is it protected by a binding at the point where the arrow rests? Note the special arrangement for holding pellets in pellet-bows. Describe all knots and loops, and especially the process of stringing the bow.

Grasp and release should be noted fully, and the observer himself should learn the whole use of each kind of bow. As to the *left hand*, note how the bow is held—perpendicular, horizontal, etc.; the position of the fingers in shooting; on which side the arrow rests; the distance of the arrow-head from the bow when fully drawn. As to the *right hand*, note carefully the exact position of the fingers in drawing the bow.

In the so-called *primitive release* the bow is drawn by the arrow being held between the thumb and first finger, which do not press on the string; in the *secondary release*, the arrow is held between the thumb and second joint of the first finger, the tip of that and any other finger resting on the string; in the *Mediterranean release*, three fingers are placed across the string, the arrow is held lightly between the first and second, the thumb not being used; in the *Oriental release*, the thumb is laid across the string, with the first finger across the thumbnail.

Accessories. Where the Oriental release is used, note whether any ring or other object is worn on the thumb to protect it and to assist in drawing the bow; describe the ring and how it is used. Sometimes a bow is drawn with the assistance of the feet, or of a ring-handled dagger, or other appliance; in all cases note at what point on the string the arrow rests. *Gloves* are sometimes worn on the right hand, to prevent the fingers from being chafed by the bowstring; any other form of protection which has the same object may be described under this heading. *Bracers* are shields or bracelets worn on the left arm or wrist for protection.

(*f*) *Firearms* are just as much a part of the equipment of the people who use them as any older or simpler kind of weapon. Firearms not of familiar European makes should be examined: note whether the people either make or can repair firearms; whether they make their own bullets, or shot, or gunpowder. Where do they obtain the materials? Describe the processes and obtain samples, if possible. What kinds of firearms are used—pistols for one hand only; guns requiring both hands; cannon, on portable carriages or stationary mountings? Are the barrels rifled? Is anything done, or attempted, to diminish the noise of the explosion? What kinds of lock-mechanism are used—matchlock, flintlock, percussion-lock? Note the outline of the stock, grip, butt, and trigger-guard, form of the device for sighting, the posture and gestures of loading and shooting, the range and accuracy of fire. Are there hand-grenades, bombs, or other explosive missiles or projectiles? Is a rest used to support the longer guns? Note, and, if possible, collect, accessories, such as powder-horns, tinder-boxes, measured charges, cartridge-cases. Is there any kind of bayonet or other muzzle-spear?

(*g*) *Capturing Weapons.* The *lasso* and *bolas* are missiles intended to entangle rather than to injure. The *lasso* is a running noose thrown so as to encircle the quarry. In some instances the noose is attached to the end of a long pole. Describe the material of the line; its preparation and manufacture; all knots, splices, and loops; how it is attached to the thrower; the method of casting. At what quarry is the lasso thrown? The *bolas* consist of one or more weighted cords. When thrown, the cords become wound round the quarry and so prevent escape. Describe as above, and add particulars of the weights, their number, their attachment, and the action of throwing. Are feathers attached to the opposite end to act as a drag? Against what quarry are the bolas used? Other capturing weapons may be in the form of a cane loop at the end of a rod, or of a forked spear with hinged barbs or of nets (*v.* Fishing, p. 256).

II. *Self-acting Weapons*

(1) *Stakes* and *spikes* are pointed instruments fixed in the ground with the object of wounding an advancing foe or their horses; *calthrops* are objects with a number of spikes or points so arranged that, when thrown upon the ground, one or more is always upwards. Describe the material; form and construction; method of use; where they are usually set; are they used with poison?

(2) *Traps.* The trap mostly used in war is the pitfall, with sharpened stakes at the bottom; others correspond with one or other of the patterns mentioned (*v.* Hunting, p. 252).

III. *Weapons of Defence*

Weapons of defence are usually designed for the protection of the individual; but occasionally they offer cover for more than one person. Most defensive equipment has developed out of the ornaments or clothing of the individual; but a distinction may be drawn between those articles of *body-armour* which are assumed on entering a fight with the express intention of protecting the body, and those which are worn habitually and yet are a protection. Note also separately those ornaments which are used with a view to dazzling the enemy and disconcerting his aim, and accoutrements, such as masks, calculated to produce a terrifying effect, but sometimes combined with protection. Among the defensive ornaments may be mentioned: forms of hair-dressing, neck ornaments, armlets, and belts which protect either by their number or size. All these should be described in terms of Ornament and Clothing (p. 234).

The principal categories of *body-armour* are helmets or head-armour, with or without face protection; cuirasses or body-armour; brassards and gauntlets, to protect arms and hands; greaves and other leg-armour; special protection for the feet; panoplies or complete suits of armour.

Parrying weapons are to protect the user by deflecting the offensive weapons of the foe. Considerable variety is found, such as parrying-sticks or clubs held in the hand; parrying-shields; shield-like weapons, the primary function of which is to deflect an offensive weapon rather than to afford shelter from it; pieces of wood bound to the forearm. Parrying weapons often resemble offensive weapons, and should then be described in terms of the weapon they most resemble. In the same category may be included weapons which are employed to render the enemy's weapon useless.

Shields are intended to afford portable shelter either to the whole body or part of it. Observe the material or materials; shape and size, outline and section at different points (with diagrams); construction, solid or composite; the method of manufacture, especially in the case of composite shields; the method of holding; the nature of the handle (loop or bar); its position, horizontal, perpendicular, etc.; in which hand is it held? How is it manipulated in action? Note any design or ornament on the face and back, and native accounts of their meaning.

For the defence of groups of individuals, there are: (*a*) *Movable screens* and large shields, which should be described, with diagrams, and notes of the method of transport, and of the number of men which each screen can protect. (*b*) *Fixed defences*, including earthworks, entanglements, moats, breastwork and palisades, forts and the like, and pile-dwellings or tree-huts occasionally. The materials and method of construction of any of these should be described fully; plans and sections and elevations should be given, and some idea of the localities where they are erected, and their access to water; buildings, such as forts, may be described in terms of Habitations and Earthworks (pp. 237, 204).

Receptacles for Food, Drink, Etc.

Gourds, Coconuts, etc. The use of natural objects, with little or no modification in form, as receptacles for food, drink, and other substances, is widespread amongst backward peoples. Mention may be made of the rinds of gourds, the shells of coconuts, and the like; lengths of bamboo stem may be used, and some bark-vessels are made with very little constructive effort. Bottles, bowls, beakers, ladles, and other forms may be noted and in all cases the provision of handles, necks, spouts, basal supports, slings, and mounts should be recorded. Decoration by carving, engraving, burning, painting, inlaying may be present, and the designs as well as the technique are important. Observe what plants are made use of, and whether they are cultivated or not. Gourds intended for use are often given an artificial shape by bindings applied during growth, and the methods employed deserve study. In all cases it is important to note how the fruit or other product is obtained and prepared for use, and what tools are used.

Wood. Similar considerations apply to the use of wood for making into receptacles, and here the processes of shaping are necessarily more arduous, calling for the application of better tools. Where two or more pieces of wood are fastened together to form a dish, bucket or beaker, close attention should be given to the fitting together of the parts, and their attachment by means of pegs or stitches.

Shells, Horns, Skin, etc. Animal products, mollusc shells, horns of sheep and oxen, egg-shells of the ostrich, need little or no modification, though horn spoons and ladles call for carving out and, perhaps, for shaping by means of heat. Skin-vessels may be made from practically the whole skin of an animal, and here the details of skinning, stopping up apertures in the skin, and preserving or tanning, are important (*v.* Skin-dressing, p. 285). Skin-vessels may be made from the intestine or bladder of an animal, and even fish-skin may be used. Bags and cases of tanned hide or of leather are sometimes met with.

Stone. Receptacles of stone are not common, though stone mortars are widely distributed. Nature and origin of the stone, how quarried and how shaped, together with the forms and decoration are the important points for record.

In the case of all receptacles care should be taken to distinguish between those which are used for food or drink, or for special varieties of food or drink, and those used for other purposes. Some vessels may be used for cooking food over the fire, others only on ceremonial occasions, others only by special persons or classes, whilst yet others are mainly or entirely of decorative value.

Basketry

Basket-work is a convenient, though ill-defined term, including not only actual baskets, but also wattlework, matting (often indistinguishable from woven fabric), and ornamental plaitwork. Basket-work is linked on the one hand to netting and knitting, but differs from these in the absence of mesh or pins, and (except in

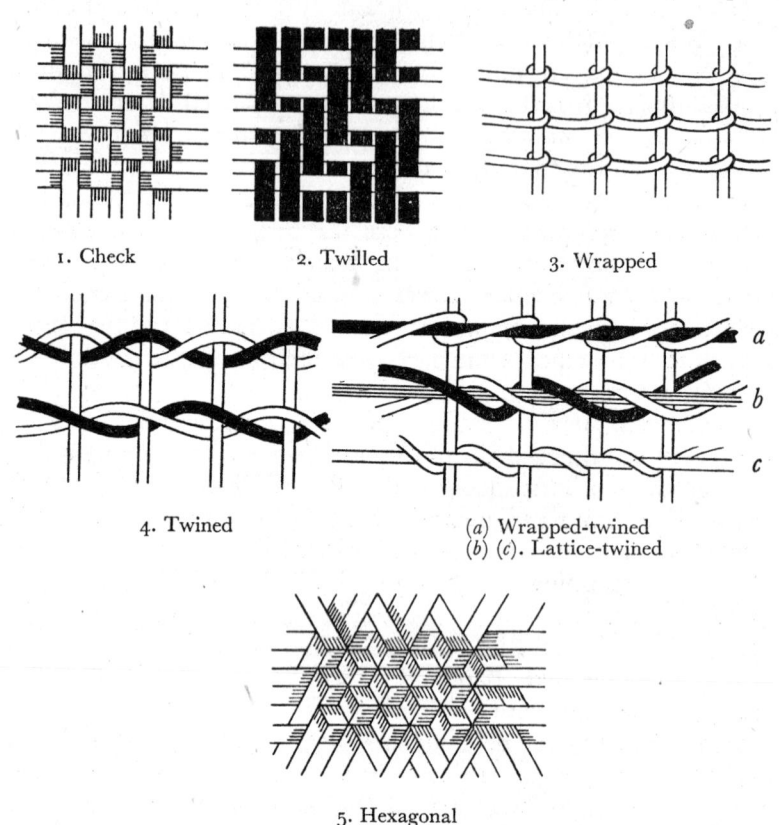

1. Check
2. Twilled
3. Wrapped

4. Twined
(*a*) Wrapped-twined
(*b*) (*c*). Lattice-twined

5. Hexagonal

single-element work), in the use of two or more sets of interlacing elements or wefts. It differs from weaving by the more general use of unspun material, and by the absence of a frame or loom. Basket-work may be (A) *Plaited* (also termed Woven), or (B) *Coiled*.

(A) *Plaited* basket-work is made by the crossing of two or more sets of elements, called, by analogy with weaving, warps and wefts, although, when the warps are indistinguishable by rigidity or direction, both sets of elements may be called wefts.

The main varieties of plaited basket-work are (1) *Check*, in which the warp and weft pass over and under each other singly, as in woven cloth. (2) *Twilled*, in which each weft passes over and then under two or more warps, producing by varying width and colour contrasts an endless variety of effects. (3) *Wrapped*, in which flexible wefts are wrapped round (take a circular bend right round) each warp in passing. (4) *Twined*, when two or more wefts pass alternately in front of and behind each of the warps, crossing them obliquely. Twining with two or three wefts is technically termed *Fitching* and *Waling* respectively. There are many varieties in twined work, plain-twined (just described) twilled-twined, when two warps are passed over each time; while warps may be upright, crossed or split. In wrapped-twined, "bird-cage" or lattice-work, the foundation consists of both horizontal and vertical elements, often rigid, at the crossings of which the weft or wefts may be twined, or wrapped. In finished specimens wrapping and twining are often indistinguishable on the outer surface, though usually distinguishable on the reverse side. (5) In *Hexagonal* work the wefts, instead of being horizontal and vertical, are worked in three directions, forming in open work hexagonal spaces, in close work six-pointed stars. In *Wickerwork* the warps or stakes are rigid, and the more flexible wefts or rods bend in and out. Check, twilled, and twined strokes may be used, but this ancient craft has its own peculiar vocabulary. In *Wattle-work* the stakes are planted in the ground.

(B) *Coiled* work is not linked with weaving, but with sewing, and it is usually done with a pointed implement which makes a hole through which the weft is passed. The technique consists in sewing together in a flat or ascending coil a spiral foundation of cane, grass, leaf strips, fibres or other materials, and the following are the commonest stitches:

(1) *Simple Oversewn Coil*. Each stitch passes over the new portion of the foundation coil, and pierces a portion of the coil below. (*a*) *Furcate coil*. If the new stitch splits the stitch in the preceding coil, a forked effect is produced, having a superficial suggestion of chain-stitch or crochet. The stitches usually lie closely side by side, covering the foundation, but with the same technique an entirely different effect can be produced as in (*b*) *Bee-skep coil*, when the stitches are spaced widely apart, connecting the coil at

intervals, each stitch passing just behind, and appearing to emerge from, the stitch in the coil below.

(2) *Figure of Eight.* The surface shows the same effect as simple oversewn coiling, but each stitch actually encloses two coils in a figure of eight. The stitch passes behind, up and over, and down in front of the new coil, then behind, down and out under the preceding coil. Also called "Navaho."

(3) *"Lazy Squaw."* The conspicuous feature of this is the long stitch passing over two coils at once. The sewing passes in front, up and over the new coil, winding right round it once, twice, or more times as desired; then it passes behind and down under the preceding coil, and right up over the new coil, making the characteristic long stitch.

(4) *Crossed Figure of Eight* or "knot stitch". The stitch passes in front, up and over the new coil, and behind, down and under the preceding coil, as in the long stitch of "lazy squaw", but the sewing is brought out between the two coils, to the right of the last long stitch, which it crosses, giving the appearance of a row of knots between the successive coils.

(5) *Cycloid* or single-element work may be grouped with coiled work, but there is no foundation, the coils, usually of cane or similarly independent material, being coiled or looped into each other. This is especially characteristic of the Malay area.

Matting forms a link between basket-work and weaving, and it is often impossible to tell from a finished specimen whether it has been plaited by hand or woven in a frame. If the mat is worked diagonally it is started at one corner and plaited like a basket in the same technique: check, twilled, etc. If it is worked horizontally, the warp elements may be arranged parallel to each other, and the weft elements inserted as the work proceeds. Both mats and baskets are also made from flat plaits the edges of which are attached side by side. When the warp for a mat is stretched on a frame or hung from a horizontal bar, the process is the same as that of weaving (p. 290), though the threading, as in primitive looms, is usually done by hand, not by a shuttle.

Specimens of every kind of basket and mat should be obtained; with materials, raw, and in various stages of preparation; and tools; photographs of the processes; half-finished specimens to illustrate technique; and all native names. Baskets which have contained food, etc., should be carefully cleaned. Notes should be made on the materials, whence obtained, whether wild or cul-

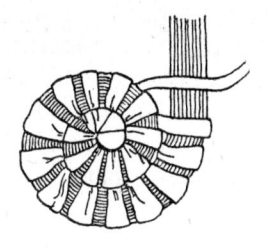

1. Simple Oversewn Coil

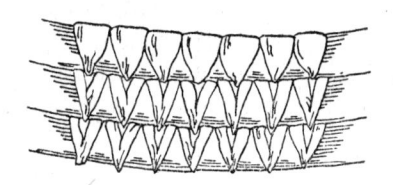

(*a*) Furcate

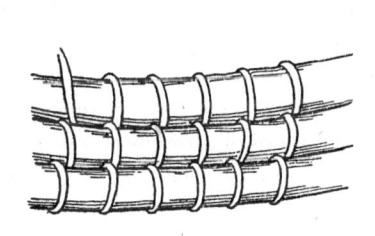

(*b*) Bee-skep

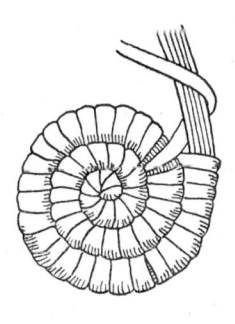

2. Figure of eight

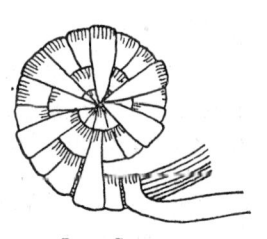

3. Lazy Squaw

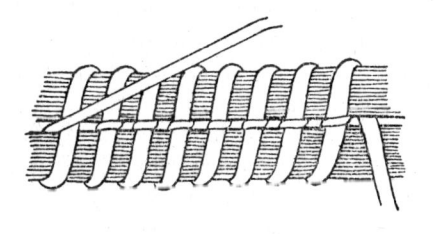

4. Crossed figure eight

5. Cycloid

tivated; the preparation of the materials by storing, drying, peeling, splitting, twisting, soaking, gauging, colouring, etc.; the processes and apparatus of manufacture; the sex and status of the makers; the designs and symbolism with their names and meanings, and uses of the finished work. Also any rites or observances connected either with manufacture or use of baskets or mats.

Pottery

Many primitive peoples make pottery. Some, though they cannot, or do not, make pots themselves, are acquainted with pottery, and import pots ready-made. Others export pots of their own making. In all cases of importation or exportation note the source of supply (or the destination), the mode of transport and traffic, and the names of each kind of vessel. As in the case of other arts and crafts, there are often traditions or legends as to the origin of pottery, the first potter, decoration of pottery; and beliefs regarding the use or abuse of pottery. Instances of absence of knowledge of pottery-making should be noted and the reasons ascertained.

Among some peoples, each family, or a member of it, makes its own pottery; in others, pot-making is reserved to particular families, or to a caste, class or sex.

The kind of clay, and its source, should be noted, and samples obtained both of the raw clay and of any substances, whether mineral or organic, which are mixed with clay or substituted for it in any kind of pot-making; of each variety of prepared clay when ready for the potter, and of all slips, paints, varnishes, glazes, or other substances which are applied to the surface to close the pores of the clay, or by way of ornament. Note the reasons given for each process and for the use of each ingredient and—though such effects do not necessarily tally with the reasons given—its observed effects.

The process of making the pots should be described in detail. The principal methods of pot-making are (*a*) *lining or coating a mould* with clay; the mould may be made specially, or it may be some other object, such as a basket or gourd. Some primitive potters begin all their pottery in a basal support of this kind; others, though this is rare, keep the clay in a perishable mould until this is burned away in the firing; (*b*) *modelling* a single lump of clay by hand, often with the aid of simple tools, for part or the whole of a vessel; (*c*) *building-up* the pot in one or more of a variety of ways, the most widely distributed of which is the *coiling* method, which involves the use of long or short rolls or pencils of clay, which are ranged upon each other, and pressed into union, in such a way as to shape the pot from the base upwards, all traces of the coiling being usually smoothed out before firing; other building methods involve the use of thick rings of clay, or slabs bent round to the form of a cylinder, or of various combinations of the moulding and modelling methods; (*d*) *throwing* a lump of clay on the *potter's wheel*, which may range from a simple disc or spoked wheel (*tournette*), to a double wheel consisting of one disc above another, on a common axis; such wheels may be spun by the hand, by the foot, or by an assistant, and the more advanced types lead up to the wheel which is driven by transmitted power. Examples, or models, or at least drawings, should be obtained, of all such wheels; of all tools and gauges used in shaping the pots; and photographs of the potter in the act of using each of them. If possible, the different stages in the manufacture of a given type of vessel should be collected. They should be baked in their incomplete stage in order to preserve them from damage. Note in such cases whether the native potter allows any interval to elapse between successive processes, and for what reason. Note whether wheel-made and hand-made fabrics coexist in the same community; if so, who makes each of them, and what does each kind of potter think of the other's method?

Most pots after shaping are subject to smoothing, polishing, or burnishing; or at first covered with a *slip* of creamy fluid clay or other surface covering, poured, smeared, or painted upon them, either all over or partially. Sometimes the pots themselves are immersed in a bath of slip. The materials, tools, and effects of these subsequent processes should be noted, and also the reasons assigned for them.

The *decoration* of all kinds of pottery should be noted, and especially the methods by which it is produced. It may be applied (either before or after firing, as the case may be): (*a*) by polishing or burnishing; (*b*) by smoking after firing, and burnishing; (*c*) by varnishing with resin or other vegetable product whilst the vessel is still hot; (*d*) by impressing with the finger, with a cord, or other material, or with a prepared stamp, by incising (note whether this is done before or after firing), scoring, carving, sometimes with the addition of white or coloured substances to fill the depressions; (*e*) by attaching surface ornament in clay before firing, such as knobs, bands, animal or human figures, etc.; (*f*) by applying foreign materials, inlaid or incrusted; (*g*) by applying white or coloured slip, or a wash of haematite or ochre, or a coating of graphite; (*h*) by applying a true mineral glaze or an opaque enamel, coloured or not; (*i*) by painting the vessel with panels or designs in slip, or with paints of vegetable or mineral origin. It may be noted that the spiral structure of coiled pottery is occasionally preserved as an ornamental feature.

In all cases the materials and tools should be described and collected, and the native names and significance of the designs should be ascertained. Designs are sometimes employed as trade-marks. A chemical analysis of clays, slips, glazes, and paints, if it is obtainable, will form a very important addition to the record.

Firing of the pots is preceded by drying, in the open air, in the sun, or under cover. The firing may be effected in an ordinary open fire, or in an open fire designed specially for pot-firing; or in a hole in the ground; or in some form of *kiln* or *pot-oven*. Photographs and drawings, and, if necessary, measurements, should be made of such kilns. Some potters are careful to saturate their pottery with smoke-stains, or to imitate smoke-stains in various ways; others exclude the smoke, or avoid it by keeping a clear fire.

Record the kind of fuel used, and any method of increasing or diminishing the quantity of air that has access to the fire or the pots; the final colour of the ware may depend in part on the extent to which the iron in the clay, or in the external coating of slip or wash, is left in a reduced (black) or oxidized (red) condition, respectively. The several methods by which black pottery is produced are especially worthy of study.

Every shape of pot should be recorded, with its name, purpose, customary dimensions, and special mode of manufacture. Special note should be taken of all forms of pottery which imitate, or are intended to imitate, any other kind of vessel, such as a shell, a gourd, bamboo, or vessel of other materials, such as skin or wood, also conversely of any vessels in other materials which seem to imitate pottery; and all substitutes for pottery, among people who use few pots or none. Note if the supposed model exists in the locality. Special varieties of pottery, such as porous water-coolers, puzzle-jugs, and purely ceremonial or ornamental pottery should be noted; and also all other objects besides actual pots, which are made of any kind of clay, such as figures of men and animals, musical instruments and toys, with their composition, mode of manufacture, and other particulars.

All occasions on which pottery is used or broken ceremonially, and the attached significance, should be recorded. Note all uses of broken pottery, as for scrapers, for games, or to make cement or pavement.

Glass

Glass is little used except in the form of *beads*, which may be either manufactured locally, or trade-beads brought from afar; both kinds may be used as currency. The materials, mode of manufacture, processes of blowing, modelling, or moulding, nature and uses of the glass objects, should be noted, and specimens, tools, and photographs obtained. Are objects of other materials covered or decorated with vitreous glaze or enamels? Note any materials used after the manner of glass for making windows or other translucent objects. Are lenses or mirrors employed, and for what purposes? Is the glass-making industry an ancient one in the region? Are arrow-heads or other implements made from fragments of imported glass bottles, etc.?

Stone, Wood and Metal Work

Stone

The processes employed in making stone implements, by those who still use them, and the precise functions which they serve, are worth special study, since they throw much light on the manufacture and function of stone implements in earlier ages. Note in detail how the raw material is obtained and shaped, whether by hammering, pecking, flaking, chipping, grinding, cutting, drilling; is fire used at any stage? Different kinds of stone or rock differ greatly in their reaction to blows or pressure, and many kinds cannot be shaped by striking off thin slices or flakes. Note whether anvils, punches, or pads are used. Pay attention to any method of preparing striking platforms before flaking and to any method of flaking by pressure. Describe any method and appliances employed for grinding and polishing stone implements, and note whether these are done in whole or in part. Distinguish between intentional polishing and that which may be produced incidentally during use in hoeing, sago-chipping, etc. Record any use made of rough chips of stone or waste flakes for implements. The manufacture of implements, such as axe-blades, of shell, should be studied in connection with stone-working.

Sketch or photograph all essential or characteristic attitudes and motions of the worker, and secure examples of all tools or accessories. This is of particular importance in the making of long thin flakes, and in the fine or ingenious flaking of masterpieces. Are natural forms used, or adapted? Are the different forms of implements well marked? Have they names? How are accidental deviations of type regarded? Is there much wastage in the manufacture? What kinds of stone are used? How are they obtained? Are they carefully selected, and by what signs of quality? Are there special stone-workers or does each individual make his own implements?

For what purposes are stone implements now used? Are substitutes available, and, if so, what reason is given for persisting with stone? How are they used? Simply held in the hand? Or hafted? If hafted, how are they secured to the handle? Obtain specimens, if possible, of every kind of hafting which is in use. How long does it take to make objects or cut down trees with stone implements? A simple test is to match two workmen against each other, one with a stone implement, the other with its nearest European counterpart. How long does a stone implement continue in use? Is it resharpened when dull; and how? Note the effects of wear, including abrasion by handles or lashings. Are there traditions of the former use of stone implements? Or ceremonials involving the use of stone where other materials are used in daily life? Do the people ever find stone implements in the ground, and if so, what account do they give of them? Are they regarded as having magical or protective qualities, or made use of medicinally? Are they termed "thunderbolts" or "thunderstones" and used as a protection against being struck by lightning? Ancient stone implements should be described and drawn, or collected, and if they are found below the surface of the ground careful record should be made of the nature of the soil, of the levels, and of animal remains, or artefacts that are associated with the stone implements (*v.* Field Antiquities, p. 345).

The Working of Wood, Bone, Ivory, etc.

Wood. In spite of the general similarity of the results, the kinds of wood and the methods of woodworking appropriate to them vary greatly, and deserve careful study. Note the different timbers which are in use, and their respective purposes. If possible obtain a botanical description of the living tree, with photographs or drawings, and specimens of leaves, flowers and fruit. How is the tree cut down, dressed, transported, seasoned, subdivided into logs and planks? What tools are used in woodworking? Is either *fire* or any hot object used to fashion wooden objects or to hollow them? Describe the process. Is the form of a manufactured wooden article determined, in part at any rate, by the direction of the grain, or is it entirely independent of it? Is the warping of wood practised? How is it effected?

How is woodwork joined together? By lashings? Splices? Sockets and morticed joints? Wooden pegs? Metal nails? Rivets or screws? Or by adhesives, gum or glue?

Note all methods of decorating woodwork, by carving, engraving, inlaying, veneering, polishing, staining, branding, painting, varnishing. Is anything done to preserve woodwork from damp, heat, or the ravages of animals or insects?

Stone, Bone, Ivory, etc. Few peoples are so copiously provided with metals, or other hard substances of artificial origin, as to neglect the primitive craft of fashioning stone, bone, ivory, and shell. Even where substitutes are at hand, and are cheaper, the rarer, more durable, or more beautiful materials are preferred by those who can afford them; rock-crystal, ivory, and mother-of-pearl are examples of this among ourselves.

The processes of manufacture are few and simple, but they vary in detail. Note how hard substances are subdivided, smoothed, made spherical or cylindrical, bored (with solid or tubular drill), hollowed, joined, engraved, polished. What tools and what accessories (such as sand, emery, leather) are employed. What materials in this class are in use? For what purposes? Where and how are they obtained? Is the raw material valued, or only the finished product, such as beads?

Metal-working

The following points require no special scientific training in the observer. Specimens, a few ounces in weight, should be obtained of all ores and fuels, of the finished product, and, if there is more than one stage in the process of smelting or refining, of the material which results at each stage; and also of all slags, dross, and other impurities which are removed, since these too may give useful information to an expert metallurgist.

All specimens should be labelled. They will travel safely in a wrapping of paper or dry grass.

What metals are in use? What are their native names? Are they produced in the country or obtained from elsewhere? To what uses are the metals put? Are any metal articles—whether plain bars, or ornaments, or useful objects or imitations of these—made for the purposes of barter, or used as currency? From what ores are the metals obtained? Are the ores found on the surface, in the beds of streams, or dug out of the ground? (*v.* Mining, p. 283). What is the native name of each ore? Is there a general word for all ores?

If ores are *smelted*, is the ore submitted to any preparatory treatment? What kind of fuel is used? Is any material mixed with the ore and fuel for smelting? Note the shape and size of the furnace or fireplace used in smelting, and make sketches. Describe the process of smelting, noting the quantities of the materials put into the furnace, and of the metals and products obtained, and the time required for the operation. What kind of bellows is used? Make sketches of the bellows and diagrams of the working of all valves. Also of all tools used in smelting. Is the product of the smelting process a metal ready for use, or is it subjected to any subsequent treatment before use? Describe all processes of purifying or refining.

Is the art of *casting* practised? Describe the furnace used for the operation? Are crucibles used; if so, describe them, and also the moulds into which the metal is poured. Are the moulds made of sand, clay, stone or metal? Are wax cores employed, to be replaced by the molten metal (*cire perdue* process). Describe in detail the method of construction of the wax model of the mould. Occasionally an easily burnt material is used for the core. Are the castings subjected to hammering or chiselling, or to both?

Forging. Is any process of hardening practised in making cutting implements, or weapons, or sonorous instruments? In forging, is *welding* known? What kinds of forge, bellows and tools are used by smiths? Is soldering practised? Is wire of any metal in use? Is it made by the people themselves? If so, is it made by simple hammering, or by drawing the metal through apertures?

The following questions refer to particular metals.

Gold. Do the people wash the sands of rivers or of the beach to obtain gold? Is it obtained in the form of "dust" (coarse or flaky powder), or of "nuggets" (small irregular pieces)? Are articles made by melting the gold dust? Or by hammering the nuggets? Are there any beliefs as to the origin of gold dust?

Silver. Is silver found native, in metallic form? Or is it obtained by smelting ores? Are lead ores smelted to obtain silver? If so, describe the process used for separating the silver from the lead.

Copper. Is copper found native, or is it always obtained by smelting ores? Is the ore burnt in heaps or treated in any way before being smelted? Is native copper formed into implements by mere cold-hammering?

Iron. Is iron found native? Some meteorites consist of iron, and such masses of iron are occasionally discovered and used. If so, what account do the people give of these iron masses? Is cast-iron made by the natives? Do they make any distinction between wrought-iron, cast-iron, and steel? Describe the process by which each is prepared.

Bronze includes all alloys of copper and tin. If it is made by melting the two metals together, what are the proportions? Are any other ores or other substances added? What is the source of the tin?

Brass includes all alloys of copper and zinc. Is it made in the country or imported? Is it made by heating metallic copper with zinc ores?

Are copper ores and lead ores smelted together, and is any use made of the alloy of copper and lead so produced?

What other metal alloys are made?

Describe any methods practised for mending broken metal articles. Describe the processes and tools used for ornamenting metal utensils, weapons, personal ornaments, etc.

Descriptions of furnaces, tools, etc., should be accompanied by sketches and dimensions.

Discover, if possible, the relative values of metals which are known to any given community, and the uses to which they are put.

Mining and Quarrying

In this section are intended to be included all operations by which mineral substances are won from the earth for human uses.

They range from simple collection of stones and fragments of ore from the surface, or from the beds of streams, by *hand-picking, sifting* or *washing*, to *quarrying* or open mining of exposed surfaces of solid rock, and *mining* proper, in which a bed or vein of the mineral is excavated underground by means of *galleries, stalls* and *workings,* approached by vertical *shafts* or horizontal *drifts.* Diagrams and plans should be made, and, if possible, supplemented by examples of any diagrams which are in use among the miners themselves. Obtain samples of the minerals, and note the uses to which they (or their ultimate products) are put.

Note all implements, receptacles for the mineral, lifting tackle, provision for light, ventilation, draining (including pumping or baling); employment of women, children, or animals in or about the works. Is the mineral prepared for use at the quarry or mine? Or is it sent away in the rough state? (For the preparation of metals from ores, *v.* Metal-working, p. 282, and for the removal of minerals, *v.* Transport, p. 299).

Salt

Salt is obtained either in solid masses of *rock-salt* by quarrying or mining; or by boiling down the *brine* from salt springs; or by enclosing sea-water in natural or artificial *salt-pans,* and collecting the deposit of salt which remains, after evaporation; or from the ashes of certain plants. Full details of each process should be recorded, with notes of all raw materials (including the names of plants if these are used), appliances, ceremonies, restrictions, and other customs connected with it. Is vegetable salt preferred to mineral salt? How is salt owned, stored, sold and distributed? Is it made up into cakes or blocks of standard size or weight? For salt as a medium of exchange *v.* Trade, Exchange and Currency, p. 170.

Are any other soluble minerals obtained in the same ways as salt? Examples are soda, potash, nitre, saltpetre, alum, borax. For what purposes are they used? Are they exported? Describe the source of supply, the mode of extraction and purification, with all implements and apparatus. Obtain samples of the crude and finished products.

Skins and Fabrics

The preparation of the skins of animals, so as to render them fit for clothing and for other purposes is an important industry.

Preparation of Skins and Leather

Skins with the hair on are frequently merely dried, the inner part being dressed with some antiseptic preparation, and sometimes curried or shaved. *Hide* is a convenient term for those raw or dried skins which are used for their toughness or pliability. *Leather*, properly so called, is either *tanned* with bark, like shoe-leather, or *tawed* with alum, etc., like kid-leather for gloves; or *dressed* with oil, like chamois, or wash-leather. For each kind the skins pass through several processes, one of the principal of these being usually the steeping of the hides in lime-water, so as to loosen the hair and prepare the substance of the skin for the final dressings.

Record the names and sex of all animals the skins of which are prepared for use; is any fish-skin used? How are the skins removed from the animals, and what instruments are employed for the purpose? What skins have the hair left on them? And how are they prepared or dressed—is the hairy side dressed or treated in any particular manner? If the hair is removed, how is this effected? And how are the hides tanned or prepared? What are the substances used for dressing them? And how are they administered? Obtain samples for analysis, if possible. Are the skins pegged out or otherwise fixed and tautened during the process? Are the inner sides of the skins scraped or curried? And if so, with what kinds of instruments? Is any beating process or manipulation employed so as to render the leather supple? Is the leather dyed, or its surface in any way ornamented by peeling, incising, dyeing or varnishing? In *Morocco-leather*, the grain is produced by crumpling.

Uses of Skins and Leather. The uses to which skins and leather may be applied should be enumerated. Are any skins, or bladder-like internal organs, used for holding liquids, like wine-skins? If so, what are they? And how prepared? Are raw hides used, either whole, or in pieces, or strips? And how applied? For what purposes are the different kinds of leather chiefly used? And are they dressed in any way for the sake of preserving them while in use? Are any portions of the human skin, such as scalps, prepared in any way by drying or otherwise? If so, how, and under what circumstances?

Bark Cloth, etc.

Felt, Bark-cloth, Bast, and other materials used instead of

textiles should be noted and collected, together with the instruments used in preparing and decorating them, and full notes of the processes employed. Add specimens intercepted at each stage, from the raw bark and the foliage of the tree, for example, to the finished garment.

Strings, Nets, Knots, Needlework, etc.

Strings are made of strips of materials, animal (skin, hair, wool, tendons, or insect product), or vegetable (stems, leaves, roots, bark, fibres), which are prepared in various ways (split, shredded, soaked, chewed, dried, spun, twisted, or plaited). Unworked strands and thongs are mere strips of the material, and, if a longer piece is required, these strips are knotted, spliced, or otherwise fastened end to end. Worked strings are made of more than one strip of the material, *twisted, plaited,* or *braided* together to form a cord or plait of two or more "ply". The twist may be simply made between the fingers, or by rolling on the thigh, with or without a spindle, spinning-wheel, or other mechanical aid. *Sennit*, or *sinnet*, is a braided cordage, formed by rope-yarns or spun yarn, of which three or more threads are plaited in various forms; common sennit is flat, but there are round and square forms. In Oceania the term sennit is applied to a braid of coconut fibre, which is used for innumerable purposes (twisted coconut fibre is string and not sennit). *Cord* and *rope* are convenient words for thick strings, and *thread* for slender ones. *Twine* should be used only for two-ply twisted threads. Ascertain the purposes for which the various kinds of string are employed. Inquire what substances are employed for the manufacture of the various kinds of string, and if they are subject to any preparation before or after manufacture. Are any animals or insects kept for the purpose of supplying materials to be spun? Are any plants cultivated for the sake of their fibre? If dyes are employed, how are they obtained and applied? Describe all methods of joining ends. Describe the methods of making worked strings. Is string made by women or men, or by both? How long does each process take to produce a given length?

Knots. Collect examples of all the different kinds of knots, with the native names for each, and the purpose for which it is used.

Sewing and Embroidery. How are articles sewn? Are needles, awls, or tweezers employed to pass the thread through the material? Note the direction in which they are used, to or from the body? Obtain as many varieties as possible of every kind of stitch, with their names. Embroidery is the decoration of surfaces with needlework. Note on what materials it is executed, with what kinds of fibre or thread, with silk, gut, porcupine-quills. *Appliqué* is one material, plain or ornamented, sewn on to another.

Netting, Knitting, and Lace. Describe the process of netting, whether by the fingers or with a netting-needle, and make drawings of the meshes if specimens cannot be obtained. How is uniformity of the mesh obtained? Knitting includes all varieties of network in which the meshes are drawn so tight that a coherent fabric results. It is made either with movable needles (*knitting* proper), or hooks (*crochet*), or fixed pins which are afterwards withdrawn. When network is coloured, note how the colouring is done, and how patterns in different colours are effected.

Other Uses of String. Are strings used in any way as measures of length? Are knotted strings used as aids to the memory? Are any special kinds of string, or strings, made of special materials, used magically or ceremonially? (*v.* String Figures, p. 335).

Spinning and Weaving

Spinning

Although spinning may play a part in string-making, it has greater importance in the preparation of yarn for weaving. The *spindle* is an ancient appliance which has survived until modern times, and there is no great variation in its form and construction. It may consist of a single piece of wood, thickened for part of its length ("spindle-shaped"), and it usually has a nick or notch or hook at one end. More usual is the type in which a cylindrical shaft is fitted with a perforated flywheel, or *spindle-whorl*, of wood, bone, stone, pottery, etc., which may bear some form of decoration. Spinning is usually, but by no means invariably, done by women, and the work may be carried on whilst the operator is standing, walking, or sitting. As already mentioned under String, the spindle may be rotated by rolling it along the thigh, but it may also be held suspended, either free in the air, or with its lower end resting on the ground, or in a shallow bowl or cup. Occasionally (as amongst some South American tribes) a support is provided for the spindle, of such a nature that it may be said to revolve in bearings. The fibre that is being spun is often supported on the end of a stick or staff, which is called a *distaff*, but it cannot be too carefully remembered that this plays no part in the actual twisting of the fibre. The observer should record all details of the material, form, and construction of the spindle and its whorl (if present), the sex of the spinner and the conditions under which the work is done, the exact method of operation, the material (wool, cotton, etc.) that is spun, and how it is prepared, the way in which the spun yarn is wound on and off the spindle, the time taken to spin a given length of yarn, and the uses to which it is put.

Spinning-wheels have a restricted distribution, occurring only in the Old World. In all but the European "Saxon wheel" (which is continuous in its action, spinning and winding the yarn at the same time), the spinning-wheel is so constructed that a short period of twisting the fibre is followed by one in which the length of spun yarn is wound on the spindle. In this intermittent action it agrees with the spindle, and indeed the Asiatic and the older European spinning-wheels are essentially spindles fitted horizontally into bearings, and rotated by means of a cord passing round both wheel and spindle. The wheel in the simpler forms is revolved by the hand alone, but in some types a treadle is fitted. The mechanism of the Saxon wheel, in which the place of the spindle has been taken by a structure which bears no resemblance to a spindle, cannot be described here.

In the description of the spinning-wheel, it is necessary to note the construction of the wheel itself, the form of the spindle, the nature of the spindle-holder and its bearings, the manner in which the cord connects the wheel with the spindle, the construction of the support or frame of the whole structure, and the method of working.

The observer should record any special relationships or observances of a social or religious character that are associated with the use of the spindle or the spinning-wheel.

Weaving

In weaving, two series of pliable elements are interlaced at right angles to form a fabric more or less dense according to the materials and method. The elements are usually of spun thread or *yarn*, sometimes of thin strips of unspun fibre; strips of leather may also be used. As the process is capable of great elaboration and complication, and the mechanical appliances by which it

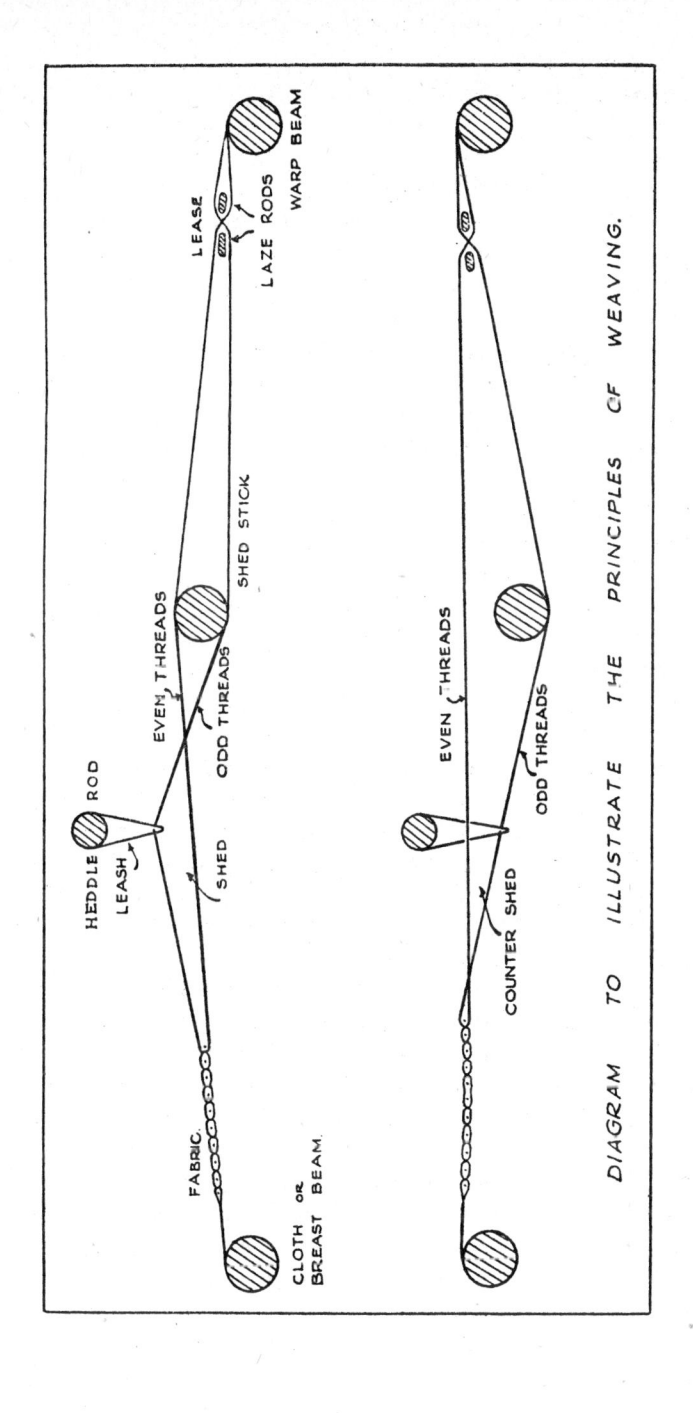

DIAGRAM TO ILLUSTRATE THE PRINCIPLES OF WEAVING.

can be furthered are numerous, a short introduction follows with the explanation of a few technical terms.

(1) *Looms and their Parts*

True cloth is usually made of material which is so soft and pliant that one series of elements has to be kept extended by artificial means, while the other series is interwoven with it. The series thus extended, which remains relatively passive during the operation, is called the *warp*; the active series, which is intertwined or *woven* into it, is called the *weft* or *woof*.

The *warp* may be attached at one end to the person of the operator, and at the other to some firm object, such as a tree or post; or it may be extended on a frame or *loom*.

The *weft*, which may consist either of a number of separate elements (as in matting), or of a single continuous element, is then guided between the threads of the warp either simply by the fingers, or by a *shuttle*, to which it is attached or on which it is wound, and each successive weft-element is called a *pick*.

If the fabric is to be of any length, a *frame* of some kind is almost indispensable. The simplest frame consists of a single *beam*, supported horizontally by its ends, at a convenient height. From this beam the threads of the warp hang down. For short pieces of cloth, the warp may be set up with a single thread, passed in long loops over and over the beam; whence these hanging loops are kept stretched by a second beam which passes through them all and is either supported by them or held apart from the first beam by side-pieces. Sometimes a special frame of parallel bars is used for setting up the warp, which is afterwards transferred from them to the actual loom.

For *continuous weaving* of longer pieces of cloth, each thread of the warp is fastened separately to the beam. The free end of each thread is coiled on a reel or *bobbin*, the weight of which, sometimes aided by a separate *loom-weight*, keeps the thread extended and vertical, with the bobbins or weights hanging clear of the floor. To prevent the warp threads from becoming entangled, one or more *lease-rods* or *laze-rods* are interlaced over and under alternate threads, near the beam. These are especially necessary when the loom is a portable one, and when, as sometimes happens, the work has to be rolled up and put away several times before it is finished. In vertical looms of this type, weaving begins at the top and proceeds downwards till it approaches the bobbins. Then the completed part is rolled up on the beam, which can revolve on its supports like a roller, and may now be described as the *cloth-beam*, and a fresh length of warp is released from the bobbins, till they hang once more just clear of the floor.

In more elaborate looms, the bobbins are replaced by a second roller, called the *yarn-beam*, on which all the warp-threads are wound and unwound as required. In looms with two beams or rollers, weaving often proceeds from the bottom upwards. But the loom need not be vertical, and when the warp is once set in a horizontal plane, space is saved by bringing the rollers nearer together and exposing a shorter length of warp. The operator now works forward from the cloth-beam, and rolls up the finished cloth upon it at frequent intervals.

In the absence of any special contrivance, each alternate warp-strand must be raised (or brought forward, if the warp is vertical) with the fingers or by aid of a needle or a shuttle, for the passage of the weft; but the process is greatly facilitated by appliances which raise or depress a number of warp-threads at the same time, so that the shuttle may be *shot*, or passed rapidly with a single motion, between these threads and the rest. The instrument which serves to raise or *float* a set of warp-threads is called a *heddle*, or *heald*; the opening so produced is called the *shed*; the operation of producing the shed is *shedding* the warp; the process of passing the shuttle through the shed is *shooting* the weft. Heddles may be of two kinds; the simpler form, called a *bar-heddle*, or *rod-heald*, consists of a number of loops usually fastened to a rod, each loop encircling one warp-thread, so that when the rod is raised these warp-threads rise with it and a *shed* is formed.

Another type, called a *frame-heddle*, may consist of two rigid bars connected by a series of loops, each of which has midway between the bars an *eye* for the passage of a warp-thread; the flexible loops may be replaced by narrow bars each pierced at the centre with a hole which forms the eye. The warp-threads which the heddle is to raise and lower pass through the eyes; those which are to remain stationary pass through the spaces between the loops or bars. The mechanism for raising and lowering the heddles to produce *shed* and *counter-shed* for the passage of the weft is discussed below.

When one heddle (bar-heddle) only is used, as in a great number of primitive looms, there is invariably a special lease-rod or *shed-stick* of considerable diameter immediately behind the heddle below the free threads and above those which pass through the eyes of the heddle. This, when the loom is at rest, sheds the warp to produce a *counter-shed*; thus the heddle, when raised, changes the relative position of the two sets of threads and so sheds in the other direction. When the heddle is released, the tension produced by the *shed-stick* reasserts itself, and the warp is again in the original disposition. Each time the heddle is raised or released, one or more picks of the weft through the warp are shot. The term "tension-loom" is suggested for this primitive type.

To put less strain on the warp, two or more heddles (frame-heddles) are used instead of the single heddle and shed-stick, each raising a different set of warp-threads. For the same reason, mechanism is added whereby when one heddle is raised the other or others are depressed, giving a wider shed with less strain upon the warp. Such looms are said to have *reciprocating heddles*.

In weaving, each *pick* of the weft, after being shot through the *shed*, is "beaten up" by the shuttle itself if it is long enough, or more commonly by a flat sharp-edged bar, called a *sword*, or by a pronged instrument called a *comb*, or by an appliance called a *reed*, shaped like a solid frame-heddle without eyes, between the bars of which the warp passes. Sometimes this is fixed in a swinging *batten* to give weight to its blows. The *reed* further serves in place of *lease-rods* (described above) to keep the warp-threads apart. In some cases, this is its sole function, and it is not used to beat up the weft at all. The whole apparatus of which these appliances form part is called the *loom*. In working a loom furnished with a heddle or heddles there are three distinct motions. First the shed is formed by raising the heddle; second, the weft is shot through the shed by means of the shuttle; third, the weft is beaten up with the sword, comb, or reed.

(2) *Points to be Observed*

The following points should be noted with regard to the appliances used in weaving and their action. Diagrams should be given where possible throughout the whole of this section, and the observer should for himself practise each mode of weaving if possible.

Warp. Describe the nature, preparation, and mounting of the warp; is it in simple strands or spun? How is it mounted? Is it of limited length, as in matting and carpet-weaving, or is a reserve of thread rolled on a yarn beam, or coiled on bobbins? Is it arranged horizontally or perpendicularly? Is it attached to the person of the weaver, or extended on a frame, or by means of weights? Is the frame portable or fixed? Describe the materials and construction of all frames and their parts. Give material, shape, number, and position of lease-rods if present.

Weft. Describe its nature and preparation; is it a simple strand or strands, or spun? Is each pick formed of a separate strand or is the weft one continuous thread?

Shuttle. Describe the material, shape, and method of making the shuttle, with especial reference to the attachment of the weft. Obtain examples of shuttles charged with thread. Describe how it is manipulated. Is it shot with the hand or with any mechanism? Is the shuttle used to beat up the weft?

Heddles. How is the shed formed? By means of one or more heddles? State their material, number, construction, noting whether bar- or frame-heddles, and whether rigid or not. How are they connected with the warp; how raised or depressed (by hand, treadles, or other mechanism); have they a reciprocating action? Give diagrams of all parts and mechanism.

Sword, Comb, and Reed. Describe their material, shape, method of manufacture, and use. How is the warp passed through the reed? Is the reed suspended in any way; if so, how? Is the reed used to beat up the warp?

Textile Patterns

Tied Cloth, or *Twined Weaving.* In some primitive weaving, where the warp is stretched on a frame, and the weft is threaded in and out by hand, two or more weft-elements pass together across the warp, twining round each other and at the same time enclosing one or more warp-elements between them; these weft-elements are usually spaced apart from one another. The commonest form of textile weaving is *Plain* or *Chequer.* In this the first weft-element passes alternately over and under each warp-element; the second passing over those warp-elements under which the first passed, and under those over which the first passed, and so on. This pattern of textile can be woven either by the single-heddle tension-loom, or better by a loom with two heddles. Varieties of this pattern may be produced by making the weft pass alternately over and under an equal number of warp-threads. *Twill* is a more elaborate pattern and requires a greater number of heddles. There are two varieties, the *Diagonal* twill and *Sateen* twill. Each of these is divided into a number of sub-varieties according to the number of heddles used.

In all cases the native names of the pattern should be ascertained, and it should be discovered whether they have any traditional meaning, whether patterns have any individual history, and are considered to be private, family, clan or tribal property, whether patterns may be borrowed, inherited or sold.

Textile Design. With a large number of heddles innumerable patterns can be woven on a ground of plain weaving or twill by floating definite warp- or weft-elements out of their turn; such patterns, when warp and weft are both of the same colour, are known as *diaper*. When colour is introduced, as in *brocade*, the question becomes more complicated. It need only be mentioned here that there are two principal methods of producing colour patterns: (*a*) By dyeing the warp piecemeal so as to reserve a design upon it (*v.* Dyeing, p. 296), and then weaving a plain chequer cloth with weft of any one colour; (*b*) by introducing threads of various colours in warp or weft, or both, and employing heddles, so as to show the required colour on the surface of the cloth where the pattern requires it: sometimes this is done, not with the shuttle, but with a needle held in the hand.

Pile-weaving. A more elaborate form of ornament is produced by introducing into the texture of the fabric particular weft-threads which form upstanding loops; these loops, when cut, form a *pile*. The fabrics so produced include *pile-carpets*, *plush*, and *velvet*. If the pile is not uniform, but in pattern, it is a *velvet brocade*. Describe the method of inserting and cutting the pile, and any mechanism employed for either purpose. Sometimes the pile is inserted in the finished textile by hand with a needle, but this process approximates to embroidery.

Embroidery is the working of patterns with sewn thread on the finished textile. Describe the method of sewing, the thread, needle, and other instruments; the various stitches and accessory ornaments, if any (beads and sequins, plaques, and patches), and obtain examples, if possible, of both of them and of the finished work (*v.* String and Needlework, pp. 286, 287).

Collection of Textiles

Textiles are neither difficult to collect nor costly, especially as it is only necessary to preserve a fragment sufficiently large to show the commencement of the "repeat" of the design. A collection of samples which can be submitted to experts gives far less trouble to the traveller, unless he wishes to study the subject thoroughly (in which case he will find it necessary to have recourse to the regular literature dealing with the subject), than a tedious and minute description of the design. Each sample should, if possible, be accompanied by particulars explaining which is the warp and which is the weft; which is the face or front of the cloth and which the reverse. In the case of diaper patterns, are they caused by floating the warp, or the weft?

With regard to design: collect the names of all designs, with translations and any particulars which can be discovered relative to their meanings. Are any patterns called by different names by the two sexes? Ascertain the names of particular patterns from several individuals, to see if they differ.

Is the design worked from a pattern or from memory? Is it suggested to the mind of the weaver by any song sung during the operation? Are particular patterns woven or worn by either sex alone? Have any patterns magical or protective qualities? Are flaws in the pattern purposely introduced? Are particular patterns reserved for any particular rank, caste, or occupation? Are any makers' marks inserted into the design?

If it is desired to record a certain design, and it is found impossible to collect a sample or to describe it easily in words, the simplest method is to describe it constructively so that a European weaver can repeat it, as follows:

Take sufficient of the warp-threads to show one repeat of the pattern, and number them from left to right; name the heddles from front to back, A, B, C, etc.; state which warp-threads are raised by each heddle; and, finally, state the order in which the heddles are raised for one repeat of the pattern.

Observances connected with Weaving. Is weaving the task of the men or of the women, or both? If of one sex only, may the other sex approach or touch any of the apparatus? Is the craft inherited? How is it learnt? How long does it take to weave a given quantity of cloth?

Is weaving regarded as a sacred operation? Must the weaver observe any particular prohibitions before or during his or her task? Are there any charms or incantations which must be uttered or sung in order that the work may be successfully performed?

Dyeing and Painting

The use of colour for ornamental purposes is almost universal but the number of colours, their nature, and the purposes for which they are applied vary greatly. There are, broadly speaking, two distinct methods by which colour is applied: (*a*) *Dyeing* proper, when the colouring-matter is used in a state of solution, and penetrates the pores of the object to be dyed; and (*b*) *Painting*, when the pigment is mixed with some fluid medium or vehicle, and is applied wet by a brush or some equivalent.

Dyeing is usually practised in connection with animal and vegetable fibre and tissue, such as leather, thread, and cloth; but the same process applied to wood and bone is commonly called *Staining*. Painting is more commonly applied to wood and other hard surfaces. Is any method of "stopping out" or "reserving" a design, in the natural colour of the material, employed in dyeing, either by means of a temporary coating of wax (as in *batik* and so-called "lost-colour" pottery), which is subsequently removed by heat; or by covering parts of the warp or weft strands, or of the cloth-ties with protective binding before applying the dye (*tie-dyeing*)?

What kinds of wax or other materials are used, and how are they applied? Procure specimens both of the "reserving" materials, and of any instruments used in applying them; also samples of cloth or other substances so treated at various stages of the process.

Dyes. Note how dyes are prepared, and from what substances, and how they are applied. If possible, procure specimens of the raw materials, and of the plants or other substances which yield them, as well as objects dyed with them. What are the favourite colours? (For their arrangement and proportions, *v.* Art, p. 310.) What are the articles usually dyed—skins with the hair on, leather, twine, cloth, etc.? Is any mordant or solution used, either to prepare the object for receiving the dye, or for rendering it permanent after it has been applied?

Are any portions of the human body dyed, such as the teeth, nails, hair, or skin? With what dyes? For what purpose? (*v.* Personal Care and Decoration, p. 227.)

Painting. What are the principal pigments, and how are they prepared? How are they ground or precipitated? What medium, if any, is employed to fix them? Is any subsequent process of varnishing or lacquering employed, and, if so, how is the varnish made? What kinds of brushes are used? What are the objects usually painted? What colours are most in vogue, and are they transparent, like *dyes* and *stains*, or opaque, like *body-colours*? Is *gilding* practised? Or any allied process? In painting patterns, in what order are the colours applied? Is the outline first painted and the colours subsequently filled in, or is each mass of colour roughly sketched, and its precise outline developed last? Is a given colour always applied in patches of uniform density throughout, or is the practice of "shading off" observed? Note if natural pigments are ever mixed to produce tints not otherwise obtainable. Are any mechanical means, such as compasses or stencilling-plates, used, or any process of stamping or printing? If so, describe the processes and appliances in detail. Is paint employed on the human body or hair, and, if so, on what occasions, and in what manner is it applied? (*v.* Personal Appearance, p. 229). Note all beliefs and observances about colour, or its use, or about particular colours. Is any colour regarded as especially sacred, and reserved for sacred purposes? If so, what reason is given?

Travel and Transport—By Land

Path-finding. Some people are credited with an almost instinctive power of finding their way; statements of this kind should be noted, and also carefully tested by experiment. Note all natural objects which serve as *landmarks* and their names. Are marks made upon natural objects for guidance? Are they private or public marks? Is there any customary code of such signs? (*v.* Recording, p. 195.) Are there regular signposts, or marks of distance, like mile-posts? How are *distances* estimated?

Paths and Roads. Are there customary paths from place to place? If so, how are they established, maintained, repaired? Is there any right of way? If so, is it the right of all, or of limited classes of persons? May a customary path be closed? If so, by whom, or under what authority, and for what purpose? What signs are used to indicate the closure? Are single paths used, or are there several paths running in and out of each other? Are artificial roads made? With what materials, by whom, under what authority? Are there "corduroy" roads of transverse logs? Are there walls along the roads? If so, how are they made and maintained? Are there any regulations for their use? Is any rite practised when going along a road for the first time? Are paths or roads under the special care of a particular spirit or god? If so, investigate this.

Halting Places. Are there regular halting places between villages or settlements? How are they selected and maintained? Are they under the protection of some spirit or deity? Is there any indication of distance from one place to another? How are travellers accommodated in villages and towns? Are there inns or public-houses? If so, to whom do they belong, and how are they maintained? Who may use them? Do they provide food as well as shelter? (*v.* Rules of Hospitality, p. 100.)

Mode of Travel. In what order do people travel—in single file, two or more abreast, or irregularly? Is there a relative position of the sexes or of rank? Do people travel alone? If not, do they arrange regular companies for travel? Are there periodical *caravans* for long journeys? If so, describe their size, organization, leadership, daily routine, and provision for defence. (For tents and other portable shelters, *v.* Habitations, p. 237). What ceremonies or salutes are made, on a journey, by passengers on entering villages or houses? Is any permission to enter required by travellers? Are any passes or complimentary introductions given from one community to another? If so, are they private or public introductions? Who may give them? What religious observances are made during travel, on passing sacred or notable spots, or on entering a new territory?

Swamps, Fords, and Ferries. How are *swamps* passed? Is anything sunk to solidify and preserve the roadway? Are *fords* marked, or any measures taken to preserve and improve them? Is the natural line of fords from salient to salient banks understood? How are *ferries* worked? Are any rafts or boats kept for the purpose? If so, how are they maintained? What payment is made? Is there any understanding respecting them between neighbouring tribes? What happens if the raft is on the wrong side of the river?

Bridges. The following are the principal types of construction: (*a*) Single trees, or trees from opposite sides crossed and fastened in the middle; (*b*) bridges of piles and beams; (*c*) trestle bridges; (*d*) lattice bridges; (*e*) swing bridges; (*f*) rope bridges; (*g*) suspension bridges; (*h*) bridges of upright jambs and lintels of large stones; (*i*) sloping jambs resting against each at top; (*j*) cantilever bridges or "false arches" of horizontal slabs overlapping and converging, with a large slab at the apex; (*k*) bridges of "true arches", constructed of radiating stones or bricks (with a *key-stone* at the top; note the mode of construction, and the kind of scaffold or *centering*, on which the arch is built up); (*l*) boat bridges; (*m*) raft bridges; (*n*) flying boat bridges; (*o*) cradles travelling along wires or ropes.

Transport. How are goods transported and by whom? Note the weights which are transported, the size, shape, packing and handling of the bales. Describe the contrivances employed to assist the carrying of loads; head-pads, head-straps, shoulder-bands, sticks and the like, and whether they permanently affect the head or body in any manner. Are there regular porters or carriers? What beasts of burden are used? Note the greatest work which can be performed habitually on a journey, either by man or beast. On what food is this work done? Describe all pack-saddles for animals, cradles or knapsacks worn by porters, and other methods of carrying burdens. Are there litters or palanquins for passengers? If so, are they used by anyone who can afford their use, or for people of special rank? What is the rank and sex of the occupant? Are litters used by persons entitled to use them always, or only on special occasions? What are these occasions—seasonal migrations, for war, or for religious purposes? Describe all decorations of litters, ceremonial or otherwise. How are very large weights carried? On crossed beams or poles or in cradles, supported by many men? By traction on sledges, or over rollers? If so, give details of construction, organization of the team, distance traversed between halts, devices (such as song, music, shouts) for keeping step or securing simultaneous effort (*v*. Songs, p. 206). Are levers employed? How are obstacles or steep places traversed? What are the largest weights or masses so dealt with? Observations of these modes of transport to-day may help to explain how the stones of ancient monuments were moved and erected.

Trailers and Sledges. Are poles fastened to beasts of burden so as to trail behind for the transport of goods, or is any form of *travois* constructed? Are sledges used? If so, describe the construction of base-board runners (if any), superstructure, and harness. Are there any devices to reduce friction with the ground? Are sledges or toboggans used by themselves on steep slopes? Are snow-shoes, skis, or skates used either for travel or for amusement? Are there regular slides or tracks for such travel? Are timber-slides used to bring tree-trunks down from high forests? How are they constructed, maintained, and lubricated? How are tree-trunks hauled on level ground?

Wheeled Vehicles. Describe in detail the construction of the vehicle and of the wheel, which may be either solid, or a composite disc built of planks, or a built-up rim supported by a rectangular *frame*, or a true wheel with radial *spokes* issuing from a central *nave* (state the number of the spokes). Is there a distinct *tyre* to hold the rim together and to resist wear and tear? Are the wheels fixed to their axle (as in the wheelbarrow) or do they revolve freely on an axle fixed to the vehicle? If so, how is the wheel secured? There are many varieties of lynch-pin and axle-box. Are the wheels inside of the shafts or outside? If four-wheeled vehicles are used, how are the fore-wheels and hind-wheels connected? Can the distance between them be adjusted so as to lengthen the vehicle? Describe the construction of the fore-carriage, the provision for turning to right or left, and the shafts, pole, or other means by which the draught-animal or animals are attached. How is the vehicle checked on a down-slope? By a brake, by a clog, by tying the wheels, or how? Describe also all harness, and obtain specimens or models (*v.* Domestication of Animals, p. 251).

Travel and Transport—By Water

Navigation is extremely complicated, and only a short sketch of the chief points to be observed can be indicated here. If the observer has any experience of boating or yachting he will know what to record.

Craft

In all cases the description should be constructive; that is, it should follow the process of manufacture from inception to completion. Obtain native names (and whenever possible the botanical names, or, at least, the distinctive English names) of the woods employed, and of the materials used for lashings, ropes, sails, caulking, and so forth. Diagrammatic sketches are valuable, and should illustrate details of construction, outline, plan, and section; the native name of every portion, including the ropes, should be recorded, as these are usually of great interest.

Floats and *Rafts* are the most primitive forms of water transport. Floats may consist of any buoyant material or object; light wood logs; bamboos; bundles of bark, light sticks and reeds; inflated skins and bladders; gourds and earthenware pots. These may be used either singly or grouped and lashed together to form a simple raft. Note dimensions, nature, and provenance of the material, how obtained, method of employment and of propulsion. Are gourds and pots, when used, closed or open, inverted mouth downward or upright? How are raft units connected? Do they carry a platform for cargo and passengers? Are all apertures of inflated skins ligatured or is one left open? Describe inflation; are the skins tanned or raw? Give local conditions governing usage; traditions, beliefs, taboos, etc.

Shaped rafts of logs or of bamboos, with definite fore and after ends, are *catamarans*. Some are rigged to sail. Describe the rig and all accessory gear and any variations in use for specific purposes. Are the timbers kept permanently connected or taken apart to dry after use? Other shaped rafts are formed of light, pithy stalks of plants, chiefly marsh species. These grade into the canoe class and are named according to the material used—reed canoes, ambatch canoes, etc. Note the material employed, if it undergoes any preliminary treatment; dimensions and shape of the component parts, manner of lashing them together, various sizes in use, burdens carried and method of propulsion; time taken to make, the cost and length of life. Close-up photographs are desirable.

Bark Canoes. Note the number and arrangement of pieces used in construction; if of a single strip, how the ends are closed; if of several, note particularly the mechanical means used to obtain the desired shape; whether the pieces are joined by sewing, lacing or pinning with splints. If a framework be present, detail the parts— *gunwale rods* or *poles* to reinforce the free margins of the sides; *spreaders* of cross-rods from gunwale to gunwale; cord *ties* similarly fitted and semicircular *frames* of withies or slats to keep bottom and sides in shape. Are there thwarts? Is the canoe built bottom up or gunwales up? Is the frame inserted after the bark skin has been shaped, or is it wholly or partially pre-erected? How are the seams made watertight? How is the canoe propelled? Give dimensions (length, beam and depth at mid-length, height at each end); plan, elevation and transverse section to scale, using squared graph paper.

What trees furnish the bark? Describe method of detachment and preparation. Which side becomes the outer side of the canoe?

Is fire or steam used to soften it? At what season is the bark stripped? Record ceremonies, beliefs, and taboos attending the stripping?

Skin Canoes. In these craft a framework is covered with hide or hide-substitute. Included are kayaks, umiaks, coracles, and curraghs. Note whether the gunwales are formed first and if the craft is built bottom up or not. If possible give details of the stages in the construction of the framework, its material and dimensions. Is the cover of hide or a substitute? Is the hide tanned or raw? How are seams made watertight? If a substitute be used for hide, is it painted, tarred, or otherwise treated? Are thwarts present? How is the craft propelled and steered? Is sail ever used? How is it conveyed to and from the water? How housed? For what purposes is it used—ferrying, fishing, cargo transport? Dimensions, plan, elevation, transverse section and local terms and usages are necessary details.

Dug-out Canoes are made by hollowing out tree-trunks and shaping them into boat form. Give dimensions and plans as for bark canoes. Enumerate the internal fittings and all accessory gear. Note how thwarts are secured and if strengthening ridges across the bottom have been left when hewing out. Describe mast, sail, and cordage when present, also size, shape, and number of paddles. How are they steered?

What species of trees are utilized? When possible describe methods of felling and transport to the coast or river. What tools are used? How and to what extent is fire employed in hollowing out the hull? Are the sides "spread" to increase the beam? Describe the means employed.

Five-piece Canoes are dug-outs with each side heightened by a plank on edge, the *wash-strake*, connected at stem and stern by an *end-piece*, usually of >-shape. Describe the dug-out base and these added parts as for a dug-out and note how the added parts are attached.

Outrigger Canoes. In these craft, usually narrow in beam, a counterpoise or balance fitting minimizes the danger of capsize. A number of spars, the outrigger *booms*, cross the hull and extend outward on one side (*single outrigger canoes*), or on both sides (*double outrigger canoes*), to connect with a light log of wood, the *float*. Describe the construction of the canoe hull, a dug-out or a plank-built boat, its dimensions, whether equal-ended, capable of sailing either end forward or having a permanent fore end;

if the latter, is there any arrangement for shifting the outrigger from one side to the other? Note any carved ornament or if the hull be decorated with feathers, or paint. Describe minutely the way the booms are secured to the hull and how their outer ends are connected with the float, whether *directly* by lashing, insertion or a combination of the two, or *indirectly*, by intermediate fittings, the *connectives*. These are generally pegs, sticks or stanchions, inserted below into the float and lashed above to the boom. They may be vertical and parallel, overcrossed above the boom, or undercrossed below it; they may be crutched, forked or branched; some are angled, ∧-shaped or U-shaped. Note the method and pattern of each lashing; photographs are essential. In single outriggers, a *lee platform* is sometimes present on the side opposite to the outrigger; note construction, if cantilever or otherwise. The single outrigger is normally kept upon the weather side.

Double Canoes consist of two canoe hulls joined together by beams carrying a deck platform whereon a mast is stepped. Note the relative dimensions of the hulls, distance apart, number of booms and details of all deck fittings; also carrying capacity, ownership, number of crew, and how remunerated. Give plan, elevation, and transverse sections; when sails are present, note the type of mast, single or pole, bipod or tripod; give details of method of hoisting, furling, and reefing sail and ascertain whether the craft can make headway against an adverse wind. When going about, whether they tack or wear ship. Purposes for which used—trading, fishing, and warfare; are there separate types for each? Investigate all beliefs and ritual connected with the building, launching, and usage of these canoes.

Do the boys make model canoes? Is the design different from that of full-sized canoes?

Plank-built Boats. These have hulls mainly or entirely constructed of planks, sewn, pegged or nailed together. In a few a dug-out hull forms the basal portion. Is this narrow, forming a primitive keel, or broad and canoe-shaped, giving a rounded or flat bottom? Note the presence or absence of a kelson and if a framework of U-shaped transverse frames be present, with stem- and stern-posts. Is each frame in one piece, shaped from a natural bend, or is it compound? Note the parts if compound; does a floor piece connect with a rib passing up each side? Is the stern sharp, pointed, rounded, full, or lean, or is it square or transom-shaped? Is a counter present or other form of overhang?

Is the hull full-decked or open, half-decked or decked only at each end?

Boats built with planks overlapping like roof tiles are *clinker-built*; those with planks edge-to-edge, are *carvel-built*. Are the frames pre-erected or inserted after the skin planking is complete? Note the number and position of any cross-beams and thwarts; do their ends pass through the side planking? In carvel-built boats note if and how the seams are cauled, and if battens are placed over them on the inner and/or the outer side. Note number and position of the mast or masts, whether pole, bipod or tripod; their respective lengths and diameters, the way the foot is stepped, how supported. Of what material is the sail? What is its type, and its dimensions? How hoisted, furled, and can it be reefed?

Dimensions, plan, elevation and cross-sections at intervals are essential, but the measuring-up of a boat in sufficient detail to permit of accurate building plans to be made therefrom needs far more guidance than can be given here. The reader is therefore referred to an excellent article on "Taking off the Lines of a Boat" (W. M. Blake, *Mariner's Mirror*, January 1935).

The following are a few useful technical terms: *sheer*, the longitudinal curve of the gunwale; *strake*, a run of planking fore and aft; *garboard strake*, the strake next the keel on each side; *bilge strake*, that at the angle where the bottom merges into the side, either imperceptibly or acutely; *washstrake*, properly a plank added to the gunwale to keep out spray, but commonly and usefully applied to the topmost strake.

The following references will be of great use to anyone taking up the study of primitive craft:

HADDON, A. C. and HORNELL, J., *The Canoes of Oceania*, 3 vols. (Honolulu, 1936–38).

HADDON, A. C., "The Outriggers of Indonesian Canoes", *J. Roy. anthrop. Inst.*, vol. 50, pp. 69–135, 1920.

HORNELL, J., "The Outrigger Canoes of Indonesia", *Madras Fisheries Bull.*, vol. 12, pp. 43–114, 1920.

—— "The Origins and Ethnological Significance of Indian Boat Designs", *Asiat. Soc. Bengal, Mem.*, vol. 7, pp. 139–256 (Calcutta, 1920).

—— "British Coracles and Irish Curraghs" (Bernard Quaritch, London, 1938).

—— "Frameless Boats of the Middle Nile", *Mariner's Mirror*, vols. 25 and 26, 1939–40.

Methods of Propulsion

Boats may be poled, paddled, rowed, or sailed, or they may be pushed by swimmers. If the craft is poled, describe the *poles*, with especial reference to the point; state how they are held and manipulated; how many men engage in poling; their attitude and positions in the boat; whether the poles are changed from one side to the other; the approximate speed which can be attained. Do the polers keep time; who gives the time? Do they shout or sing as they pole or paddle? How is the craft steered and by whom?

Paddles may vary from a plain piece of wood or bark, or a mere stick or shaft, to an implement composed of more than one part. Describe with reference to the following points: whether cut from the solid or composite; material or materials; how put together; outline and section of butt, shaft, and blade. Is a double paddle used? Describe how paddles are held and manipulated; are the feet or legs used in any way; are the blades of the paddle removed from the water at the end of the stroke or are they "feathered" under water; are the paddles changed from side to side? Are there different paddles for the two sexes? How many paddlers form a crew; give their attitude and positions in the boat, indicating which side they paddle; do they keep time and who gives the time?

How is the craft steered and by whom? If by a paddle, state how it differs from other paddles, and whether it has any fixed attachment to the canoe; if by a *rudder*, describe with all appliances; state how attached and manipulated, and note especially the manner in which its movements are controlled by the attachments. A true rudder, whether lateral or median in position, has two attachments, which restrict its motion to one of rotation, usually effected by means of a *tiller*.

Oars. Speaking generally, while a paddle is manipulated simply by the aid of the hands (sometimes assisted by the foot), an oar rests against some kind of fulcrum on the boat; the questions relative to paddles will serve for oars, but the following additional points require notice. Describe the nature of the fulcrum, whether a hole in the topstrake; a *thole-pin* passing through the oar, or through a loop attached to the oar; a loop through which the oar passes; a *rowlock*, which may be a notch in the topstrake, or composed of two pins, or a more elaborate construction placed on an *outrigger*. In every case describe fully with diagrams. State the number of oars used by each rower, and whether the rowers face bow or stern.

Sailing presents great difficulties to an observer who has no practical experience. The following notes may, however, be found useful.

Sails. Give the material and shape of the sails, and describe how they are prepared; whether in one piece or sewn together; if the latter, how are they sewn? (*v.* String, p. 286). Give approximate sizes, and describe with diagrams any attachments for the running gear whereby they are manipulated. The following nautical terms may be used: the lower margin, the *foot*; the anterior margin, the *luff*; and the posterior margin, the *leech*; the anterior top corner, the *throat*; the posterior top corner, the *peak*; the anterior bottom corner, the *tack*; and the posterior bottom corner, the *clew*. The same terms are applied to triangular sails, with the necessary modifications.

Spars are used to support sails, and keep them expanded. In all cases note the material, whether cut from the solid or composite; in the latter case, how are they put together? Give the position, length, girth at various points, and section. The main support of the sail is the *mast*. Is the mast single, twinned (double mast), or a tripod-mast, or is there more than one mast? Describe in detail how the mast is secured to the hull, together with all fittings which may be attached to it. A spar to which the head of the sail is laced to keep it expanded is called a *yard*; how is this kept to the mast, and how is the sail laced to it? A spar to which the foot of the sail is laced is called a *boom*. A spar running from the tack to the peak of a sail to keep it expanded is called a *sprit*. A spar attached to, and projecting from, the hull at the bows is called a *bowsprit*. A similar spar projecting from the stern is called a *bumpkin*. Describe these, with attachments. A *gaff* is also a spar to which the head of a sail is laced, but differs from a yard in having jaws which keep it wholly on the after side of the mast.

Rigging is either standing rigging or running rigging. *Standing Rigging* is fixed more or less permanently, and serves only for the support of the spars. It consists mainly of various shrouds or stays; the term *shrouds*, without qualification, is usually applied to those which are attached to the mast, and fastened to the hull or deck

TRAVEL AND TRANSPORT—BY WATER

in the immediate neighbourhood, to keep the mast from falling sideways. What are the details of the shrouds' fastenings? A stay fastened nearer the bows and keeping the mast from falling backwards is called a *forestay*; a stay fastened nearer the stern and keeping the mast from falling forwards is called a *backstay*. The material, almost invariably consisting of some variety of cordage, should be described, together with all knots and methods of attachment (*v.* String, p. 286). A separate sketch-plan of the standing rigging is desirable.

Running Rigging serves to adjust and manipulate the sails and movable spars. A rope which serves to hoist a sail is called a *halliard*, and is described by the name of the sail with which it is connected, or if a sail is furnished with more than one halliard, by that portion of the sail which it serves to hoist, e.g. throat-halliards. Ropes which serve to adjust the sails to the required angle with the wind are called *sheets*, if attached to the clew of the sail, or to the boom; when attached to the yard they are known as *vangs*. The word rope has been used above, and, as a matter of fact, most running rigging consists of some variety of cordage; but in some cases it is made of strips of hide. In every case the material should be given (*v.* String, p. 286). A separate sketch-plan of the running rigging is highly desirable; also descriptions and sketches of all appliances for, and methods of, making fast the various ropes permanently or temporarily by *pegs*, or *cleats*, or through holes in the spars or gunwale. Are *blocks* or other labour-saving appliances used?

Manipulation. Describe the operation of setting and taking in the sail. Do they sail only with or across the wind, or can they sail to any extent against the wind? Describe the operation of *tacking* (sailing in a zigzag course in the direction from which the wind is blowing). Pay special attention to the tacking of outrigger-boats and double-boats. How is the craft steered, by a paddle, oar, or rudder? Describe, with all attachments, and give details as to how and by whom manipulated. Are *lee-boards* or a *centre-board* used? These are usually fixed by a pivot to the sides or centre so that they can be dropped into the water when sailing across or into the wind to prevent the boat making leeway or drifting sideways.

General Questions

Inquire whether all the craft are built locally and, if not, where they are built and the details of the trade; also be on the look-out for craft visiting from other places. Are different patterns of craft or appliances (such as paddles) used by the two sexes? What form of baler is used? How are craft moored; is any form of anchor used? How are the craft kept when not in use; are there special boat-houses, if so, are they used for other purposes? Is there provision for a fire on board? How are provisions and gear carried and where? Are the voyages coasting trips, or does the crew venture out of sight of land? How do they lay their course? Do they make use of currents, the sun, stars, etc., in navigation? Do they make maps or charts of any kind? Describe any apparatus that is used to assist in navigation or in finding the route. How are rapids or other obstacles passed? Describe portages and other breaks in the voyage. Does the journey continue during the night? Are voyages limited to any particular season? Who own the craft—individuals, a group, the tribe, or the chief? Is any part of the craft reserved for individuals of any particular rank, or of either sex? Note the positions and names of the captain, mate, and crew, and describe their several functions. Is any particular dialect used when on the water, or are there any words which are never used through fear of ill-luck or for any other reason? Describe any ornament carved, painted, or attached to the craft; what is their meaning and object? Frequently decorations or adjuncts (such as figure-heads, flags, decorative staves, etc.) of a vessel serve for good luck or to avert danger from weather, sharks, etc. Note whether these are temporary or permanent fittings. Objects such as coconuts, limes, etc., may be carried for similar purposes. Canoes and boats are of such importance in the life of a people, and are subject to so many dangers, that one frequently finds that everything appertaining to them is liable to be attended with ritual or magical practices. These may begin even before the tree is felled, and continue through all stages of making and rigging, and not only at the first launching but whenever the craft is to undertake an important voyage. All these and similar usages should be described.

Arts

Introduction

In the past anthropologists appear to have been mainly concerned with decorative art and the meanings and development of designs. The art of illiterate peoples, whether graphic, plastic, music, drama or poetry, can be regarded in the same way as the art of the great Eastern and Western civilizations.

1. Art as an aesthetic activity.
2. Art in relation to society.
3. The place of the artist in society.
4. Schools and styles of art.
5. Objects of art and history of art.

These headings are not mutually exclusive, and meaning, whether representative or symbolic, in art forms may come under all of them.

Plastic, Glyptic, and Graphic Arts

1. *Art as an Aesthetic Activity*

The attempts of children as well as of adults should be studied. The natural available material may determine the typical art of any given culture or area. The presence of suitable material—rock surfaces which may be painted or sculptured, boulders which may be hewn and carved, wood for carving, etc.—may be a powerful incentive to artistic activity, but the presence or absence of suitable material is not the only deciding factor. Peoples living in similar natural environment may show great differences in artistic achievement in quantity, scope, style, and artistic level.

It should be noted whether any, all, or only exceptional children make attempts at modelling in mud or clay, carving wood, etc., and whether such attempts are disregarded or encouraged by adults and lead to later specialization.

Where there is a general tendency to decorate buildings, utensils, etc., the feelings of the people towards these objects should be investigated, whether value is set on certain objects and why, whether there are any accepted standards of taste, and whether artists of outstanding merit are respected and their works valued.

2. *Art in Relation to Society*

The religious beliefs and ritual and the social organization may greatly influence art. Objects in any way connected with ritual (*v.* Ritual and Belief, p. 184) may be given special care and may become the outlet for great aesthetic development. Temples may become treasure houses. Where representations of deities, heroes or ancestors are made, it should be discovered what are the guiding principles—conformance to traditional patterns, expression of some outstanding quality, i.e. power, beneficence, protection, etc., or likeness to a human or animal prototype. The attitude of the artist to the ritual object should be investigated. The existence of kings or important chiefs may influence artistic development. Their houses and courts may be specially decorated and objects used ceremonially may be works of art. Wealthy men may enhance their prestige by the possession of highly decorated dwellings and objects of art. It should be noted whether only indigenous art is encouraged or whether art objects from other tribes or foreign cultures are esteemed.

3. *The Artist in Relation to Society*

Is the artist an expert, or is art a normal spare-time activity? If the former, what is his place in society, and how is he rewarded for his work? (*v.* Experts, pp. 182, 222.)

4. *Schools and Styles of Art*

Do artists of outstanding ability tend to develop schools? If so, what is the relation of the artist to his pupils? Is skill in art hereditary? Do artists tend to train some members of their family to carry on their traditions in art?

It frequently happens that whole culture areas are characterized by definite styles in art. One or more elements may be so typical that where another element is seen foreign influence may be suspected and investigation should be made. Correlation of the style in art with the prevalent religious beliefs, the types of material available and the tools in use should be investigated.

5. *Objects of Art*

In describing an object of art it should be noted: Whether it is a cult object or memorial (*v.* Ritual and Belief, p. 184), an object for ceremonial or for ordinary use, for personal adornment, or whether any object is made purely for its aesthetic value; whether it is part of a building, canoe, or other artifice. The material of the object should be described, whether wood, stone, metal, skin, or other substance. Note whether pigmentation—painting, stencilling, staining—drawing, engraining, carving, or sculpture in wood, stone, ivory, or bone is used. Designs may be geometric, using angular or curved lines, naturalistic or conventional representations of life—sculpture in high or low relief, or in the round, or by means of incised lines. Designs may be made by burning (poker work). The tendency to repetition, symmetry or balance should be noted, also the purposeful interruption of a rhythm. There may be a tendency to fill in all spaces in a design. Backgrounds may be filled in many ways or may be left plain.

Note whether a design is anthropomorphic, zoomorphic, phylomorphic, physiomorphic, or skeuomorphic (i.e representing human beings, animals, plants, phenomena of the physical world, or a design based on a feature no longer functional). In some cases the forms may be so conventionalized that they cannot be understood by the observer, but are definitely recognized by members of the culture to which they belong. In all cases the meaning of designs should be investigated and the symbols used identified.

Representative collections should be made, but it may not be possible to acquire objects of ritual or ceremonial use. In some cases copies can be made by skilled artists; any deviation from the original in detail or material should be noted. Photographs, rubbings, and squeezes (*v.* p. 365) should be made and measurements taken. If possible, notes should be made of the methods of the artist at work. Photographs of objects in various stages of production should be made. All tools, pigments, etc., should be described and if possible collected.

A Terminology of Decorative Art

A description of a decoration, more particularly of geometric patterns, must essentially be a description (1) of what the artist did; (2) as far as possible, of the order in which he did it, distinguishing *motive* from *enhancement* or *filling*; (3) if necessary, of the *effect* produced by the completed work. Thus in Fig. (*a*) we have a band containing a double series of alternate triangles, i.e.

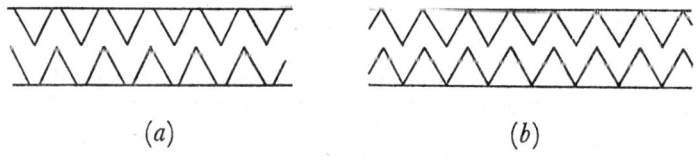

(*a*)   (*b*)

between parallel lines, a convergent series of recurrent alternate triangles, or it may be a band containing two continuous chevrons, (Fig. (*b*)), which limit a broad central zigzag band. The triangles may be enhanced by *hachure* or *hatching* as in Figs. (*c*), (*g*), (*h*), in

(c)   (d)

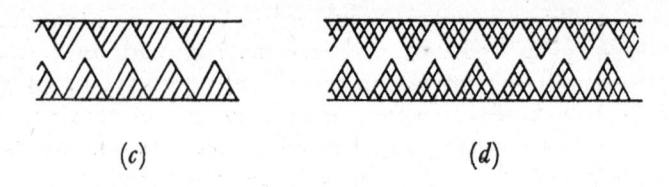

these cases the triangles are hatched *from the left*, i.e. when viewed with their *base* downwards and their *apex* upwards; by *cross-hatching* (Fig. (d)); by *internal repetition* (Fig. (e)); by *punctation*; or by other methods. The zigzag band may be enhanced by internal repetition (Fig. (f)), or by other methods, or negatively so by the enhancement of the background, in this case the triangles, as in Figs. (c), (d), (e). The description depends upon what was the primary object of the artist, i.e. either triangles or a zigzag. In Fig. (g) the recurrent triangles are *opposite*, and the effect is that of a string of lozenges, which may or may not have

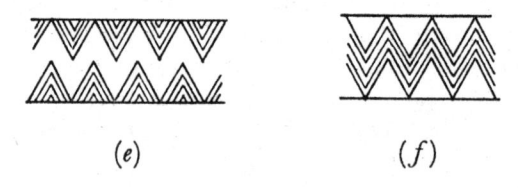

(e)   (f)

been the actual aim of the artist. In Fig. (h) the triangles are not recurrent but *intermittent* or *sparse* and opposite, the effect being

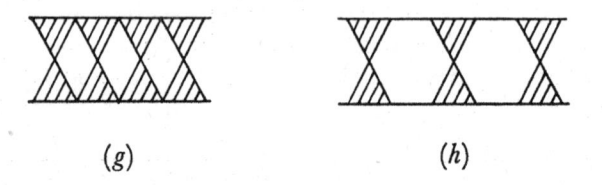

(g)   (h)

that of a hexagon pattern, which similarly may be incidental rather than intentional. Descriptions of decorations should conform to the usages of heraldry, systematic botany, and other sciences of classification, which fortunately agree in essentials.

The following are some of the principal common geometrical motives, with their names, and may be of service in describing the elements of patterns and designs:

1. Circular dots or punch-marks.
2. Contiguous and detached circles.
3. Concentric circles.
4. Elliptical punch-marks or dots.
5. Bands.
6. Hatching which may be (*a*) vertical, (*b*) horizontal, (*c*) oblique from the left, (*d*) oblique from the right; these are so described as from the base line.
7. Rectangular cross-hatching.
8. Oblique cross-hatching.
9. Chequer.
10. Chevrons.
11. Zigzag.
12. Herring-bone.
13. A cross formed by the intersection of one or two lines or bands. There are many varieties of crosses of which only six are mentioned here.
14. St. Andrew's cross or Saltire.
15. Tau-cross.
16. Swastika or Gamma-cross.
17. Triskele.
18. Croix patée.
19. Maltese-cross.
20. A star is formed by the intersection of more than two lines.
21. A rosette may be formed by a line joining the extremities of a star. Stars and rosettes, like crosses, may be conformed of other motives, for instance:
22. is a rosette of simple coils.
23. Triangles may be plain or filled in (enhanced) with any of the foregoing motives, or with simple dots.
24. Lozenges include all figures bounded by four sides, except squares.
25. The Pentagram unites five points.
26. The double triangle unites six points.
27. Plain coil or spiral.
28. Reversed coil or spiral.
29. Loop coil.
30. Continuous loop coil.
31. Frets may be four-square, triangular or hexagonal.

32. A wave-pattern is often enhanced, and is a filled-in continuous loop coil.
33. Scrolls.
34. Impressions produced by twisted coils or thongs.
35. A plait ornament or guilloche of two or more interlacing lines; this one is a guilloche of three.

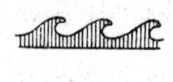

Basket-work decorations should be described according to their (1) apparent mode of construction; (2) the effect of the completed design.

Symbolic Art

Symbolic Art may be naturalistic, conventional or geometric, but these forms may be employed with meanings peculiar to the artist or to the society or caste of which he is a member. For example, the sun, the eyeball, and many flowers and fruits may be represented as circular, because of their shape. Some purely decorative designs are also circular because this form fits into a given space, or from the mere use of revolving tools. But circles may also be used symbolically to suggest the notion of Deity, the Universe, Zero, and other abstract ideas. In the same way, all intersections of lines give rise to some kind of cross or star, but many crosses and stars are found to symbolize religious or magical ideas.

Whenever any motive or design is observed to be common or conspicuous, it should be noted whether the people connect it with beliefs, observances, or restrictions. It is probable that many peculiarities in decorative art result from the continued use of symbols, or of representations, after their meaning and use have become obsolete. For this reason investigation of names of decorative motives may be valuable. Many symbolic motives do not represent the object after which they are named, but may stand for something connected with it, thus a bird's footprint stands for the bird itself, or scales for a crocodile, etc. The same simple design may have different meanings even among the same people; in such cases, without direct information, it is impossible for an observer to be sure of the significance of any particular design. Natural forces may be symbolized by figures of animals or men. A particular arrangement of symbolic designs may be dreamt by a person, and thus becomes his or her property, and often will not be explained until the object has been bought.

Most symbols have social or magico-religious meaning and are applied to persons or objects, to classify them, to secure supernatural aid, the fulfilment of a wish, or protection from harm. The meaning even of the simplest marks or additions should be investigated. But though all symbolic art is also either representative or decorative, it does not follow that all representative or decorative art is also symbolic.

The foregoing remarks apply solely to the actual designs, but it is quite as important to consider the object or surface as a whole and what relation the decoration bears to it. Taking any object, note whether it is decorated all over or only in certain areas, in the latter case, whether there is any functional or constructional reason that suggests that the decoration is peculiarly appropriate at such a spot. For example, in an object made all in one piece, a design may occur at a spot or spots where formerly there were joins which united separate parts. Lines or simple designs may be so placed as to emphasize swellings or constrictions. Note what value is placed on undecorated areas—not only may one decorated area balance another, but it may be balanced by an "area of repose". Note all *rhythm*; when in a design it usually produces a pattern, but there may be a rhythm of designs which are not in themselves identical. In any object or surface note whether the decoration is symmetrical, actually or by balance to a median line, or whether it is frankly or intentionally asymmetrical. Note to what extent and in what order the decoration was sketched in before being actually executed, for the layout of a decoration may give a clue as to what was passing in the mind of the artist.

In descriptions of methods employed in producing designs—whether realistic or conventional—it is important to state whether the effect is produced by carving, engraving, painting, printing, stencilling, staining, stopping-out, etc.

Music

The music of every people, whether vocal or instrumental, has its own characteristics, and can be estimated rightly only on the evidence supplied by accurate records. General impressions—even those of a trained European musician—are of little value unless the sounds and phrases which they describe can be reproduced. Music may be recorded either in writing, or by means of the phonograph or other recording instrument.

Musical Instruments

In recording information regarding musical instruments, special care should be taken to describe them accurately, and to avoid such slipshod and usually erroneous expressions as "a kind of rude violin", "a sort of harp", etc., unless it is certain that the instruments referred to properly belong to the "violin", "harp", etc., categories. It is essential that the mechanism, however simple, whereby the primary sounds, or tones, of an instrument are produced should be carefully diagnosed and described (with sketches where necessary), in order that the instruments may be assigned to their proper groups in the general classification. For this purpose, it is desirable to restrict the use of special names, e.g. harp, violin, zither, flute, trumpet, gong, drum, etc., to special categories of instruments. Further subdivision can be based upon variation of types within the groups.

It should be noted whether the instruments described are used merely for rhythmic purposes, or whether they are intended for producing melody (i.e. a succession of notes in ordered sequence, a "tune"); also whether two or more notes are sounded simultaneously ("harmony").

Any methods whereby the primary notes are varied deliberately, e.g. by "stops" or by "fingering" and so on, should be recorded in detail.

The simplest one-note instrument may serve for melody by associating together a number of the simple units, and sounding them in appropriate sequence. On the other hand, an instrument may be fitted with the mechanism necessary for varying the notes.

The uses to which particular instruments are put should be described; any ceremonial or other restrictions limiting their use should be noted; also the employment of instruments for signalling and conveying information should be described in detail.

Any object, whether natural or artificial, and however simple, which is employed for the purpose of producing sounds (whether "musical" in an aesthetic sense or not) should be included as a musical instrument.

The more important groups[1] and subdivisions of musical instruments are as follows:

[1] While the accepted terms "Wind Instruments" and "Stringed Instruments" serve well enough as main headings on pp. 320 and 323, examination of the groups (1)–(4) on pp. 317–19 shows that no main headline such as the usual "Percussion" or other suggested terms will fit these four groups.

(1) Instruments of inherently RESONANT MATERIALS which are caused to vibrate by percussion.

(*a*) *Clapper Series*. Instruments consisting of two *similar* objects which are struck together, e.g. two sticks or two specially shaped pieces of wood or other material (castanets, cymbals, etc.). Spring-clappers, in which the two parts are united, or hinged together, at one end, the free ends being clashed together.

(*b*) *Gong Series*, in which the instrument consists of a shaped piece of sonorous material (wood, metal, etc.) which is struck with a special striker. Subdivisions:

(i) *Simple Xylophone:* a shaped piece of hard wood giving a single note when struck.

(ii) *Compound Xylophone,* consisting of a number of the simple wooden bars, tuned to different notes and associated together (either temporarily or permanently) in graduated series from high to low notes. Note the number of the bars; how they are tuned; whether the sequence of the notes represents a scale; and whether the tones are reinforced by the addition of a resonator or a series of resonators (one to each bar). Where each bar has its own resonator, note whether the capacity of the latter varies with the length of the bar.

(iii) *Hollow Xylophones:* wooden gongs which are hollowed out (like troughs or canoes), so as to increase the resonance. Describe the shape in detail, the manner of striking, and note whether different tones are emitted when different parts of the xylophone are struck.

(iv) *Metallophones:* gongs made of metal, which, like the wooden types, may be either solid or hollow, and of very variable shape. Simple units may be combined to form compound scales of gongs, and one or more resonators may be added.

(v) Similar types in other materials (crystallophones, glass, lithophones, stone, etc.).

(vi) *Pellet-bells:* more or less globular, hollow bodies, usually of metal, containing one or more loose pellets, which rattle when shaken. There is a slit-like orifice in the casing.

(vii) *Clapper-bells:* Hollow gongs of wood, metal, pottery, etc., usually widely open below and having the striker suspended inside, so that it swings and strikes the sides when the bell is shaken or swung (alternatively, the clapper may be swung, the bell remaining stationary).

(*c*) *Jingle Series,* consisting of a number of sonorous objects, not necessarily similar, loosely attached together, so as to clash together when shaken.

(*d*) *Sistrum Series*, consisting of a number of perforated discs, rings, or cup-like bodies, loosely "threaded" upon a rod, so as to clash together when shaken.

(*e*) *Hollow Rattles*. Hollow bodies (of gourd, wood, pottery, metal, etc.) enclosing a number of loose pellets, which rattle when the instrument is shaken. Rarely, the pellets may be attached to string-work *outside* and strike upon the exterior of the hollow body.

(*f*) *Musical Rasps*. Solid or hollow instruments of wood, bone, gourd, metal, etc., having one or more edges or surfaces serrated or furnished with a number of parallel grooves, across which a stick is drawn, so as to produce a loud, harsh sound. When used singly these are mere "noise-instruments", but when several, tuned to different notes, are used in association, melody can be achieved and the effect may be quite pleasing. Various mechanical forms, e.g. the "Watchman's rattle", have been invented.

(*g*) *Stamping Tubes*, usually lengths of bamboo, closed at the lower end, which are stamped upon the ground (like pestles); or open tubes which are stamped on a man's thigh. Several differently tuned may be used in association for producing melody.

(2) FRICTION INSTRUMENTS. Instruments of sonorous material (wood, glass, metal, etc.), whose surface or surfaces are thrown into vibration by friction of the fingers; or in some evolved mechanical types, by friction of special rosined pads.

(3) MEMBRANOPHONES. Pulsatile instruments furnished with tensely strained membrances, which are caused to vibrate by percussion or by friction, and to emit sounds whose pitch varies with the degree of tension of the membrane. To these the name *Drum* properly applies. Subdivisions:

(*a*) *Single-membrane Drums*, in which a membrane (of hide, snake- or lizard-skin, parchment, etc.) is strained across the rim of a wooden or metal hoop, e.g. tambourine, or across one end of a hollow cylinder, or the orifice of a pottery vessel or other hollow body. Played usually by striking the membrane with the fingers, or with one or two beaters.

(*b*) *Double-membrane Drums*, in which membranes (or "drumheads") are strained across both ends of the hollow body, both membranes being used for production of sound. A variant consists of double-membrane drums in which only one membrane is thrown into vibration; the other serving as an attachment for bracing cords whereby the sounding-membrane is strained.

(*c*) *Friction Drums.* Single-membrane drums having a stick or string fastened to the centre of the tense membrane. In sounding these, the fingers are wetted or rosined, and are drawn along the stick or string, causing vibrations which are communicated to the membrane. Friction drums may be large or small. They are usually associated with ceremonial and mystic rituals.

Note. In describing any type of drum, note especially the material and shape of the hollow body, the material and mode of attachment, of straining and of tuning the membranes, and whether the tension of the membranes can be varied at will; how the membrane is vibrated, i.e. with the fingers, wrists, with drumsticks, or with pellets attached to the ends of strings, or by friction.

(4) LINGUAPHONES. Instruments fitted with one or more "tongues" or spars of bamboo, wood, bone, or metal, which are fixed at one end, the other end being free to vibrate when plucked or otherwise agitated.

(*a*) *Jews-harp Series.* Small instruments consisting of a single flexible tongue (or rarely, two) enclosed in a frame to which it is attached at its base. In the primitive types, the tongue and frame are in one piece and the tongue is straight; it is caused to vibrate either by jerking a short string attached to one end of the frame, or by plucking the end of the frame itself. In the more developed examples made of metal, the tongue is usually bent so that its free end stands out clear of the frame; in these the tongue is agitated by plucking the free end. The instruments are held to the lips and the vibrations of the tongue are communicated to the air in the hollow of the mouth, and by varying the capacity of the latter a variety of notes can be achieved.

(*b*) *Sansa Series.* Instruments consisting of a number of flexible tongues of bamboo, wood, or metal, whose bases are attached to a board or to a box-like resonant body; the tongues are supported by a bridge so that their free ends can vibrate freely when plucked with the fingers or thumbs. The length of the tongues can be adjusted and varied, so that a series of notes can be produced, varying with the length of the vibratile portions of the tongues. The tongues may be arranged in graduated series from the shortest to the longest, or they may be disposed irregularly in order to adapt them to a particular melody.

The "musical box" is an elaborated mechanical analogue of the *sansa*, from which it may have been derived. Essentially it consists of a steel comb with graduated vibratile tongues, plucked by short pins fixed into a mechanically revolving barrel.

(*c*) *Nail Violin Series*. A series of short rods with their bases fixed to a resonant body, the other ends being free. They are graduated in length and tuned to give the notes of the scale. They are vibrated by the friction caused by drawing an ordinary fiddle-bow across them. "Nail-violins" were sometimes fitted with sympathetic strings (p. 320).

Wind Instruments

Instruments which are caused to sound by directing a jet of air against the edge of an orifice, or through a valve, in such a way as to set up vibrations producing musical tones.

(1) FLUTE SERIES. A jet of air, directed by the lips or through a duct, is caused to impinge upon a portion of the edge of an orifice in a hollow tubular or globular instrument. This causes the jet to be cut, so to speak, the interference creating vibrations which are communicated to the air inside the hollow body.

(*a*) *Syrinx, or Pan-pipes Group*. These may be single tubes, open above and closed below, or open at both ends. The jet of air is directed against the edge of the upper opening. (A tube which is open at *both* ends will give a note an octave above that given by a tube of similar size which is closed below.) Two or more of the simple tubes of varying length may be combined to form a compound syrinx (the typical Pan-pipes). The tubes may be arranged in graduated series, or may be variously grouped to suit some particular melody or sequence of notes.

(*b*) *End-flute Group*. Cylindrical tubes, usually open at both ends, sounded by blowing across the upper edge, which is usually bevelled; a series of open stops along the lower part of the tube enables a variety of notes to be produced, since by opening or closing the stops, the vibratile length of the column of air in the tube is varied. These flutes are held nearly vertically in playing.

(*c*) *Transverse-flute Groups*. Cylindrical tubes, closed at the proximal end, near which is a sound-hole against the edge of which the jet of air is directed. The tube is usually furnished with open stops. Held horizontally in playing.

(*d*) *Notched-flute Group*. These are "end-flutes" having a notch cut in the edge of the upper opening, thus specializing the portion of the edge against which the air-jet is directed. Usually furnished with stops. Held vertically in playing.

(*e*) *Duct-flutes.* End- or transverse-flutes furnished with some kind of duct through which the air is automatically directed against the edge of the sound-hole. The form and construction of the duct varies greatly among the more primitive types and should be carefully noted. The "flute-à-bec", modern whistle, and flageolet are evolved members of this group.

(*f*) *Aeolian Flutes.* Tubes or other hollow bodies having natural or artificial perforations, which are caused to sound by the wind as it blows across the perforations.

(*g*) *Whistling Arrows,* pigeon-whistles, humming-tops, etc., are varieties of "flutes" which are driven against the air.

(2) VALVE INSTRUMENTS. Wind instruments whose primary sound is caused by the action of some kind of valve-like mechanism, which causes a very rapid succession of checks to the free passage of the air jet, and thus sets up vibrations which are communicated to the column of air inside the tube.

(*a*) *Single-beating-reed Series* (or Clarinet group). In the more primitive types, of reed, corn-stalks, etc., the tube is closed above and open below. Near the closed end a flexible tongue is cut in the side of the tube. The tongue remains attached by its base and forms a valve; the free end is caused to vibrate by air pressure, the orifice covered by the tongue or reed being closed and opened in very rapid succession, creating vibrations. The resultant notes can be varied by the use of stops. In more advanced types the tongue-valve is cut in a separate piece of reed or bamboo, which is fixed into the open upper end of the tube. The modern clarinet is a developed form of this type, having the "valve" sounding-mechanism built up, the tongue being separate from the mouthpiece.

(*b*) *Double-reed Series* (or Oboe series). In these the "valve" is in the form of a slit-like orifice having thin, elastic walls, both sides of which vibrate when blown upon, alternately closing and opening the "valve" in rapid succession and thus creating the necessary vibrations. The note is varied by use of stops. Note the material and structure of the "valve"; whether it is a natural cylinder whose sides are pinched together at the end, or whether it is built up from two or more pieces. The sounding-reeds of the modern oboe, the *cor anglais* and the bassoon are evolved forms of this type of "valve".

(c) *Free-reed Series.* In these the sounding-valve consists of a vibrating tongue enclosed in a frame to which it is attached at one end, the other end being free to oscillate on either side of the frame. Simple forms are merely cut out of a thin piece of bamboo, wood or brass; the advanced types have the tongue of springy metal screwed to the frame. The instrument may consist of a single tube fitted with a free-reed and may have a series of stops for varying the notes. Or it may consist of a number of tubes, each with a free-reed set in a common air chamber. "Mouth-organs" are comparatively simple types; concertinas, accordions and harmoniums are developed types, fitted with keyboards.

Note. Some composite instruments combine two or more of the sounding mechanisms above described, e.g. some bagpipes, church organs, etc.

(d) *Spinning-valve Series* ("Bull-roarers"). Thin, elongated wooden blades (rarely of other materials) to one end of which a string is fastened; the other end of the string is frequently attached to the end of a stick. The wooden blade is whirled round at the end of the string, so as to drive it against the air and cause it to spin very rapidly, and so to present its sharp edge and its flat surface to the air resistance in rapid alternation. An effect is produced analogous to that produced by a valve alternately closing and opening, and the intermittent checks set up vibrations creating sound which varies in pitch with the rapidity of the spin of the blade. It is important to note any ceremonial or mystic ritual attaching to the use of this instrument, and any restrictions imposed upon its use.

(e) *Slit-valve Series.* A rather rare class of simple wind instruments whose sound is produced by blowing air through a very fine slit (or series of slits) cut longitudinally in a reed-stem or grass-stalk.

(f) *Oscillating-ribbon-reed Series.* The sound is produced by blowing upon the edge of a thin blade or ribbon-like band which is strained between its ends. (This may be a blade of grass, a very thin strip of bamboo or tape strained between the ends of a small bow, or a piece of feather-quill flattened and strained with a bow-string, etc.) The pressure of the air causes the blade or ribbon to oscillate alternately one way and the other in rapid succession, setting up vibrations causing sounds. The South African *goura* is a specialized type of this group.

(g) *Trumpet Series* ("Loose-lip" instruments). These are sounded by blowing air through a relatively large aperture into which the lips are tightly pressed. The air-current in passing throws the edges of the lips into vibration (the lips being elastic membranes), the lips performing a function analogous to that of a valve which in rapid alternation allows and checks the free passage of an air-current. The notes produced can be varied within harmonic limits by the action of the lips, and the more rudimentary types of trumpets (of shell, horn, ivory, wood, gourd, pottery, etc.) usually have no mechanical means of varying the notes; the same applies to some evolved types in metal, e.g. the bugle, post-horn, etc. But the intermediate notes not obtainable by unaided lip-action may be sounded by varying the vibrating length of the air-column by means of stops, slides, e.g. trombone, or pistons (cornet). The *embouchure*, or sound-orifice, may be at the side or at the end of the tube.

Sympathetic-membranes. Many of the types of instruments referred to in the foregoing classificatory groups are fitted with a peculiar accessory in the form of a small piece of very thin membrane strained across a hole in the instrument. This membrane picks up automatically the sound vibrations set up in the instruments and, by its own vibrations, serves to modify the tones and give them a kind of "reedy" or "buzzing" quality. In Africa, drums, *sansas* and xylophones frequently have such membranes (usually obtained from the egg-capsules of spiders) added to the resonator. Among wind instruments, certain types of flutes (in China, Siam, Europe) have a membrane covering a small hole in the side of the tube. Certain instruments have two orifices, one into which the performer hums or speaks; the other covered by a thin membrane which vibrates and alters the quality of the sounds produced. (These are mere toys in Europe, but are serious instruments in parts of Africa.) The *nyastaranga* of Brahminic ritual in India is a specialized type in which a thin membrane is fixed near the embouchure, which, instead of being held to the lips, is pressed against the outside of the throat, the membrane picking up the vibrations of the larynx when the performer hums or speaks.

Stringed Instruments

Instruments furnished with one or more tightly strained strings which emit notes when thrown into vibration by (*a*) plucking, (*b*) striking, (*c*) friction, (*d*) wind. The pitch of the notes given by a vibrating string will depend upon the length of the vibrating portion of the string, the degree of tension, and the weight of the string.

(1) EXTEMPORIZED MONOCHORDS. Single strings tightly strained and bridged, either near the two ends and capable of emitting one note only when struck or plucked; or towards but not *at* the centre, so as to furnish two unequal vibrating lengths, giving two different notes.

(2) SPLIT-STRINGED INSTRUMENTS. These form a special group of instruments of reed, bamboo, or palm-leaf stalks, having one or more strings formed by prising up very narrow lengths of the cortex, leaving them attached at their extremities. These split-off strings are bridged up near their extremities (or in some types near their centres), to raise them from the surface, so that they can vibrate freely when struck or plucked. Of the polychord examples, two chief types can be differentiated: (*a*) those instruments, usually of stout bamboo or palm-leaf stalk, having several strings so formed upon the single body or rod; (*b*) those which are composed of a number of slender reeds each carrying a single string, bound together to form a compound instrument. Note any accessories, such as resonators, which may be added.

(3) MUSICAL BOWS. Simple bow-like monochords (or rarely, fitted with more than one string).

(*a*) *The ordinary shooting-bow* temporarily converted into a musical instrument; the string, which may be divided into two unequal vibrating portions by tying a string loop round the bow-string and the bow, is either struck or plucked, or, rarely, vibrated by friction. The bow is usually held with a portion touching the teeth of the performer, so as to magnify the sounds to his ears; or the bow may be rested against a gourd, pot, or other external resonator, to augment the sounds for the benefit of an audience.

(*b*) *Monochord instruments* resembling the shooting-bow in structure, but made for musical purposes only. The notes may be reinforced as before.

(*c*) *Derivatives* from the above, in which the "bow" is straight and the string is bridged up near its ends to allow of vibration.

(*d*) *Simple musical-bows* having a resonator, e.g. a gourd, attached to the bow.

(*e*) *Simple musical-bows* fitted with more than one string.

(*f*) *Compound musical-bows*, consisting of two or more "bows" united together to form a polychord instrument. The bows may be attached to a common resonator, and the plane of the line of strings is more or less parallel to the surface of the resonator.

(g) *A variant* in which a single bow is substituted for the several bows and is fixed to a resonator. All the strings are attached at their upper ends to the "bow", one above the other; at their lower ends they are attached to the resonator surface, or, after passing over a notched, vertical bridge, indirectly to the end of the "bow", which projects beyond the resonator. The plane of the strings is at right angles to the surface of the resonator.

(4) PRIMITIVE AND "ORIENTAL" HARPS. Polychord instruments, whose framework consists of a box-like resonator (usually elongated, hollowed out and covered with hide or parchment), from one end of which issues a long "neck", which is either curved or set at an angle to the resonator. The strings are attached at one end to the neck and at the other to the surface of the resonator along the median longitudinal line. The plane of the line of strings is at right angles to the resonator surface. The strings, which are plucked with the fingers, may have tuning pegs for regulating their tension, or they may be simply wound round the "neck" (as in the Burmese harp). Practically all African and Oriental harps are of this type, and as the framework lacks rigidity, the strings can at best be only approximately tuned to a scale.

(5) THE DEVELOPED, OR "WESTERN" HARP. The essential structure is similar to that of the last group, but a prop, or "fore-pillar", is fixed between the free ends of the "neck" and resonator. The framework is thus rendered triangular and rigid, and more effectively stands the strain of the strings, which can be accurately tuned, since tightening one string does not involve slackening all the others, as happens in the former group. The strings are plucked with the fingers, both hands being used.

(6) ZITHER SERIES. Instruments in which several strings are stretched across a flat board, or across the hollow of a shallow trough-like body, or across the upper surface of a box-like resonator. The line of strings is in a plane parallel to that of the resonator surface and lies close to it, the strings being either bridged up at both ends, or provided with fixed or movable bridges towards their centres. Each string is tuned to a particular note, which usually is not varied by "fingering". The strings are *plucked* with the fingers or with a plectrum. (The Spinet, Virginal and Harpsichord are highly evolved zithers, fitted with a keyboard and having a plectrum of quill for each string.)

(7) DULCIMER SERIES. Generally similar to the last series, but the strings are *struck* with light beaters, instead of being plucked. (The Clavichord and the Pianoforte are evolved keyboard types of dulcimers.)

(8) LUTE-GUITAR SERIES. Instruments having a resonant body and "neck". The strings (usually few in number) are attached at one end to the "neck", at the other to the distal end of the resonator, and they lie in a plane parallel to the surface of the latter. The notes to which the strings are tuned can be varied by "fingering"; one hand being thus occupied, while the other plucks the strings with the fingers or a plectrum. By varying at will the vibrating portion of the strings, by the action of "fingering", a very wide range of notes can be obtained with a few strings.

(9) LYRE SERIES. Instruments having a hollow resonating body from which project two upright rods whose upper ends are united by a transverse bar, forming an open, trilateral, and rigid framework. The strings, which may be numerous, are attached at one end to the transverse bar, and at the other end to the resonator, being strained across the open space between the two. They are plucked with the fingers or with a plectrum. (A very rudimentary type of lyre consists of a triangular frame formed by the V-shaped junction of two branches of a tree, with a cross-bar uniting their free ends. There is no resonator.)

(10) INSTRUMENTS WHOSE STRINGS ARE VIBRATED BY FRICTION. These usually consist of a resonator and a "neck" between the distal ends of which the strings are strained. They are caused to vibrate by drawing across them a friction appliance, which may consist merely of a wetted blade of grass, a thin strip of palm-leaf or bamboo, or of fibres or hairs stretched upon a small bow; or it may be an elaborated instrument, such as the modern fiddle-bow. Rosin, or its equivalent, is usually applied to increase the friction. The strings are usually few in number and their notes are varied by fingering, or "stopping" the strings along the "neck" so that all the notes of the scale are attainable.

In a specialized group of friction instruments (the vielle, hurdy-gurdy, etc.), a rosined wooden wheel, rotated by a crank, is substituted for the fiddle-bow, and a keyboard is added, for stopping the strings.

(11) AEOLIAN STRINGED INSTRUMENTS. Instruments, usually of zither-like structure, whose strings are thrown into vibration by wind. The instruments are placed so that the wind blows across

the strings, which are all sounded together, causing an attractive medley of soft notes.

(12) SYMPATHETIC STRINGS. These are added to a considerable variety of stringed instruments (*v.* also Nail-violin, p. 320). They consist of a supplementary series of thin wire strings which underlie the primary set of strings, and are not caused to vibrate by any of the usual methods. The "sympathetic strings" are tuned either in unison or in some harmonic relationship with the primary strings, and, when the latter are caused to vibrate (by plucking, striking or friction), the "sympathetic strings" pick up automatically the vibrations to which they are responsive. As they are not subjected to "fingering", their vibrations are sustained and die away gradually. This "background" of sustained notes gives a peculiar quality to the melodies performed on these instruments. Sympathetic strings are widely used in India. In Europe the *viola d'amore* and some other instruments of the violin family were fitted with them.

The classification of musical instruments here given is intended merely as an aid to the description of such types as may be met with. The list is by no means exhaustive and the groups could be further subdivided. The classification adopted is based mainly upon the different methods whereby the vibrations required for the production of sound are initiated. It is very important to collect all examples of rudimentary types of musical (or noise) instruments before they become entirely obsolete. There is a tendency for primitive types to be retained for ritual purposes, and the special instruments associated with ceremonial and mystical performances should particularly be sought for and placed on record.

WRITTEN RECORDS OF MUSIC

The characteristics of non-European music, and of folk music even in Europe, are often such that they cannot be accurately transcribed by means of ordinary musical notation; they require additional signs and full verbal commentary. Simple, relatively short tunes with simple intervals can be written down either in staff notation or in tonic solfa directly from the lips of the singer by any person with a moderately good musical education and an accurate ear and rhythmic sense.

Written Recording

There are certain points which even the trained musician must bear in mind when recording non-European music and the less highly trained worker should observe when using a mechanical recorder.

No attempt should be made to alter anything in the pitch rhythm or metre of the music performed, however strange or incorrect it may appear to the observer. There is a phenomenon in vocal music that might be called "pitch attraction", somewhat comparable to what is known in linguistics as "assimilation", that has barely been noticed as yet, to the study of which all such apparent aberrations contribute valuable data. Such "defects" also may be the signposts which subsequent musical analysis relies upon to indicate new patterns of melody, musical form and rhythms.

Recording of Tempo and Pitch

If the rhythm and tonality of a melody are simple the tempo should be ascertained by means of a metronome. Variations in tempo, the lengthening of particular notes and other peculiarities clearly due to expressive and emotional causes, may be noted down in strict time values and explained in the usual way as pauses, rallentandos, accelerandos, and other similar modes of expression. But many kinds of music have rhythms and metres so complicated that the metronome speed used as an indicator of strict time is useless except to give a rough idea. In writing out these irregular and constantly shifting metres, it is best to insert the usual measure bar and proper time signature at every change and before the primary accent, as this helps materially to orientate the European musician who may wish to read this music and to facilitate his grasp of what is happening. The absolute pitch of the composition should be determined by means of a pitch-pipe or tuning-fork sounded primarily to identify and fix the pitch of the ground tone (a sort of tonic) of the melody. From this ground tone as a pivot about which the melody moves the absolute pitch of the other tones may be worked out in their interval relation to it. If there are intervals of less than a semitone (or more properly half-step) they should be indicated on the staff by special signs placed over, under or through the notes. These should conform to diacritical marks already accepted and adopted by musicologists unless nothing already in use will fit the case. If new signs must be used, they must be carefully explained.

In the mechanical recording of a piece of music, the note of the pitch-pipe used should be included, preferably at the end of the piece.

Pitch

The subject of pitch in general is often given undue prominence in musical investigations to the detriment of other equally important considerations. Some aspects of pitch are exclusively musical problems; some belong more properly to the field of psychology; some are intermediate and a few are tied in with language. Only experience in the light of a full understanding of the various implications involved in pitch questions will guide the investigator in pursuing this problem.

Recording Machines

No advice can be given as to the choice or use of a recording machine. These are constantly being produced and improved, and an up-to-date machine suitable for the conditions in which the work will be undertaken must be chosen and instruction obtained.

Instruments

The use of all musical instruments should be described, not only the types (*v.* p. 316), but the ways in which they are employed. All instrumental or other accompaniment to singing and dancing, such as clapping, stamping and inarticulate cries, should be recorded. In some instances such accompaniment may seem to be at variance with the rhythm and metre of the melody. Note should be made of this and the phenomenon recorded on the mechanical recorder.

The social context of each piece of music should be noted, whether it is for general performance or restricted to a particular occasion, occupation or class of performer. It has become fairly axiomatic that music associated with special occasions, like ceremonial music, tends to follow definite melodic or other patterns of composition which may be quite obscure to the listener not familiar with them. Sometimes only the most painstaking analyses will reveal these subtle features, which may nevertheless be recognized by the people creating them even though they can seldom isolate and describe the traits in the abstract. This tendency to patterns resembles patterns in types of poetry of which there may be many in a given language. So there may be many music patterns in the music of a people, and it is the sum total of these which characterizes the music in general and distinguishes it from that of another ethnic group. Also, since these patterns are complex, their common ownership by two peoples divergent in most other respects, offers valuable evidence of borrowing.

Beliefs and Ideas

Beliefs about the origin of music or of particular pieces, recitals of experience in composing, estimates of composers' ability by themselves or by others, any data procurable about the theory of music however little may exist, any remarks about repetitions in the music with changes appropriate to each reiteration or to any accompanying dance, are extremely valuable for musicological study, while descriptions of observances connected with the music and musicians will be equally welcome to ethnologists. There may be special regulations governing the use of particular modes and instruments, or special methods of singing appropriate to special occasions like the whispering of certain sacred words, and these may apply to the different sexes or age groups when in the presence of others, or just on general principles. Some usages may hold only in the presence of strangers or of persons of rank.

Musical Vocabulary

Any musical terms in use should be recorded, i.e. words for "high", "low", "loud", "soft", "noise", "tone", and any others like the names of notes and including terms that might be elicited in discussing the material with the informants. The different styles of music, modes and rhythms may have terms to identify them. If evidence of musical notation is discovered it should be thoroughly explored, but this is rare.

Voices

The complete range of individual voices other than as revealed by the songs sung, and the range common to the songs as a whole, may prove of psychological or physiological interest and should be obtained, although analyses and summaries made after transcription in the laboratory will give the latter point. Tricks of performance, mannerisms, the use of falsetto, tremolo, glissando, and so on, and the ways in which various emotions are expressed should be noted. In more sophisticated instrumental music similar observations should be made on ways of playing.

Texts

The texts of all songs should be recorded, syllable by syllable, in writing as well as on a recording machine (*v.* Knowledge and Tradition, p. 206).

Where a tone language is involved the spoken tones of the words should be carefully written down. Comparison of the spoken tones with the sung tune is then possible and the measure of correspondence or divergence between the two can be assessed.

General Observations

Laboratory analyses and comparative studies concern themselves with all questions of form and design within the melodies, the tonal content and intervals employed, and in the general conclusions, with the predominance of certain forms over others considering the music in question as a whole. Thus a music may be characterized as predominantly vocal or instrumental or mixed in a certain proportion: it may be monophonic or polyphonic: there may or may not be harmony or other forms of part singing. It may be solo or ensemble. The styles may be simple or complex, with varying degrees of preference for preludes, refrains, postludes, verses, and so on. These will come out with study. But any field observations that can be made along these lines will be helpful in a final summary, particularly if the collection from which it is to be drawn is only a sampling rather than exhaustive.

DANCING

Dancing, like other arts, includes representations of states of feeling, or of events and occupations or imitations of objects such as animals in motion; and these representations may be naturalistic, or more or less conventional. Imitative or mimetic dances must be described with reference to the performer's motive (as elicited from native testimony, not as merely guessed at by the onlooker), and to the movements, gestures and external aids, such as costume, by which he gives expression to it. When such dancing attempts the narration of a series of incidents, it passes over into pantomimic drama (p. 333).

There are also many formal dances in which it is difficult to detect an intention to represent anything. The gestures and other movements are here mainly formal, like the patterns of decorative art; and are performed either for the pleasure which they afford, or for some symbolic reason, as part of a religious or social ceremony. There are, of course, intermediate instances, in which the meaning of a dance is not apparent but is preserved by tradition or in its name; or where a conventional but still clearly mimetic dance is executed by combining formal steps and gestures.

Records of dances are best made by the kinematograph, and by a recording instrument for the music and singing (*v.* pp. 359, 329). Observers who are themselves trained dancers may also be able to record the steps by a written notation. In any case the evolutions of individual dancers and the concerted movements of figure-dancing can be described in writing with diagrams. A descriptive notation-symbol may then be given to each, to shorten the description of more complicated movements in which they recur. All native names for steps and figures should be recorded and employed as far as possible. Any native systems of teaching dancing, with its technical terms and standards of criticisms, would be of special interest.

Photographs should be taken, showing each important moment of the dance and the appearance of the dancers. Costumes, decorations, and accessories should be minutely described; they may have ritual significance, and they are sometimes valuable evidence for migration or transmission of culture.

Dances may be held in conjunction with religious or other feasts. Inquiries should always be made as to the occasion of a dance and who come to it, noting what social or local groups are present. There may also be special seasons when dances are held which appear to be purely social functions; it should be discovered who is responsible for such dances and whether invitations are sent.

Dances are often owned by groups or individuals, note how these can be transferred (bought, or borrowed) to other groups or individuals; and what is done when these are copied without permission.

Are new dances invented, and is the inventor honoured or paid in any way? Records of the history of any dances would be of interest. What is dancing called, and what purpose is it supposed to serve? Make notes of any spontaneous, unorganized dancing.

Is there a definite age at which an individual begins and ceases to take part in dances?

Do the sexes dance together or separately, or in separate groups which meet?

Drama

Closely connected with music and dance are all the simpler forms of drama. Under this head it is convenient to include all performances, where action is used to rehearse actual, legendary, or mythical events, otherwise than as a pastime (*v.* Games, below). The action may take place in dumb-show, or may be supplemented by speech, song, instrumental music, and more or less mimetic dance (*v.* Music and Dancing, pp. 327, 331).

Note, in regard to all dramatic performances, the place, date, and occasion. Is it performed at some fixed point in the calendar year, such as Midsummer Day, etc., after a death or some other event of importance? The number, status, and organization of the performers; all provision, either temporary or permanent, for the convenience of actors or audience; all accessories, such as costumes, masks, or other make-up, manikins, puppets, or other images, stage properties, scenery, and artificial lighting, if any. Are any shadow-plays performed? Obtain specimens or models of all these, and particularly of masks. Who may perform drama? Are there temporary or permanent companies of players? If permanent, note their status, habits, and how they are maintained. Obtain a full account of each dramatic performance which is observed, including, if possible, the exact words, music, dances, and habitual gestures, and make full use of diagrams, sketches, and photographs, and, if possible, of the kinematograph and sound recorder.

It is better to avoid general expressions such as tragic and comic, and to describe rather (1) the subjects of each performance, which may be a folk-tale or legend, or an historical incident, or drawn from current events, or daily life; (2) the reasons given by the people themselves for its selection on that occasion; and (3) the expressions of emotion, if any, which are evoked by it. Is the story familiar to the audience beforehand? How are the words, business, etc., of the drama preserved, and taught to performers? How are new plays, or new features, invented, produced, popularized? Dramatic performances are frequently part of a ceremony and should be described in their appropriate context.

Games and Amusements

Games are worthy of special study, and they may be primarily for amusement, and played whenever suitable opportunity arises. It should, however, be particularly noted whether any games or pastimes are confined to one sex, to certain ages, to particular groups or ranks, to ceremonial occasions, or to particular seasons. If there is opportunity, join in a game and learn to play it yourself. Games should be recorded fully, if possible, with photographs and drawings. Collect toys, and the implements used in games. Sides may be taken in games. If so, the social significance should be investigated.

Games may be classified as: (1) *Games of movement* that incidentally or purposefully develop and exercise the bodily powers, such as games of agility, strength, endurance, skill and the like. Examples are: leaping, climbing, boxing, wrestling, sham-fights, throwing of missiles, feats of horsemanship, aquatic sports, shooting game or vermin for sport. (2) *Games of dexterity* which exercise the memory and develop manual dexterity, such as string figures and tricks (*v.* p. 335). (3) *Games of skill and calculation*, with apparatus (mancala, chess) or without (morra). (4) *Games of chance*, with particulars of the dice, other apparatus, and stakes; beliefs connected with chance. (5) *Amusements with animals* include baiting tame or captured animals, and setting animals to fight. Note the training of the animals, the conditions of the fights, and any weapons and defences which are provided. (6) *Dancing and dramatic acting* are often practised for amusement (*v.* Dancing, p. 331). (7) *Shows and professional performances.* The foregoing are mainly the diversions of adults, but most of them are to some extent played by children. More particularly played by the young are (8) *Round games* of simple amusement in which singing is a frequent accompaniment—some of these may be played without any special appliances; but where such occur, the objects, as well as the game and songs, should be described.

Toys may be simple, such as dolls, tops, balls, kites, etc.; others imitate the contrivances of grown-up people; there are mechanical toys, and musical toys. It is occasionally found that a toy is a survival of a ritual or practical object which is no longer employed as such by adults.

The spirit of emulation enters into the majority of games, and usually the contest element masks other features of the games. An important point to remember is that a game is frequently a simplified and secularized ceremony of an older culture, or it may be but the merest vestige of a rite or social custom. Note the ditties or formulae which accompany games; they often contain archaic words and phrases. Besides the sex, age, and status of the players of games, note what games are considered manly or unmanly; whether active or sedentary games, individual contests, or games played between sides or teams are preferred; if the latter, how are the sides chosen, organized, and led?

Note *betting and gambling* in connection with any amusement, or social or religious occasion; whether the betting is done by the players or bystanders; what are the stakes, the social effect of the practice, and the state of public opinion on the matter?

The partaking of narcotics and stimulants of various kinds may be regarded from the point of view of an amusement, though it usually subserves some social or religious purpose.

String Figures and Tricks

A Method of Recording String Figures and Tricks[1]

Making string figures is so widespread an amusement that it deserves special attention. Not only should the method be recorded, but the name of each figure and its alleged meaning as well as any songs and stories associated with it.

The term *string figures* is employed in those cases in which it is intended to represent certain objects or operations. The "cat's cradle" of our childhood belongs to this category. *Tricks* are generally knots or complicated arrangements of the string which run out freely when pulled. Sometimes it is difficult to decide which name should be applied.

Terminology

A string passed over a digit is termed a *loop*. A loop consists of two strings. Anatomically, anything on the thumb aspect of the hand is termed *radial*, and anything on the little finger side is called *ulnar*, thus every loop is composed of a *radial string* and an *ulnar string*. By employing the terms thumb index, middle finger, ring finger, little finger, and right and left, it is possible

---

[1] Abridged from an article by W. H. R. Rivers and A. C. Haddon in *Man*, 1902, 109.

to designate any one of the twenty strings that may extend between the two hands.

A string lying across the front of the hand is a *palmar string*, and one lying across the back of the hand is a *dorsal string*. Sometimes there are two loops on a digit, one of which is nearer the fingertip than the other. Anatomically, that which is nearer to the point of attachment is *proximal*, that which is nearer the free end is *distal*. Thus, of two loops on a digit, the one which is nearer the hand is the *proximal loop*, that which is nearer the tip of the digit is the *distal loop*; similarly we can speak of a *proximal string* and a *distal string*.

In all cases various parts of the string figures are transferred from one digit or set of digits to another or others. This is done by inserting a digit (or digits) into certain loops of the figure and then restoring the digit (or digits) back to its (or their) original position, so as to bring with it (or them) one string or both strings of the loop. This operation will be described as follows: "Pass the digit into such and such a loop, take up such and such a string, and return." In rare cases a string is taken up between thumb and index. A digit may be inserted into a loop from the proximal or distal side, and in passing to a given loop the digit may pass to the distal or proximal side of other loops. We use these expressions as a general rule instead of "over" and "under", "above" and "below", because the applicability of the latter terms depends on the way in which the figures are held. If the figures are held horizontally, "over" and "above" will correspond as a general rule to the distal side, while "under" and "below" will correspond to the proximal side.

A given string may be taken up by a digit so that it lies on the front or palmar aspect of the finger, or so that it lies on the back or dorsal aspect. In nearly all cases it will be found that when a string is taken up by inserting the digit into the distal side of a loop, the string will have been taken up by the palmar aspect, and *vice versa*. Other operations involved are those of transferring strings from one digit to another, and dropping the strings from a given digit or digits.

The *manipulation* consists of a series of movements, after each of which the figure should be extended by drawing the hands apart and separating the fingers; unless this would interfere with the formation of the figure, in which case a note should be made that the figure is not to be extended.

There are certain opening positions and movements which are common to many figures. To save trouble these may receive conventional names; but it is better to repeat descriptions than to run any risk of obscurity.

Position 1. This name may be applied to the position in which the string is placed on the hands when beginning the great majority of the figures. For example "Position 1" denotes: Place the string over the thumbs and little fingers of both hands so that on each hand the string passes from the ulnar side of the hand round the back of the little finger, then between the little and ring fingers and across the palm; then between the index and thumb and round the back of the thumb to the radial side of the hand. When the hands are drawn apart the result is a single radial thumb string and a single ulnar little finger string on each hand with a string lying across the palm. This position differs from the opening of the English cat's cradle, in which the string is wound round the hand so that one string lies across the palm and two across the back of the hand with a single radial index string and single ulnar little finger string.

Opening A. This name may be applied to the manipulation which forms the most frequent starting point of the various figures. For example, "Opening A" denotes: (1) Place strings on hands in Position 1; (2) with the back of the index of the right hand take up from proximal side (or from below) the left palmar string and return. There will now be a loop on the right index, formed by strings passing from the radial side of the little finger and the ulnar side of the thumb of the left hand, i.e. the radial little finger strings and the ulnar thumb strings respectively; (3) with the back of the index or the left hand take up from the proximal side (or from below) the right palmar string and return, keeping the index with the right index loop all the time so that the strings now joining the loop on the left index lie within the right index loop. The figure now consists of six loops on the thumb, index, and little finger of the two hands. The radial little finger string of each hand crosses in the centre of the figure to form the ulnar index strings of the other hand, and similarly the ulnar thumb string of one hand crosses and becomes the radial index string of the other hand.

Example of a string figure from Torres Straits.

Ti meta, "nest of the *Ti* bird" (Mer); *Gul*, "a canoe" (Mabuiag). Opening A. Insert each index into the little finger loop from the distal side and pass it on the proximal side of radial little finger string and bring it back to its previous position by passing it between the ulnar thumb string and the radial index string. Let go little fingers. There are now two loops on each index and a large loop passing round both thumbs. Insert the little fingers from the distal side into the index loops and pull down the two ulnar index strings. Let go both thumbs gently and insert them into the same loop in the opposite direction to which they had been previously, i.e. change the direction of the thumbs

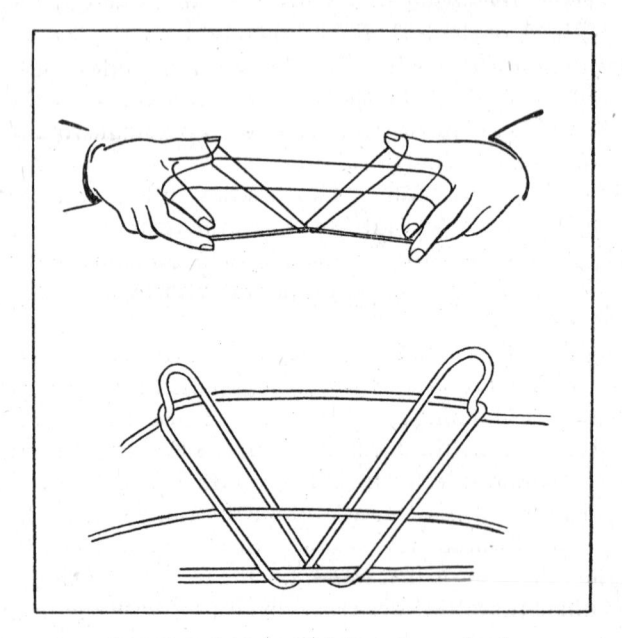

Examples of a string trick from Torres Straits

in their loop. With the dorsal aspect of the thumbs take up from the palmar side the strings passing obliquely from the radial side of the index fingers to the ulnar little finger strings and extend the figure. The inverted pyramid in the centre represents the bird's nest.

Kebe mokeis, "the mouse" (Mer). Hold left hand with the thumb uppermost and the fingers directed to the front. Put whole left hand through the string, letting the loop fall down its dorsal and palmar aspect from the radial side of the thumb. There will then be a pendent palmar and dorsal string on the left hand. Pass index of right hand beneath the palmar string, and between the thumb and index of the left hand, then pass it round the pendent dorsal string, bringing it between the thumb and index. Give the loop thus made a twist clockwise, and place it over the index of the left hand. Pull tight the pendent strings. Again pass right index beneath the pendent palmar string and between the index and middle fingers, hook it over the dorsal string as before. Bring it through as before, twist the loop clockwise, and put it over the middle finger. Repeat so as to make similar loops over the ring and little fingers. Pull all the strings tight. Remove the loop from the left thumb, hold it between the left thumb and index. With right hand pull the palmar string, and make a squeaking noise as the loop disappears from the left hand.

PART IV

FIELD ANTIQUITIES

FIELD ANTIQUITIES

Work on field antiquities is closely linked with anthropology, especially in the study of movements of peoples and cultures and of technological evolution. A lowly society of the present day may be living in a manner that in some aspects resembles that of prehistoric society. Hence an archæologist may discover remains which help in tracing early phases of present-day patterns of culture. Equally, a field worker inquiring into the social life and technological equipment of a people may contribute to the study of early man and, generally, of our cultural heritage. A researcher should therefore try to interweave social and technological studies with work on field antiquities, especially among peoples whose life may be, in some respects, described as "a living past".

Equipment and Technique

A survey of the field antiquities of a region may be merely a reconnaissance, or be of an intensive nature. We are mainly concerned here with the first type, since intensive study is a matter for trained specialists over many seasons of minute exploration and excavation. The traveller attempting such a reconnaissance should know the general sequence of phases of cultural evolution described by archaeologists in the standard works. Before going into the field he should visit museums and become familiar with various types of tools, stone implements, pottery, etc., if possible by actual touch. Equipment for collecting, for field recording and registration of material is necessary. Flints and other stone implements, potsherds, ornaments, etc., may be collected from sites of ancient settlements. Ample supplies of small and medium-sized bags of stout material are needed; paper is generally unsuitable. The bags should have tie-strings and an attached tab for numbering and labelling; a paper slip label inserted in the bag is useful as an additional check. These bags should be used only on reconnaissance, and should be repacked securely on return to camp to avoid breakages. For repacking ample newsprint is useful with cotton-wool and straw. Stiff cardboard boxes packed into wooden cases are valuable, and sometimes essential. Beads, microliths and small metal objects are best carried in cardboard boxes which can be obtained in graduated sizes. A small hammer and a pick with one end pointed and the other solid or square are useful tools, but serious digging should only be undertaken by trained archæologists.

Field recording of finds and observations is essential, and each item should be numbered and dated; the memory cannot be trusted; collections without records made on the spot have led to serious misapprehensions. Two books should be kept, one as a daily field journal, the other as a register with field numbers of objects found and brief descriptions, preferably in indelible pencil or waterproof ink. A field number and, if possible, some mark giving the place of the find should be indelibly marked on each specimen. The entry in the register should give this number, the date and place of discovery, name of the finder and a short description of the article found. Sometimes it is advisable to add a note on the possible period of the specimen, on its condition and its technological relations. Photographs, sketch maps, etc., of objects or sites should be included in the register and numbered; the number given to any find shown in the photograph should be included in registering the photograph. Sketches should be made and registered in the same way. It is important to place a scale (preferably wooden metre-scale with black-and-white decimal segments) in position for photography. Large-scale maps for marking positions of finds are important, so that a return to the exact sites may be possible. If air-photographs of the area to be surveyed are available, they should be carefully studied, as they often disclose archæological features not easily visible from the ground.

Finds of Movable Objects

Surface collections often indicate the existence of important sites, and therefore if registered as above can provide valuable clues for subsequent research. A broad description of the neighbourhood of a find is necessary. A surface find may be derived from an adjacent deposit; a stream may cut into its bank and re-expose objects formerly embedded in its lateral terraces. The condition of the surface find should be noted. It should be noted whether the flint implements are wind-worn, patinated, or water-rolled. Patination and physical condition may be points of primary importance, especially for estimating the period to which the find belongs.

The main interest of the field observer is in the objects and data representing the mode of life and the ways in which needs are met, such as stone implements, pottery, ornaments and miscellaneous material in bone and metal.

Stone Implements. These are of special interest. Their occurrence normally, but not necessarily, suggests prehistoric date; certain types of stone implements have remained in use in some places to the present day. In collecting and classifying implements the type and variety of stone must be noted—whether flint, chert, quartzite, obsidian or other stone. If a factory site is discovered, i.e. a place where implements were made, then the "provenance" of the raw material may be easy to determine and noted on the map together with the site itself. Flint mines and obsidian quarries can often be discovered by the use of the geological map. Flint may be either nodular or tabular; in many cases it forms bands in cretaceous formations. Pebbles of flint found lying on the surface were often used, but "fresh" flint that has not been thus exposed to changes of temperature can be flaked more easily. Early men, like modern men, knew this and therefore frequently mined to obtain good nodules. This is not the place to give a detailed account of the various types of stone tools: that can be found in any standard work. But the distinction between tools made on cores and tools made on flakes or blades should be noted, and also that between tools flaked on both faces (so-called bi-facials) and those flaked or retouched on one side only (unifacials). Stress is generally laid on the method used to give a sharp edge to chopping or cutting tools, whether the edge has been produced by simple flaking or the intersection of one or more flake-scars, or by grinding and polishing. The use of the latter method was, till recently, considered the essential feature of the "neolithic stage". Finally patination, general condition, wind and water effects, and traces of utilization should be noted. Utilization can be shown in the form of battering, blunting or breakages along the working edge or end of a tool. In certain cases, such as sickle-blades used in cutting silica-bearing grass (wheat for example), utilization results in shiny lustre; this may be an indication of cultivation. Types of tools (used in agriculture or in hunting, etc.) and their proportions to each other may indicate the prevailing occupations of a certain people or community. During the later phases of the Stone Age, specialization in tools became well marked; and we can clearly distinguish between tools used in agriculture (hoes, sickles, etc.), hunting (arrow-heads, javelin-points, spear-heads, bolas, staves, etc.), fighting (mace-heads, arrow-heads, etc.), or domestic work (scrapers, knives, chisels, gravers, borers, adzes, etc.) Chopping tools sharpened by polishing or grinding are often termed celts—a convenient word that covers equally axes and adzes and even chisels or gouges; an adze can be distinguished from an axe if the plane of the blade is asymmetric.

Pottery. Use of pottery began in the neolithic stage or somewhat earlier. It is frequently found that early pottery forms are exact copies of vessels previously made in other materials, i.e. leather, gourds, wood or basketry. Details essential to the older material but unnecessary in the pottery forms are frequently retained as decoration. Sir John Myres has given the name "skeuomorph" to such imitations in one material of objects originally made in another, and calls patterns that enhance the resemblance "skeuomorphic". For the classification of pottery the method of manufacture—building up by hand or throwing on the wheel —is of major importance; but slip, burnishing, firing, painted, incised or impressed decoration, glazing, etc., are significant, as are also shape, including handles, lugs, spout, feet, and bases. Certain vessels have property-marks on them, and some even bear inscriptions cut either before or after firing.

Ornaments. Beads and shells, sometimes in strings, ear-rings, bracelets, anklets, armlets, rings, pins of metal, bone, ivory, etc., pendants, escutcheons, amulets, and other charms should be sought. In some communities one may find stone or other palettes, sometimes ornamented or coloured, sometimes used for painting with ochre and malachite.

Miscellaneous Objects. Bone and horn have been used from early times for harpoons, fish-spears, fish-hooks, borers, needles; also for polishing, decoration, and other decorative purposes. Wood has been in use, too, but, being perishable, little has survived. Metal has long been used. In some parts of the world copper or bronze gradually took the place of stone for the manufacture of cutting or piercing tools and weapons, and initiated what archæologists call a Bronze Age (or Stage). Some archæologists recognize an intermediate Chalcolithic Stage when both copper and stone were used for knives, daggers and axes; in such finds, stone implements may often be imitations of copper forms. Eventually iron, being cheaper, superseded copper and bronze for industrial purposes, or in some parts of the old world replaced stone for these purposes without the intervention of a Bronze Stage. In general, when metal tools begin to compete with stone ones, the technique of stone-working declines, but it takes a long time for metal to replace stone and bone altogether, even for the manufacture of axes or daggers. Stone is exclusively used by most backward societies for rough tools like hammers, and for things like hand-mills. The field worker should be on the alert for all evidences of cultural and technological contacts. It is almost a general rule that contact of cultures which are not too widely divergent is a fruitful source of initiative, and may bring about new techniques.

Preservation of Metal Objects. This demands special care. Metals used are copper, copper alloys (including bronze), iron, gold, silver, electrum (gold and silver), etc. Cleaning should never be attempted in the field; it demands specialist attention to avoid damage and obliteration of important features. Thin wires, pins, rings, hooks, etc., may be so corroded or rusted that hardly any metal core is left; in fact sometimes only a discoloured outline of the original object remains. This should be drawn with minute accuracy. Sometimes one may find unfinished objects, or lumps of metal or ore, or moulds; these can all be valuable in shedding light on sources of metal and on metallurgical processes. Ore deposits should also be sought and specimens of the ore collected for analysis by experts.

Settlements, Cemeteries and Surface Sites

Habitations, ritual centres and burial-grounds may yield finds; those from the first two are likely to give the truer picture of the habits, activities and general life of the people concerned. Objects in graves may have been selected for ritual purposes; the finds from both habitation and burial places should be kept distinct.

Cemeteries need special training and experience for their exploration. An exposed grave should be drawn, accurately measured, and photographed. If a skeleton is exposed in the grave a line drawing should be made to show its attitude (contracted with knee up to the chin, flexed with legs making more than 90° with the spinal column or extended, lying on right or left side or on the back, etc.). These facts have yielded clues concerning ritual and belief. Large pots may have been used either to contain the ashes of the cremated dead, or for food and drink for the dead man's use. Skeletons that have long been in the earth should not be exposed to the air as they are apt to break up quickly. Shellac solution painted immediately on bones hardens and protects them, and should be allowed to set before the bones are removed to be cleaned and examined in the laboratory.

Domestic Sites, Caves and Rock Shelters, factory and fabrication sites, surface stations, mounds, mines and other workings and ritual centres connected with them, localities with rock drawings and other spots may be found. Deposits containing implements and other relics may fill a cave almost up to the roof. A traveller should inquire about the occurrence of caves and shelters under overhanging rocks and should visit them for traces of former occupation. Ashes, midden, broken bones, flints, sherds, smoke on the roof, drawings on the walls, impressions of hands and feet are all possible occurrences. Surface stations in the open may be recognized by deposits of ash, or by potsherds and other relics strewn about the surface, but may yield little or no deposit of material made by man. Everything possible should be observed and collected, including any indications of a succession of occupations of the site, e.g. finds at a lower level separated from surface finds by a barren layer, however thin. On such sites one is liable to find objects of different periods. In certain cases, careful collection and registration of material and exact large-scale mapping of the position of each object will reveal that parts of the site have been differently occupied, probably at different times. Under certain conditions, however, if a site has been occupied for a long time the repeated reconstruction of buildings on the same spot produces a mound or *tell*, rising above the level of the surrounding ground. Pottery and other relics of different ages are likely to be exposed on its surface, but a cutting into the mound, made for instance by a water-course, may disclose successive "floors" marked by layers of ash or stamped clay, the relics from each of which should be carefully separated. On the other hand, mounds or barrows may be heaped up over graves. Superficially a *tell* and a large barrow are hardly distinguishable, except by size, and all artificial mounds are archæologically significant: the traveller should note the height and the diameter of each and record its position on a map.

SETTLEMENTS, CEMETERIES AND SURFACE SITES 349

Fabrication or Chipping Sites for Stone Implements are usually near a source of raw material, a flint, obsidian, diorite or other mine or quarry, or a beach with flint or other suitable pebbles.

Discarded or unfinished tools are likely to be abundant, and a good sample should be taken. Mistakes have been made in the past through a collection of only the best pieces. Completed tools are most helpful for dating; but unfinished tools should be collected. However, care must be taken to distinguish unfinished unpolished neolithic implements from earlier palaeolithic types which were never intended to be polished. It must also be remembered that, even in Europe, flint implements have remained in use in some places until recent times.

Rock Paintings and Drawings on a rock face or in a cave may be naturalistic or schematic and may belong to periods from the late Palaeolithic to modern times, e.g. in South Africa and elsewhere. Figures may be indicated by line-grooves, by pecking a whole surface, by painting in outline or over the whole surface, or by cross-lining. If the picture has faded or weathered, spraying or damping the surface gently with clear water may help to make it clearer, but this is an emergency treatment which may cause damage, and, if repeated, lead to evaporation and the deposit of a crust. Some workers touch up the engraving with chalk or other colouring matter before photographing; this is dangerous and is seldom justified. A photograph, a hand drawing, and, if possible, a tracing of the drawing, should be secured, and each should have a scale shown on it. Figures often overlap and overlie one another, and this may be helpful in determining chronology. Indication of climatic changes may be detected in rock drawings by the representation of animals no longer extant in the environment. It is therefore important to try to identify animals shown in rock drawings. One should also look carefully for cases in which a deliberate imperfection is to be seen, e.g. in a South African drawing on stone a very finely drawn elephant is shown with the hind legs of a rhinoceros. This may be related to a taboo against perfect representations or may represent a composite mythological animal. Thus a study of rock drawings may give indications of both environmental and cultural features.

Environment

Prehistoric sites often provide valuable evidence for the physiography, climate, fauna, and flora of the time of occupation. Such evidence must not be neglected.

Fauna and Flora. Bones, teeth, shells, pieces of wood, charcoal, leaves and other organic remains found in association with remains of prehistoric man are always worth collecting, even though they might not bear signs of having been utilized by man. They help in the reconstruction of the ancient environment and, incidentally, in the dating of the sites. The specimens should be treated like archaeological objects, i.e. treated with preservative if necessary, carefully packed and labelled.

Stratigraphical Sequence. If the site provides a vertical section, for instance of river gravels in a pit, samples should be taken of all the distinguishable strata, and their thickness, depth below surface, and other characters recorded. The samples should, if possible, cover the *entire* section, beginning well below the archaeological layer and continuing up to the modern surface. They are easily packed in canvas bags or tins, and their size need not normally exceed that of a fist. But they should be taken in such a manner that contamination from other deposits is avoided. If the samples have to be stored for some time before investigation, it is best to allow them to dry. If they can be submitted to a geologist within a few weeks, loams and clays will keep damp in a tin without turning mouldy and may be welcome in this condition. This applies particularly to weathering soils.

A set of stratigraphical samples is of value, even if it cannot be studied by an expert. It will always provide the factual record of the embedding material which it is often difficult accurately to describe in words.

Physiographic Features. Implements dropped by hunters are often found along river terraces, thus where the river has made a cut is often a good place at which to search. Implements may have weathered out and may lie at the foot of the scree if there is one. Finds in scree will suggest that it may be worth while to examine the material of the terrace itself. Terrace implements should be examined to see whether they are rolled (i.e. waterworn and perhaps water-borne from older formations), or wind-worn, or relatively fresh. Quarries and gravel pits are often useful places for special study especially when they occur in river terraces. Terraces often occur in series and should be measured and distinguished according to their height in metres above the adjacent flood-plain of the present-day river. In doing so it is important to give the figures both for the surface of the terrace and, if available, for the rock-bench on which the gravel rests. If this cannot be obtained, a figure for the maximum thickness of the gravel observed is often useful. The height of the level at which implements occur in the gravel should also be determined. Higher terraces are usually older than lower ones, but, especially at river bends, they tend to merge into one another. In some areas, loess or other non-fluviatile material may have covered or masked the terrace.

In warm and tropical belts pluvial periods seem to have occurred which correspond to the Ice Ages of the temperate zones. In such areas remains of lakes may be detected or present lakes may have covered wider areas. Their outlines may be traced by following beaches near which men lived and dropped tools. Shells, etc., from these beaches should also be collected and carefully registered, storm beaches being differentiated from normal ones. The height of each beach above the level of the present lake or its dried-up floor should be noted.

Implements are found also in raised sea-beaches, but more often occur on the surfaces. Shells, etc., associated with the beach or with its surface may give important clues, e.g. the common winkle (*Litorina*) in Scandinavian beach deposits of a certain post-Pleistocene phase suggests conditions warmer than those of the present day. The same types of shells in Portuguese beach material suggest conditions cooler than those now prevailing. Physiographic work on raised beaches in a complicated matter, but the field worker should note the height and thickness of the deposit and look out for ancient cliffs and, if possible, determine the height of their base.

Moraines may be noticed in regions now quite temperate or even warm, but implements are not likely to be found in them. Stones in a moraine are apt to have been chipped as the ice ground its way along, and observers should be on their guard against too credulous interpretations of such chippings as indications of human workmanship. Mapping of moraines and ascertainment of their height above sea level and distance from the probable source of the ice may all be helpful.

Former sand dunes, sometimes now quite fixed and transformed (fossilized) into sandrock or sandstone, may exist in areas now covered with vegetation, and may thus show a former extension of desert conditions. Other fossil dunes are now covered with vegetation and others still contain fossil soils. Palaeolithic and Mesolithic industries are often associated with dunes.

Loess deposits need careful study. They are fine yellow or brown silts of aeolian origin, and apt to form vertical cliff-like faces where attacked by erosion. They often contain fossil soils, usually indicated by a browner hue and loamy texture. It is important to state whether implements are found in fresh loess, the fossil weathering horizon or soil, or on the surface of the fossil soil. Loess seems to have accumulated chiefly around regions of persistent high pressure, e.g. ice sheets or cold continental interiors; it has lent itself to early cultivation by its friable character and its organic content, and incidentally to the excavation of shelters and habitations. Palaeolithic habitation sites in the loess are sometimes concentrated into well-defined round or oval areas of dark soil which probably represent the bases of former huts. Volcanic flows may cover implementiferous deposits such as terraces and beaches.

Soils. All deposits containing remains and artefacts of early man are subject to the process of weathering. Thus, soils are formed. The structure of a modern soil may easily be inspected in any pit. Familiarity with modern soils will help the field worker to recognize buried or fossil soils. These represent ancient land surfaces on which man may have left his artefacts, and they are important evidence for the type of climate in which early man was living.

APPENDICES

Photography

A GOOD photographic record is an essential part of every kind of anthropological field work.

It is not the purpose of these notes to give instruction in photographic technique. The intending field worker who has had no previous experience with a camera should consult someone with a good knowledge of photography before starting, and under his guidance should practise both taking and developing photographs until he understands the technical processes involved and can recognize the results of correct and incorrect exposure and development. There are, however, some problems arising in anthropological field work which would not come within the purview of the photographer working only at home, and it is these which will be briefly considered in the following paragraphs.

Apparatus

It does not greatly matter what kind of apparatus is used, provided the operator is thoroughly familiar with it. At least two cameras should be taken, to guard against risk of the failure or loss of one of them. If they are the same size it simplifies the provision of films and developing apparatus, but there are some advantages in having cameras of different kinds.

If architectural work is to be undertaken, it will be well to use a camera not smaller than quarter-plate. It must have a rising front and should be fitted with a wide-angle lens. Telephoto lenses are expensive and heavy, and are seldom required. Some kinds of archaeological work also require this type of camera, which will, in general, be used with a stand. But for most anthropological purposes a hand camera is the most suitable, and there is little if any advantage in taking one whose negative size is larger than $3\frac{1}{2}$ in. by $2\frac{1}{2}$ in. This size is large enough to give clear prints by contact, and lends itself well to the making of lantern slides.

Miniature cameras such as Leica, Contax, and the like have the great advantage that a large number of exposures can be made without changing the film. They also stand up well to tropical conditions. On the other hand, the need for raising the camera to eye level for focusing most cameras of this type sometimes makes its presence undesirably obtrusive. It is essential that pictures taken with miniature cameras should be perfectly sharp in focus, and free from dust specks (sometimes difficult to ensure under field conditions), otherwise the results will not be satisfactory when enlarged to a reasonable size, or as lantern slides. Fine-grain film and a fine-grain developer should be used with miniature cameras.

The majority of subjects photographed by the anthropologist require exposure speeds at which the camera can safely be held in the hand. A good durable stand should be taken in case of need, but if the camera is one with which focusing is done on a ground-glass screen with the aid of a dark cloth, great care must be exercised in using it among people unaccustomed to photography. The manipulations under the dark cloth may suggest black magic, with disastrous results, and if, as will certainly happen, onlookers wish to look in the focusing screen, they will probably be frightened by seeing people and objects upside down. For the field worker who depends upon the goodwill of those among whom he works, cameras of the Reflex or Rolleicord type have the great advantage that people can look into the reflector and will there see what is being taken in a way which they can understand. Special lenses can be obtained which fit in front of the permanent lens, for taking portraits or very small objects with this type of camera, and these should be carried.

Exposure

The field worker is likely to meet with light conditions varying considerably from those under which he works at home. Some form of exposure meter should therefore be used. The most satisfactory kind is that which works by means of a photo-electric cell, but these do not always continue to function accurately after prolonged use in the wet tropics. Under the worst conditions it may be necessary to use an exposure calculator instead; these can be obtained with tables for use in different latitudes. If no mechanical aid is available, when taking photographs under difficult conditions, or when the subject is one of particular importance, it is worth while to make three negatives, one with what is judged to be the correct stop and exposure, one at half and one at double this, altering either stop or exposure as the subject may require. Unless the photographer is particularly unlucky one of the three should be a useful negative. (The writer has found this method satisfactory.) With using film of wide latitude, some photographers recommend eight times more and eight times less than the "best guess".

Development

It may be possible to send films to a base for development, and this is, at present, necessary if colour film is used. But it is advisable to get the film (whether coloured or not) developed as soon as possible after exposure. In the wet tropics even twenty-four hours' delay may affect the quality of the negative. Except when exposed film can be sent away quickly, it is advisable for the photographer to do his own developing. Tanks suitable for daylight loading now do away with dark-room difficulties, but if these are not available, it is quite possible, by working at night, to develop by the old-fashioned "dish" methods, even if the "darkroom" is no more than a hut walled and roofed with leaves or even a light-weight tent carried expressly for the purpose. But bright moonlight, which will fog very sensitive film, is sometimes hard to exclude entirely. A large-sized electric torch, with a dry battery, can be fitted with a red shield to provide a dark-room lamp, and if used sparingly will last for several months.

Under tropical conditions an extra hardener is essential. With it water may be used up to a temperature of 80°. Formalin is useful, in the proportion of 1 part formalin to 20 parts water. Chrome alum may be used instead, in the same proportion.

Standard makes of tabloid chemicals will keep good for an indefinite time if the container remains unopened; it is therefore possible to take a good supply of these, packed in small quantities, and also of acid hypo, packed in tins.

The water supply may be a problem. If necessary, sea water can be used for washing negatives, but great care must be taken to see that it is free from sand, etc., and subsequent washing in fresh water is necessary. If the worker is making a long stay in the field, batches of developed negatives could be sent home with instructions that they should be placed in a fixing bath and then well washed before being stored, as a safety measure.

Care of Apparatus

Special precautions are desirable to ensure that cameras and other photographic apparatus are kept in good order under various climatic conditions, especially those involving extremes of heat and cold, dryness and damp. Considerable advance has been made in dealing with these problems by researches carried out during the war. It is not possible to deal with this subject in detail within the compass of these notes, but attention is called to the facilities for instruction in such matters offered by the larger firms specializing in photographic equipment, and the field worker would do well to consult the firm whose products he intends to use. Those intending to work in the tropics, especially the wet tropics, must take only film with adequate tropical packing. Tins sealed with adhesive tape, though sometimes offered as "tropical packing", are not suitable. Roll-film must be hermetically sealed in a metal container which is broken open when the film is required for use.

Storage of Negatives

Negative albums containing numbered envelopes bound into a stiff cover are probably the best means of storing negatives in the field, and for reference later. Do not attempt to economize in the use of these by putting more than one negative into each envelope. It is important that exposed films should never be allowed to remain in contact with one another.

The subject of each negative with date of exposure should be recorded on the numbered list included in the book. At the outset of work in unfamiliar climatic and lighting conditions, it is useful to record the stop and exposure used, until the photographer has become familiar with the range appropriate to his surroundings. Even if an exposure meter is available, there will be occasions when there is no time to use it, and experience is the only guide. All negatives, even very poor ones, should be retained as they can sometimes be made to serve if an opportunity for repeating the picture does not present itself (which happens exasperatingly often) or may be useful for making outline drawings.

Printing

It is not usually necessary to make any prints in the field. Considerable saving in the amount of equipment carried can be effected if printing is not attempted, and it is difficult to keep the paper in good condition. If possible, avoid promising prints to anyone; such promises are sometimes hard to keep after returning home, and failure to do so leads to disappointment.

Films not good enough to give a useful print can, if they show any details which are valuable, serve as a basis for line-drawings. A print is made on matt paper, the details required are then drawn over the image in waterproof Indian ink, and the photograph is then bleached out (formula for bleaching solution is given in the *British Journal Photographic Almanac*). The result is a sketch in black lines on a white background, which is relatively cheap to reproduce for publication, and probably more accurate than a freehand drawing.

Indexing of Prints

A useful method of recording the photographs on returning home is to make a card index of prints. A contact print (or small enlargement if miniature camera is used) is made from each negative (rough prints will serve), and stuck on a card leaving space for writing in the number of the negative, subject, place, date, and any other relevant data. Size 8 in. by 4 in. cards are suitable, unless a large camera is used. These cards can then be filed in a Card Index Storage Box with guide cards giving subject headings or any other method of classifying. Although the making of such an index involves the expenditure of a considerable amount of time and of some money, it will be found well worth while if much use is to be made of the photographic record, as it enables the worker to ascertain quickly what photographs are available to illustrate any part of his material for publication, or for making lantern slides.

Subjects

The type of subject selected for photography will, of course, vary with the interest of the field worker. But it should be remembered that he will be asked for details of aspects of the life of the people studied by him which may fall outside his own particular line of study, and he should therefore make the range of his photographs as wide as possible, having regard to the time and resources at his disposal. This is especially important if the people among whom he is working are little known. Typical local scenery should not be forgotten, and it is well to take these soon after arrival, before familiarity has dulled appreciation of characteristic features of the landscape.

Portraits should be of individuals, though some groups may be included showing people sitting or standing about in characteristic postures. Representative individuals should be photographed full face and profile, head and shoulders, close enough to show the features in detail; a full length may conveniently be added to show proportions of limbs. If the photographs are accompanied by measurements it is well to include a disc bearing a number large enough to be visible on the print; this must correspond with the number on the measurement record sheet, to facilitate identification.

For portraits, and some other subjects, a plain background is almost essential; as this is sometimes difficult to obtain, a roll of neutral-coloured cloth, which could be stretched between two poles should be provided. It must be kept at a sufficient distance from the subject to be out of focus, but near enough to cover the whole field shown. A plain wall, especially of mud, makes a good background, and an open doorway can sometimes be effectively used. In open country the sky above the horizon serves well if there is no danger of halation.

In photographing technical processes care should be taken that the worker is looking at what he is doing, and not at the camera. The photographer should take the same stage of the operation from several different positions, if possible. Close-ups of the hands showing the method of manipulating the object being made or of handling a tool add greatly to the value of such studies. Stills exposed at frequent intervals will, if taken carefully and numbered in sequence, show the details of technological processes, such as weaving, pottery-making, etc., as well as, and sometimes better than, cinematograph pictures.

It cannot be too strongly emphasized that no photographs of a sacred place or building, or of a ceremony, dance or ritual performance, should be taken unless the permission of the people concerned has been obtained. It will usually be given if the field worker has been able to win their confidence. When permission has been given the photographer should do his work as inconspicuously as possible. It is far better to do without such pictures than to run the risk of incurring ill-will or even hostility by tactless disregard of local feeling which we should not tolerate in our own society. To enlist the co-operative interest of the group being studied it is usually sufficient to explain that the pictures will enable your own people, who live a long way off, to see how their people live and what they do. In addition a few photographs of people and of easily comprehensible activities from home, and from other communities may be shown; they will be examined with the greatest interest and will help to establish good relations with the camera.

Cinematography

Although a most useful, and in some circumstances essential, adjunct to anthropological field work, the cine-camera should not be looked upon as a substitute for the ordinary camera. Cinematograph films are not often satisfactory for making enlargements or lantern slides, as the image is not sharp enough, though they can be used for these purposes in case of necessity. Some cine-cameras have a switch which enables stills to be taken in the middle of a film, but the individual frame is not easy to pick out afterwards. There are many occasions where the results obtained with an ordinary camera are preferable to those given by the cine-camera. Both can advantageously be used in conjunction in many cases, but if there must be a choice, it is probably better to concentrate on a good ordinary camera for most kinds of field work.

There are several types of portable cinematograph apparatus which are suitable for anthropological field work. Full size (35 mm.) apparatus is too expensive, too heavy and too bulky for most workers, unless cinematograph records are to be made a special feature of the expedition in charge of a whole-time photographer. For most purposes apparatus using 16 mm. film is adequate. The 8 mm. size should not be used as the pictures cannot easily be enlarged sufficiently to be satisfactory for lecture purposes, though good enough in a small room. Coloured cinematography should be attempted only if the film can be sent for development within a short time of being exposed.

As with ordinary photography, those who are taking motion pictures for the first time should consult an expert before starting, and should take at least one trial film and see it on the screen. They will thus learn many small technical points which make all the difference between an interesting film and a dull one, without detracting from the accuracy of the film as a record. A little experiment under skilled supervision is worth far more than printed instructions, but a few points may be noted.

The first essential in producing good cinematograph pictures is to remember that *moving pictures must move*. This may seem obvious, but judging from considerable experience of amateur films, it is a rule too often ignored. Do not waste cinematograph film on subjects which can be shown as well, if not better, by means of stills. Most architectural and many archaeological subjects come under this head. So do photographs intended to illustrate physical type, unless special features, such as gait, are being studied. Scenery is not usually very interesting unless something is happening all the time, though a short panorama may be effectively used as an introduction. People should not, as they so often do, come into the picture only to stand still staring at the camera. Faults of this kind can, of course, be eliminated by judicious cutting later, but it is better to switch off and save the film.

It should be remembered that to the audience who will see the film at home, both place and people will be unfamiliar. Enough general shots should therefore be taken to give a background, say of village life, which can be shown before films of particular activities.

Long sequences, such as those illustrating technological processes, are best shown by a number of fairly brief scenes taken from different positions. Shots including the surroundings, and the onlookers, if any, should be interspersed between nearer ones showing the position of the worker in relation to that of the object he is making, and close-ups showing technological details as clearly as possible. Similarly, pictures of a ceremony might be varied with close-ups of the reactions of the audience, but not so frequently as to distract attention from the performance.

Shots should be varied in length as well as in position. The time required on the screen will be the same as that occupied in taking the shot. It is a good general rule to start with an introductory shot, say of not less than 10 seconds, to enable the audience to get their bearings. For the others, not less than 5 or more than 15 seconds is a good working rule. Close-ups should be kept brief. A long shot after a number of short ones heightens its effect by contrast. No shot should be prolonged after the scene has conveyed its point, when the attention of the spectator begins to wander. This applies in particular to such scenes as native dances and ceremonies, which can become extremely monotonous on the screen, as they often consist of simple actions or movements repeated almost indefinitely. The repetitions should be omitted from the film, though the number of times they occur should, if likely to be significant, be recorded in the fieldworker's notes.

If taking pictures when yourself on the move, as in a boat or an aeroplane, include a piece of the boat or plane in a corner of the picture to emphasize the nature of the shot. Avoid taking shots of people or objects moving parallel with you in the same direction; the result will be to show them as if they were not moving at all. Such shots should be taken at an angle of 45°.

Notes should always be made of the scenes taken, and of the order in which they should be shown. This is particularly important when filming ceremonies or technological processes, as the order of the different parts is not always apparent from the film and may become inverted during editing. The filming of different parts of a long process, such as pottery-making, which may need to be taken on different days, should not prevent the worker from taking other shots in the meantime, as the relevant portions of the film can easily be brought together when it is prepared for showing. For this work the help of an expert will probably be required, and as this can easily be obtained no notes on the subsequent treatment of films will be given here.

Collecting and Packing

The purpose of collecting ethnographical specimens in the field is to provide material for the study of the material life of primitive people, and, to a lesser extent, the paraphernalia associated with religious or magico-religious ceremonies. Isolated objects are of little value in themselves and undocumented specimens are useless. Therefore all collections should be well labelled, and selected with a view to showing as completely as possible an industry of a people. For example, a hoe by itself may not perhaps be of great interest except for comparison with specimens from other areas. Each implement should form part of a series illustrating all aspects of the agriculture of a particular people, such as other cultivating instruments, reaping knives, winnowing trays, grinding and milling apparatus and so on, and each instrument should if possible be accompanied by photographs showing method of use. Wood carvings, however interesting they may be as examples of the art of a people, are considerably enhanced in value by being accompanied by the tools used, and, if possible, by unfinished specimens showing various stages of their manufacture.

Care should be taken to see that the collection is as complete as possible for any particular aspect of culture. The negative evidence inferred by the absence of a type of object from a collection is sometimes of great value, but the inference can only be safely used by the student if he is reasonably certain that the collection is comprehensive. It is better to make a complete collection of objects illustrating one aspect of a culture than a miscellaneous collection depending purely on the fancy of the collector.

A question of some difficulty may arise in the case of cult objects—especially if they are unique and their removal would be liable to destroy a culture. Apart from any question of offending native susceptibilities the anthropologist will seldom be justified in removing unique specimens of a ritual character except where there is evidence that they are neglected and will probably be destroyed. In such cases not only will he be justified if he preserves them but it is obviously his duty to do so.

In such cases advice might be sought of local officials, if accessible, or from the authorities of the national museums if communications permit. In any case the anthropologist should be at pains to consult one or more of the National or University Museums before setting out on an expedition. They will be most willing to advise him what to collect.

For general convenience a specimen label with appropriate heads is appended.

It must, however, be emphasized that labels, however detailed, are not a substitute for, but complementary to, the investigator's notebook.

Packing

Good materials for packing are the following: textiles, paper, wood-shavings, straw; sawdust is extremely bad. If there is sufficient paper, wrap each specimen up separately, so as to prevent chafing and the obliteration of painted designs. Perishable specimens ought to be packed in a tin case and soldered, to protect them against tropical damp and sea air. Delicate objects must be protected by being enclosed within a small box or the cavity of a strong specimen, but in the latter case make sure that they will not be shaken out in transit. It is sometimes possible to get a cheap fabric, such as hessian, with which various specimens can be wrapped up, the loose sides of which, and if necessary the

OBVERSE   REVERSE

COLLECTOR'S NAME:

No.   Date

COLLECTED FROM:
Name:
Tribe:
Place:

REMARKS:
*e.g. material,
any peculiarities,
method of manufacture.*

DESCRIPTION AND USE:

BY WHOM USED, CARRIED OR WORN:

NATIVE NAME:

Made by?
Locally?
Obtained by trade?
From where?

ends, should be coarsely sewn by means of a packing-needle. See that all specimens are quite dry and free from beetle and other insect pests before packing, and sprinkle the packing freely with paradichlorbenzine, naphthalin, pyrethrum, or other deterrents. Hair, fur, feather, and perishable fabrics should be sprinkled well with paradichlorbenzine before being packed.

In packing pottery the interior of each pot should be filled tightly with paper or straw, so that in case of breakage the space occupied by the vessel shall not be left vacant. It is a wise precaution to paste narrow strips of paper inside or outside (or both) the pot. These should pass from rim to rim across the bottom of the pot and fairly close together; they tend to prevent vibration and consequent breakage, and to hold pieces together should fracture occur. Pottery is best packed in plenty of crumpled paper which is springy. Ropes of twisted paper may be rolled tightly around necks, handles and other projecting parts. Hay and twisted straw, dry grass, and even leaves may be used in an emergency but they are more or less liable to shrink and contract, leaving the specimens loose and unprotected. Each pot should be completely wrapped in paper or textile, so that if it is broken the fragments will be kept together and apart from those of others. When heavy specimens are packed in a case with fragile ones, it is necessary to secure the heavy specimens to the walls of the case, either by wiring or by battens running across the case and nailed to the sides. It is essential to remember that all objects should be firmly packed; the cases or packages will inevitably be subjected to rough handling and constant joltings and jarrings—any movement of the specimens is bound to act to their detriment. Metal (especially iron) objects should be well greased to prevent oxidation by sea-air and damp. Labels should not be fixed with paste or gum to any iron pieces.

Human Remains

It is often difficult to collect recent human skulls and bones, but every effort should be made to do so. These are generally collected in a state fit to be packed at once, and they should be fully labelled with identification marks written on each with lead pencil. Where the people are head-hunters it is important to endeavour to ascertain the name of the tribe from whom heads were taken. In every case care should be taken to learn the history of each skull; sometimes the name of the individual can be obtained and thus the sex can be recovered. It would be worth while tracing that individual in the genealogies collected.

PRESERVATION OF BONES

Bones of animals and of human beings are often found *in situ* under conditions which render their preservation a matter of great importance, yet lack of knowledge on the part of the finder prevents their preservation.

The most important things to remember are (1) that the edges of broken bones must be kept intact, and not allowed to rub together or crumble. (2) That it does not matter into how many pieces a bone breaks up, provided all pieces, however small, be kept, and the edges of the pieces are intact so as to give direct points of contact. (3) That if a bone is broken or cracked it is usually fatal to attempt to keep the parts in position when packing, as usually the edges rub together. (4) That if, therefore, a bone is broken or cracked it should be taken to pieces, care being taken that all fragments of a single bone are either marked with an index number, or else packed in the same box so that they can be reassembled. (5) That if the bone is at all fragile the edges, if not the whole of each fragment, should be treated with some preservative.

The question of the preservative used must necessarily depend on circumstances. If available, a mixture of commercial shellac and methylated spirit in the proportion of $\frac{1}{2}$ lb. of shellac to $1\frac{1}{4}$ pints of methylated spirit is perhaps the best for hardening crumbling bone. Liquid shellac or varnish will serve in the place of flake shellac, and if methylated spirit is not available, most other spirits, but not petrol, will serve.

Failing these, a thin solution of glue or of gum arabic may be used to harden the edges of the bone, but in no circumstances must a thick one be used. Another method is to use some form of wax, melted and applied liquid, and then allowed to harden. If this method is used it is not absolutely necessary to separate the broken bone, as the wax holds each piece firmly in position, and prevents any chance of the edges rubbing together.

Only bones that have become soft or fragile and liable to crumble need be treated at all; bones that have retained a high percentage of animal fats do not need it.

Paper Squeezes

Squeezes are copies of sculptured or engraved objects, made by pressing moist paper on the carved surface, of which, when dry, the paper mould retains the exact form and texture.

Requisites for taking squeezes are (1) tough paper unsized; (2) a stiff bristle-brush; (3) a scrubbing brush; (4) a vessel of water.

Special paper is made for this purpose and is strongly recommended. It is supplied in rolls, like wall-paper, and is best carried in cylindrical tin boxes, which serve also to transport finished squeezes; note, however, that the squeezes will be much more bulky than the paper was. But any tough, thick, wrapping paper will do, such as plant-collectors or orange-packers use; and newspaper may be used, though it needs care and patience. The best forms of brush are (*a*) a clothes brush with thickset bristles and long handle; (*b*) less expensive, the long narrow one made for washing carriage wheels; (*c*) in emergency, an old nail-brush or stiff hairbrush will serve. Squeeze-making wears out some brushes very quickly; but with a long handle, labour is saved, and the brush itself lasts longer. For deep carvings it is good to have several brushes of different widths and shapes. A shallow bath, large enough to hold a whole sheet of paper, is convenient, but not necessary; good paper can bear some unfolding after it is wetted, and can always be refreshed by sprinkling with the brush or a handful of water.

The process is a simple one. Sweep off all loose debris and dust. Then wash the sculptured surface, removing every trace of sticky clay, etc. This is essential, as a dirty squeeze is difficult to decipher and of little value for publication. But use the scrubbing brush gently until it is certain that the surface of the stone is in good condition. Especial care should be taken with surfaces recently excavated, and with all sandstones and absorbent limestones. Leave enough water on to hold the paper by capillary attraction. Squeezes of large monuments must be made in sections, taking care that the edges of each section overlap their neighbours, to allow for trimming and fitting. As a rule, let the paper remain on till quite dry; it will then be easy to release without using force, and without distortion. Before a squeeze is detached from the stone, write on it a serial number and enter this in your notebook; locality and date should be added, as a squeeze looks very different from the original.

Difficulties. Limestone is often too absorbent, and will not leave water on the surface to hold the paper. If the face is upright the paper must be held up along the top edge, beaten on quickly, and then lifted off the stone and laid to dry on a flat surface, such as smooth sand. If the sculpture is too polished and smooth for the paper to get any hold, it draws out the hollows when it begins to contract in drying, and fails to obtain an impression from low relief. If this happens, lift it off (as soon as it begins to dry) by peeling it from one edge (holding the other corners), and lay it flat till it is dry. Wind is the greatest trouble; on this account, never use larger paper than is necessary, and sometimes beat it on temporarily at the edges, that it may hold until you have done the rest and can lift and re-beat the edge finally. It may be necessary to plaster the edges down with a little mud. It is always well to beat the edges on very thoroughly in finishing, to prevent the paper being blown up while drying. After the paper has been beaten in, stamp-edgings or adhesive binding-strips are useful to prevent it from peeling off at the edges. If the surface overhangs, strings across it are useful. Cracks which disfigure the design, or deep under-cutting which is not essential to it, may be partly filled up with clay or sodden paper before the squeeze is begun. Full measurements should be taken of the monument itself, in case the squeeze should be distorted accidentally, either in taking it off the original or afterwards. Torn squeezes may be mended with strips of the squeeze-paper, and paste or barley-gruel; if the paper splits over prominences, or in deep grooves, patches of squeeze-paper, well soaked, may be applied and beaten in till they adhere. These patches necessarily take longer to dry.

Packing. When dry, squeezes may safely be packed tight together in a box or crate, or rolled up if they represent flat surfaces. Such rolls are conveniently stored in cylindrical tin boxes, which any tinsmith can make.

Squeezes must be kept dry, and are liable to absorb moisture in damp atmosphere, especially if the stone has yielded any salt during the process. To extract salt, a squeeze may be soaked in a tray of water, but this risks disfigurement. In a very moist climate, squeezes may be waterproofed by giving them, when once dried, a coat of boiled linseed oil; apply this hot, and use a soft brush.

Reproductions. When a cast is required, heat the squeeze on a stove and brush over with beeswax, or solid paraffin (or candle-grease in an emergency), enough to choke the grain, but not to efface it. Then plaster may be cast on it, taking care that it does not run round the edges. Support the mould well, lest the weight of the plaster distort it.

A photograph, tracing, or drawing usefully supplements a squeeze; and if the surface of the stone be smooth enough, a *rubbing* with heelball may be made. The simplest and quickest method of making a rubbing is to use moist grass or leaves in place of heelball. For rubbings, common Japanese paper is the best; it is very strong, and not being sized it is not much affected by damp.

Another method of making incised ornaments more distinct is to dab the smooth surface, while it is drying upon the stone, with water-colour mixed with a little gum or size, applied with a porous pad (a sponge, rolled sock, or rag stuffed with leaves). The colour must not be so copious as to flood the incisions, which should remain white on a coloured ground.

For publication, squeezes should be photographed on the side which was next the stone; the negative thus produced is, of course, reversed, but this can be corrected in making the print. The light must fall on the squeeze so as to give the relief correctly on the print. It is customary to arrange that the lighting should be from the upper left-hand corner. When this normal lighting does not show all essential details, the illumination used should be specially recorded.

BOOK LIST

GENERAL WORKS

Boas, Franz, *General Anthropology* (Heath, Boston, Mass.; Harrap, London, 1938).
Firth, R. W., *Human Types* (Nelson, 1941).
Goldenweiser, A. A., *Anthropology* (Harrap, 1937).
Haddon, A. C., *A History of Anthropology* (Watts, 1910).
Herskovits, M. J., *Man and his Works* (Knopf, 1948).
Kluckholn, C., *Mirror for Man* (McGraw-Hill Book Co., 1949; Harrap, London, 1949).
Kroeber, A. L., *Anthropology* (1949).
Linton, R., *The Cultural Background of Personality* (Kegan Paul, 1947).
—— (ed.), *The Science of Man in the World Crisis* (Columbia University Press, 1945).
Marett, R. R., *Anthropology* (Oxford University Press, 1912).
—— *Man in the Making* (Nelson, 1937). Out of print.
Smith, G. Elliot, *The Diffusion of Culture* (Watts, 1933).
Tylor, E. B., *Primitive Culture* (Murray, 1903). 2 vols.

Physical Anthropology

Barzun, J., *Race: A Study in Modern Superstition* (Methuen, 1938).
Benedict, Ruth, *Race and Racism* (Routledge, 1942).
Boule, M., *Fossil Men* (Oliver and Boyd, 1923).
Boyd, W. C., "Blood Groups," *Tabul. Biol.*, *17*, Part 2, 1939.
—— "Critique of the Methods of Classifying Mankind," *American Journal of Physical Anthropology*, 27, 1940.
—— *The Determination of Blood Groups* (H.M. Stationery Office, 1943).
Buxton, L. H. D., *Peoples of Asia* (Kegan Paul, 1925).
Chambers, E. G., *Statistical Calculation for Beginners* (Cambridge University Press, 1943).

CLARK, W. E. LE GROS, *Early Forerunners of Man* (Baillière, Tindal and Cox, 1934).
COON, C. S., *The Races of Europe* (Macmillan, 1939).
DAHLBERG, G., *Race, Reason and Rubbish* (George Allen and Unwin, 1942).
DOBZHANSKY, T., *Genetics and the Origin of Species* (2nd ed., Columbia University Press, 1941).
FISHBERG, A. M., *The Jews: A Study of Race and Environment* (Scott, 1889). Out of print.
FLEURE, H. J., *The Peoples of Europe* (Oxford University Press, 1922).
GREGORY, W. K., *Man's Place among the Anthropoids* (Oxford University Press, 1934).
GRIEVE, S. W., and MORANT, G. M., "Records of Eye-Colour for a British Population and Description of a new Eye-colour Scale," London, *Ann. Eugenics*, *13*, 1946-7).
GUHA, B. S., *Outline of the Racial Ethnology of India* (Calcutta, 1937).
HADDON, A. C., *The Races of Man and their Distribution* (Cambridge University Press, 1924).
—— *The Wanderings of Peoples* (Cambridge University Press, 1911).
HARRIS, J. A., JACKSON, C. M., PATERSON, D. G., and SCAMMON, R. E., *The Measurement of Man* (University of Minnesota Press, Minnesota; Oxford University Press, London, 1930).
HOOTON, E. A., *Up from the Ape* (New York; Macmillan, 1947).
HOWELLS, W., *Mankind So Far* (Sigma Books, 1947).
HRDLIČKA, ALES, *Practical Anthropometry* (Wister Institute of Anatomy and Biology, Philadelphia, 1947).
—— *The Skeletal Remains of Early Man* (Smithsonian Institution, 1930, Miscellaneous Collections, v. 83).
HUXLEY, J. S., HADDON, A. C., and CARR-SAUNDERS, A. M., *We Europeans* (Cape, 1935). Temporarily out of print.
KEANE, A. H., *Man, Past and Present* (revised by A. Hingston Quiggin and A. C. Haddon) (Cambridge University Press, 1920).
KEITH, A., *Antiquity of Man* (Williams and Norgate, 1925). 2 vols.
—— *Ethnos, or the Problem of Race* (Kegan Paul, 1931).
KLINEBERG, O., *Race Differences* (Harper, New York, 1935).
KROGMAN, W. M., "Anthropometric Instruments," *American Journal of Physical Anthropology*, N.S. 6, 1948.
MARTIN, R., *Lehrbuch der Anthropologie* (Jena, Fischer, 1928).

MATHER, K., *Statistical Analysis in Biology* (2nd ed., Methuen, 1946).
MOLLISON, P. L., and MOURANT, E. E., "The Rh Blood Groups and their Clinical Effects," *Medical Research Council Memorandum*, No. 19 (H.M.S.O., 1948).
MONTAGU, M. F. ASHLEY, "The Location of the Nasion on the Living," *American Journal of Physical Anthropology*, 20, 1935.
—— *Introduction to Physical Anthropology* (C. C. Thomas, Springfield, Illinois, 1946).
—— *Man's Most Dangerous Myth: The Fallacy of Race* (Columbia University Press, 1942).
MORANT, G. M., *The Races of Central Europe* (Allen and Unwin, 1939). Out of print.
PITTARD, E., *Race and History* (Kegan Paul, 1926).
RACE, R. R., and SANGER, R., *Blood Groups in Man* (Oxford, Blackwell Scientific Publications, 1950).
SELIGMAN, C. G., *The Races of Africa* (Oxford University Press, 1930).
SMITH, G. ELLIOT, *Human History* (Cape, 1930).
SNEDECOR, G. W., *Statistical Methods* (Iowa State College Press, 1940).
TANNER, J. M., and WEINER, J. S., "The Reliability of the Photogrammatric Method of Anthropometry", *American Journal of Physical Anthropology*, N.S. 7, 1949.
TILDESLEY, M. L., "Choice of the Unit of Measurement in Anthropology." (London, Royal Anthropological Institute, *Man*, 72, 1947).
—— "The Relative Usefulness of Various Characters on the Living for Racial Comparison" (*Man*, 14, 1950).
WEIDENREICH, F., *Giant Early Man from Java and South China* (Anthropological Papers, American Museum of Natural History, New York, 1945).
—— *Apes, Giants and Man* (Chicago University Press and Oxford University Press, 1946).
WIENER, A. S., *Blood Groups and Transfusion* (3rd ed., Thomas, Springfield, Ill., 1943).
WILDER, H. H., *The Pedigree of the Human Race* (Holt, New York, 1926).
—— *A Laboratory Manual of Anthropometry* (Pitman, 1921). Out of print.

SOCIAL ANTHROPOLOGY

General Theory

BARTLETT, F. C., *Psychology and Primitive Culture* (Cambridge University Press, 1923). Out of print.
BARTLETT, F. C., GINSBERG, M., LINDGREN, E. J., and THOULESS, R. H. (editors), *The Study of Society: Methods and Problems* (Kegan Paul, 1939).
BENEDICT, R., *Patterns of Culture* (Routledge, 1935).
CHAPPLE, E. D., and COON, C. S., *Principles of Anthropology* (Cape, 1948).
FORDE, D., *Habitat, Economy and Society* (Methuen, 1946).
GINSBERG, M., *Sociology* (Home University Library) (Oxford University Press).
HOCART, A. M., *Kingship* (Watts, abridged edition, 1941).
LANDTMANN, G., *The Origin of the Inequality of the Social Classes* (Kegan Paul, 1938).
LOWIE, R. H., *An Introduction to Cultural Anthropology* (Harrap, 1936).
—— *Primitive Society* (Routledge, 1949; 2nd edition).
—— *Social Organization* (Rinehart, 1948).
MALINOWSKI, B., *A Scientific Theory of Culture* (1947).
—— *Magic, Science and Religion* (Beacon Press, Boston, Mass., 1949).
RADCLIFFE-BROWN, A. R., *Social Anthropology* (Home University Library) (In press).
RIVERS, W. H. R., *Social Organization* (edited by W. J. Perry), (Kegan Paul, 1924).
WEBSTER, H., *Taboo: a Sociological Study* (Stanford University Press; Oxford University Press, 1942).
WESTERMARCK, D., *A Short History of Marriage* (Macmillan, 1926).
WILSON, G. and M., *Analysis of Social Change* (Cambridge University Press, 1946.)

Social Systems

BATESON, G., *Naven* (Cambridge University Press, 1936).
COLSON, E. D. (ed.), *Seven Tribes of British Central Africa* (Oxford University Press). Forthcoming.
EMBREE, J. F., *A Japanese Village, Surye Mura* (Kegan Paul, 1946).
EVANS-PRITCHARD, E. E., *The Nuer* (Oxford University Press, 1940).

FIRTH, R. W., *We, the Tikopia* (Allen and Unwin, 1936).
FORDE, D., *Marriage and the Family among the Yako in South-Western Nigeria* (London School of Economics Monographs on Social Anthropology, No. 5, 1941).
FORTES, M., *The Dynamics of Clanship among the Tallensi* (Oxford University Press, 1945).
—— *The Web of Kinship among the Tallensi* (Oxford University Press, 1949).
KRIGE, E. J. and J. D., *The Realm of the Rain Queen* (Oxford University Press, for Int. African Institute, 1943).
KUPER, H., *An African Aristocracy: The Swazi* (Oxford University Press, for Int. African Institute, 1947).
LANG, O., *Chinese Family and Society* (Yale University Press and Oxford University Press, 1946).
MEAD, M., *Sex and Temperament in Three Primitive Societies* (Routledge, 1935).
NADEL, S. F., *A Black Byzantium* (Oxford University Press, for Int. African Institute, 1942).
RADCLIFFE-BROWN, A. R., *Social Organization of Australian Tribes* (Oceania Monographs, No. 1, reprinted from *Oceania*, vol. I, 1931).
RADCLIFFE-BROWN, A. R., and FORDE, D. (eds.), *African Kinship Systems* (Oxford University Press, for Int. African Institute, 1949).
RATTRAY, R. S., *Ashanti* (Oxford University Press, 1923).
SCHAPERA, I., *Married Life in an African Tribe* (Faber and Faber, 1940).

POLITICAL AND LEGAL SYSTEMS

BARTON, R. F., *Ifugao Law* (Berkeley, University of California Press, 1919).
FORTES, M., and EVANS-PRITCHARD, E. E. (eds.), *African Political Systems* (Oxford University Press, 1940).
GLUCKMAN, M., *Essays on Lozi Land and Royal Property* (Livingstone, Rhodes-Livingstone Institute, 1943).
HOGBIN, H. I., *Law and Order in Polynesia* (Christophers, 1934).
LLEWELLYN, K. N., and HOEBEL, E. A., *The Cheyenne Way* (Oklahoma University Press, 1943).
MAINE, SIR H., *Ancient Law*.
—— *Early Law and Custom*.

MALINOWSKI, B., *Crime and Custom in Savage Society* (Kegan Paul, 1926).
MEEK, C. K., *Law and Authority in a Nigerian Tribe* (Oxford University Press, 1937).
RATTRAY, R. S., *Ashanti Law and Constitution* (Oxford University Press, 1938).
SCHAPERA, I., *Handbook of Tswana Law and Custom* (Oxford University Press, 1938).
SEAGLE, W., *The Quest for Law* (Knopf, New York, 1941).

ECONOMIC SYSTEMS

FEI, H. T., and CHANG, C. I., *Earthbound China* (Kegan Paul, 1945).
FIRTH, R. W., *Primitive Economics of the New Zealand Maori* (Routledge, 1929).
—— *Primitive Polynesian Economy* (Routledge, 1939).
—— *Malay Fishermen* (Kegan Paul, 1946).
GOODFELLOW, B. M., *Principles of Economic Sociology (as illustrated from the Bantu Tribes)* (Routledge, 1939).
HERSKOVITS, M. J., *The Economic Life of Primitive Peoples* (Knopf, New York, 1940).
RICHARDS, AUDREY I., *Land, Labour and Diet in Northern Rhodesia* (Oxford University Press for Int. African Institute, 1939).
SCHAPERA, I., *Native Land Tenure in the Bechuanaland Protectorate* (Lovedale Press, Lovedale, South Africa, 1943).
SHIH, K. H., and T'IEN, J. K. (ed. by Fei, H. T., and Hsu, F. L.K.) *China enters the Machine Age* (Harvard University Press, 1944).
STEVENSON, H. N. C., *Economics of the Central Chin Tribes* (Government of Burma, 1944; Times of India Press, Bombay, 1943).
THURNWALD, R., *Economics in Primitive Communities* (Oxford University Press, for Int. African Institute, 1932).

RITUAL AND RELIGION

DURKHEIM, E., *Elementary Forms of Religious Life* (trans. by Swain) (Allen and Unwin, 1915).
EVANS-PRITCHARD, E. E., *Witchcraft, Oracles and Magic among the Azande* (Oxford University Press, 1937).
FIELD, M. J., *Religion and Medicine of the Ga People* (Oxford University Press, 1937).

FIRTH, R. W., *The Work of the Gods in Tikopia* (Lund Humphries, 1940; 2 vols.).
—— "Religious Belief and Personal Adjustment" (*J. Roy. Anthrop. Inst.*, vol. LXXVIII, 1948).
FORTUNE, R. F., *Sorcerers of Dobu* (Routledge, 1932).
—— *Manus Religion* (American Philosophical Society, 1935).
FRAZER, J. G., *The Golden Bough* (Macmillan, 1922; abridged edition).
GREENBERG, J., *The Influence of Islam on a Sudanese Religion* (Monographs, American Eth. Society, 1946).
GRIAULE, M., *Les Masques des Dogons* (Trav. et Mém. Inst. de l'Inst. d'Eth., Paris, 1938).
JAMES, E. O., *Comparative Religion* (Methuen, 1938).
KARSTEN, R., *Origins of Religions* (Kegan Paul, 1935).
LEENHARDT, M., *Do Kamo: La Personne et le Mythe dans le Monde Mélanésien* (Gallimard, 1947).
MARRETT, R. R., *The Threshold of Religion* (1909).
RADCLIFFE-BROWN, A. R., *Taboo* (Cambridge University Press, 1939).
—— "Religion and Society" (*J. Roy. Anthrop. Inst.*, vol. LXXV, 1945).
RATTRAY, R. S., *Religion and Art in Ashanti* (Oxford University Press, 1927).
READ, CARVETH, *Man and His Superstitions* (Cambridge University Press, 1925).
SKEAT, W. W., *Malay Magic* (Macmillan, 1900). Out of print.
SMITH, W. ROBERTSON, *Lectures on the Religion of the Semites* (A. and C. Black, 1894).

LANGUAGE

ARMFIELD, N., *General Phonetics* (Heffer, Cambridge, 1918).
BLOOMFIELD, L., *Introduction to the Science of Language* (New York, 1933).
BOAS, F. (ed.), *Handbook of American Indian Languages* (Bureau of American Ethnology, Washington, Bulletin No. 40, parts 1–2; J. J. Augustin Inc., part 3, 1911–22).
FIRTH, J. R., *The Tongues of Men* (Watts, 1937).
GRAFF, W. L., *Language and Languages* (New York, 1932).
GRAY, LOUIS H., *The Foundations of Language* (New York, 1939).
JESPERSEN, O., *Language: Its Nature, Development and Origin* (London, 1922).

—— *Mankind, Nation and Individual from a Linguistic Point of View* (London, 1946).
MALINOWSKI, B., *Coral Gardens and their Magic*, vol. II (Allen and Unwin, 1938).
MAROUZEAU, J., *Lexique de la Terminologie Linguistique* (Paris, 1944).
—— *La Linguistique ou Science du Langage* (Paris, 1944).
MEILLET, A., *Linguistique Historique et Linguistique Générale* (Paris, 1921, 1936; 2 vols.).
MEILLET, A., and COHEN, M., *Les Langues du Monde* (Paris, 1924).
NIDA, E., *Learning a Foreign Language* (Foreign Missions Conf., New York, 1950).
OGDEN, C. K., and RICHARDS, I. A., *The Meaning of Meaning* (London, 1923).
PEDERSEN, H., *Linguistic Science in the Nineteenth Century* (trans, by J. W. Spargo) (Harvard, 1931).
SAPIR, E., *Language* (Oxford University Press, 1922 (out of print), New York (reprint), 1947).
STEBBING, S., *Introduction to Modern Logic* (Methuen, 1950).
WARD, I. C., *Practical Suggestions for Learning an African Language in the Field* (Oxford University Press for Int. African Inst).
WESTERMANN, D., and WARD, I. C., *Practical Phonetics for Students of African Languages* (Oxford University Press for Int. African Inst., 1933).
Handbook of African Languages (Int. African Inst.). In course of publication.

MATERIAL CULTURE

BALFOUR, H., *The Evolution of Decorative Art* (Percival, 1893). Out of print.
BEST, E., *Maori Agriculture* (Bull. Dominion Museum, No. 9, Wellington, 1925).
—— *The Maori Canoe* (Bull. Dominion Museum, No. 7, Wellington, 1920).
BLAKE, W. M., "Taking off the Lines of a Boat" (*Mariner's Mirror*, Jan., 1935).
BRITISH MUSEUM, *A Guide to the Ethnographic Collections* (Oxford University Press, 1925).
BUCK, P. H. (TE RANGI HIROA), *Material Culture of the Cook Islands: Ethnology of Tongareva: Ethnology of Mangareva* (Bernice P. Bishop Museum Bulletins 75, 92, 157, The Museum, Honolulu, 1902, 1938).

—— *Samoan Material Culture* (Bernice P. Bishop Museum Bulletin 75, The Museum, Honolulu, 1930).
Dixon, R. B., *The Building of Cultures* (Scribner, 1928).
Geiringer, K., *Musical Instruments: Their History in Western Culture from the Stone Age to the Present* (trans. by Bernard Miall) (Oxford University Press, 1945).
Haddon, A. C., *Evolution in Art* (Scott, 1895). Out of print.
—— "The Outriggers of Indonesian Canoes," (*J. Roy. Anthrop. Inst.*, vol. L, 1920).
Haddon, A. C., and Hornell, J., *Canoes of Oceania* (Bernice P. Bishop Museum, Special Publication No. 27).
Haddon, K., *Artists in String* (Methuen, 1930).
Hornell, J., *Water Transport: Origins and Early Evolution* (Cambridge University Press, 1946).
Horniman Museum, *The Evolution of the Domestic Arts*, by H. S. Harrison: Part I, *Agriculture, the Preparation of Food and Fire-Making* (1925); Part II, *Basketry, Pottery, Spinning and Weaving* (1924).
—— *Travel and Transport by Land and Water*, by H. S. Harrison (1925).
Leroi-Gourhan, A., *Milieu et Technique* (Paris, 1945).
—— *L'Homme et la Matière* (Paris).
Mason, Otis T., *Origins of Invention* (Scott, 1895).
Nordenskiöld, E., *Origin of the Indian Civilizations in South America* (Comparative Eth. Studies, 9, Goteborg, 1931).
Pitt Rivers, A. L. F., *The Evolution of Culture* (Oxford University Press, 1906).
Rickard, T. A., *Man and Metals* (McGraw-Hill, New York, 1932). 2 vols.
Roth, H. Ling, *Studies in Primitive Looms* (reprinted from the *Journal of the Royal Anthropological Institute*, vol. XLVI, 1916; Bankfield Museum, Halifax).
Sayce, R. U., *Primitive Arts and Crafts* (Cambridge University Press, 1933).

FIELD ANTIQUITIES

British Museum, *Flints* (Oxford University Press, 1928).
—— *A Guide to the Antiquities of the Stone Age* (Oxford University Press, 1926).
Burkitt, M. C., *The Old Stone Age* (Cambridge University Press, 1933).

CHILDE, V. G., *The Bronze Age* (Cambridge University Press, 1930).
—— *The Dawn of European Civilization* (Kegan Paul, 1948).
—— *New Light on the Most Ancient East* (Kegan Paul, 1934).
—— *Prehistoric Communities of the British Isles* (Chambers, 1946).
CLARK, GRAHAME, *Archaeology and Society* (Methuen, 1939).
—— *From Savagery to Civilization* (Cobbett Press, 1946).
GOODWIN, A. J. H., and LOWE, C. VAN RIET, "The Stone Age Cultures of South Africa," *Annals of the South African Museum*, vol. XXVII, 1929 (The Museum, Cape Town).
HAWKES, C. F. C., *The Prehistoric Foundations of Europe: to the Mycenaean Age* (Methuen, 1939).
HAWKES, C. F. C., and HAWKES, J., *Prehistoric Britain* (Chatto and Windus, 1947).
HORNIMAN MUSEUM, *Handbooks* (1920–37).
—— *From Stone to Steel: The Ages of Stone, Bronze and Iron*, by H. S. Harrison.
HUZAYYIN, S. A., "The Place of Egypt in Prehistory" (*Mémoires Inst. d'Egypte*, 43, Cairo, 1941).
LEAKEY, L. S. B., *Stone Age Africa* (Oxford University Press, 1930).
—— *The Stone Age Cultures of Kenya Colony* (Cambridge University Press, 1931).
—— *Adam's Ancestors* (Methuen, 1934).
MORLEY, S. G., *The Ancient Maya* (1934).
MOVIUS, H. L., *Early Man and Pleistocene Stratigraphy in Southern and Eastern Asia* (Peabody Museum Papers, XIX, 3, and Cambridge University Press, 1944).
O'BRIEN, T. P., *The Prehistory of the Uganda Protectorate* (Cambridge University Press, 1939).
DE PRADENNE, A. V. VAYSON, *Prehistory* (Harrap, 1940).
SHAW, T., *Field Archaeology* (Oxford University Press for Int. African Institute, 1946).
DE TERRA, H., and PATERSON, T. T., *Studies in the Ice Age in India and Associated Cultures* (Carnegie Inst. Pub. No. 493, Washington, 1939).
THOMPSON, J. E., *Archaeology of South America*, Anthropology Leaflet No. 33 (Field Museum of Nat. Hist., Chicago, 1936).
—— *Mexico before Cortes* (Scribner, 1933).
VAILLANT, G. G., *The Aztecs of Mexico* (New York, 1944).
WRIGHT, W. B., *Tools and the Man* (Bell, 1939).
ZEUNER, F. E., *Dating the Past* (Methuen, 1946).

REGIONAL STUDIES
(see also previous sections)

Africa

Brown, G. G., and Hutt, A. McB., *Anthropology in Action* (Oxford University Press, for Int. African Institute, 1935).
Culwick, A. T. and G. M., *Ubena of the Rivers* (Allen and Unwin, 1935).
Hailey, Lord, *An African Survey* (Oxford University Press, 1938).
Hayley, T. T. S., *The Anatomy of Lango Religion and Groups* (Cambridge University Press, 1947).
Hunter, Monica, *Reaction to Conquest* (Oxford University Press for Int. African Institute, 1936).
Junod, H. A., *The Life of a South African Tribe* (Macmillan, 1927), 2 vols.
Lindblom, K. G., *The Akamba in British East Africa* (Uppsala, 1920).
Mair, Lucy P., *An African People in the Twentieth Century* (Routledge, 1934).
Meek, C. K., *A Sudanese Kingdom* (Kegan Paul, 1931).
Nadel, S. F., *The Nuba* (Oxford University Press, 1947).
Peristiany, J. G., *The Social Institutions of the Kipsigis* (Routledge, 1939).
Raum, O. F., *Chaga Childhood* (Oxford University Press, 1940).
Schapera, I., *The Khoisan Peoples of South Africa* (Routledge, 1930). Out of print.
—— (ed.), *The Bantu-Speaking Tribes of South Africa* (Routledge, 1937).
—— *Western Civilization and the Natives of South Africa* (Routledge, 1934). Out of print.
Seligman, C. G., and Brenda Z., *Pagan Tribes of the Nilotic Sudan* (Routledge, 1933).
Smith, E. W., and Dale, A. M., *The Ila-Speaking Peoples of Northern Rhodesia* (Macmillan, 1920). 2 vols. Out of print.
Stayt, H. A., *The BaVenda* (Oxford University Press, 1931).

America

Birket-Smith, K., *The Eskimos* (Methuen, 1936).
Ford, C. S., *Smoke from their Fires: The Life of a Kwakiutl Chief* (Yale University Press and Oxford University Press, 1941).

FORDE, D., *Ethnography of the Yuma Indians* (University of California Press, 1931).
HERSKOVITS, M. J., *Life in a Haitian Valley* (Knopf, New York, 1937).
JENNESS, D., *The Indians of Canada*. (Department of Mines Bulletin, 65, Anthropological Series, No. 15. National Museum of Canada, Toronto, 1934).
KARSTEN, R., *The Civilization of the South American Indians* (Kegan Paul, 1926).
—— *The Headhunters of the Western Amazonas* (Societas Scientarium et Litterarum: Commentationes Humanarum Litterarum, Helsingfors, 1935).
KROEBER, A. L., *Handbook of the Indians of California* (Bureau of American Ethnology, Bulletin 78, 1925).
—— *Native Cultural Areas of North America* (American Archaeology and Ethnology, vol. XXXVIII, University of California Press, 1939).
LANDES, R., *Ojibwa Sociology* (Contributions to Anthropology, vol. XXIX, Columbia University Press, 1937).
LOWIE, R. H., *The Crow Indians* (New York, 1935).
NIBLACK, A. P., *The Coast Indians of Southern Alaska and Northern British Columbia* (Annual Report, Smithsonian Institution, 1888, pp. 225–386, Washington, 1890).
OPLER, M. E., *Apache Life Way* (University of Chicago Publications in Anthropology: Ethnological Series, University of Chicago Press, 1941).
PARSONS, E. G., *Pueblo Indian Religion* (Publications in Anthropology: Ethnological Series, University of Chicago, Press, 1939). 2 vols.
RADIN, P., *The Winnebago Tribe* (Bureau of American Ethnology, Washington, 37th Annual Report, 1915–16).
REDFIELD, R. (ed.), *Social Organization of North American Indian Tribes* (University of Chicago Press).
—— *Folk Culture of Yucatan* (Chicago University Press, 1941).
SPECK, F. G., *Naskapi* (Norman, University of Oklahoma Press, 1935).
UNDERHILL, R., *Papago Indian Religion* (Columbia University Press, New York, 1946).
WISSLER, C., *The American Indian* (Oxford University Press, 1938).
See also other publications of the Universities of Chicago, California and Yale, and of the Bureau of American Ethnology,

Asia

Anderson, J. D., *Peoples of India* (Cambridge University Press, 1913).
Barton, R. F., *Pagans: The Autobiographies of Three Ifugaos* (Routledge, 1938).
Bateson, G., and Mead, M., *Balinese Character* (Academy of Sciences, New York, 1942).
Benedict, R., *The Chrysanthemum and the Sword* (Secker and Warburg, 1947).
Du Bois, C., *The Peoples of Alor* (University of Minnesota Press, 1944).
Cole, Fay-Cooper, *The Peoples of Malaysia* (D. van Nostrand Co., 1945.)
Cuisinier, J., *Danses Magiques de Kelantan* (Institut d'Ethnologie, Paris, 1936).
Czaplicka, M. A., *Aboriginal Siberia* (Oxford University Press, 1914).
Elwin, V., *The Baiga* (Murray, 1939).
—— *The Agaria* (Oxford University Press, 1942).
—— *Maria Murder and Suicide* (Oxford University Press, 1943).
Evans, I. H. N., *Negritos of Malaya* (Cambridge University Press, 1937).
Ferrars, M. and B., *Burma* (Sampson Low, Marston, 1900).
von Fürer-Haimendorf, C., *The Reddis of the Bison Hills* (Macmillan, 1945).
Gilbert, W. H., junr., *Peoples of India* (Smithsonian Institution's War Background Series, No. 18, 1944).
Grigson, W. V., *The Aboriginal Problem in the Central Provinces and Berar* (Government Printing, Nagpur, 1944).
Hose, C., and McDougall, W., *The Pagan Tribes of Borneo* (Macmillan, 1912). 2 vols. Out of print.
Hutton, J. H., *The Sema Nagas* (Macmillan, 1921). Out of print.
—— *Caste in India* (Cambridge University Press, 1946).
Jochelson, W., *Peoples of Asiatic Russia* (American Museum of Natural History, 1928).
Kroeber, A. L., *Peoples of the Philippines* (American Museum of Natural History, 1928).

MARSHALL, H. I., *The Karen People of Burma* (Columbia University Press, 1922).
MILLS, J. P., *The Ao Nagas* (Macmillan, 1926).
―― *The Rengma Nagas* (Macmillan, 1937).
MOWBRAY, G. H. DE C. DE, *Matriarchy in the Malay Peninsula* (Routledge, 1931).
PARRY, N. E., *The Lakhers* (Macmillan, 1932).
RADCLIFFE-BROWN, A. R., *The Andaman Islanders* (Cambridge University Press, 1922).
READ, M., *The Indian Peasant Uprooted* (Longmans Green, 1931). Out of print.
RIVERS, W. H. R., *The Todas* (Macmillan, 1906). Out of print.
SELIGMAN, C. G. and BRENDA Z., *The Veddas* (Cambridge University Press, 1911).
SKEAT, W. W., and BLAGDEN, C. O., *Pagan Races of the Malay Peninsula* (Macmillan, 1906). 2 vols. Out of print.
YEE, CHING, *A Chinese Childhood* (Methuen, 1940).
YEO, SHWAY (SIR J. G. SCOTT), *The Burman: His Life and Notions* (Macmillan, 1882.)
DE ZOETE, B., and SPIES, W., *Dance and Drama in Bali* (Faber, 1938).

OCEANIA—AUSTRALASIA

BEAGLEHOLE, E., and P., *Some Modern Maoris* (Whitcomb and Tombs; Oxford University Press, 1946).
ELKIN, A. P., *The Australian Aborigines* (Angus and Robertson, 1938).
―― *Studies in Australian Totemism*, Oceania Monographs No. 2 (Reprinted from *Oceania*, vol. III, March 1933, No. 3, and Vol. IV, No. 1. Australian National Research Council, Sydney).
HOGBIN, H. I., *Experiments in Civilization* (Routledge, 1939).
KABERRY, P., *Aboriginal Women* (Routledge, 1939).
SPENCER, B., *Native Tribes of the Northern Territories* (Macmillan, 1914). Out of print.
SPENCER, B., and GILLEN, F. J., *The Arunta* (Macmillan, 1927). 2 vols.
―― *Native Tribes of Central Australia* (Macmillan, 1900).
SUTHERLAND, I. L. G. (ed.), *The Maori People To-day* (Whitcomb and Tombs; Oxford University Press, 1940).

WHITING, J. W. M., *Becoming a Kwoma* (Yale University Press, 1941).
WILLIAMSON, R. W. (ed. Piddington, R. O'R.), *Essays in Polynesian Ethnology* (Cambridge University Press, 1939).

OCEANIA—PACIFIC

ARMSTRONG, H. E., *Rossel Island* (Cambridge University Press, 1928).
BEST, E., *The Maori*, Memoirs of the Polynesian Society, Vol. V, 1924 (The Society, Wellington, New Zealand).
—— *An Introduction to Polynesian Anthropology.*
BLACKWOOD, BEATRICE, *Both Sides of Buka Passage* (Oxford University Press, 1935).
BUCK, P. H. (TE RANGI HIROA), *Vikings of the Sunrise* (Stokes, New York, 1938).
—— *An Introduction to Polynesian Anthropology* (Bernice P. Bishop Museum Bulletin, No. 187, The Museum, 1945).
CODRINGTON, R. H., *The Melanesians* (Oxford University Press, 1891). Out of print.
DEACON, A. B., *Malekula* (Routledge, 1934).
FIRTH, R. W., *Art and Life in New Guinea* (Studio, 1936). Out of print.
HADDON, A. C., *The Decorative Art of British New Guinea*. Cunningham Memoirs, No. 10 (Royal Irish Academy, Dublin, 1894).
KEESING, F. M., *South Seas in the Modern World* (John Day; Allen and Unwin, 1941).
—— *Native Peoples of the Pacific World* (Macmillan, New York, 1945).
LANDTMANN, G., *The Kiwai Papuans of British New Guinea* (Macmillan, 1927). Out of print.
LAYARD, J., *Stone Men of Malekula: The Small Island of Vao* (Chatto and Windus, 1942).
LINTON, R., *Arts of the South Seas* (Museum of Modern Art, New York, 1946).
MALINOWSKI, B., *Argonauts of the Western Pacific* (Routledge, 1922).
—— *The Sexual Life of Savages in North-Western Melanesia* (Routledge, 1932).
—— *Coral Gardens and Their Magic* (Allen and Unwin, 1938). 2 vols.

MEAD, MARGARET, *Coming of Age in Samoa* (Cape, 1925). Out of print.
―― *Growing up in New Guinea* (Routledge, 1931).
POWDERMAKER, HORTENSE, *Life in Lesu* (Williams and Norgate, 1933).
RIVERS, W. H. R. (ed.), *Essays on the Depopulation of Melanesia* (Cambridge University Press, 1922).
ROTH, H. LING, *Aborigines of Tasmania* (King, Halifax, 1899). Out of print.
SELIGMAN, C. G., *The Melanesians of British New Guinea* (Cambridge University Press, 1910).
WILLIAMS, F. E., *Papuans of the Trans-Fly* (Oxford University Press, 1936.
―― *Orokaiva Society* (Oxford University Press, 1930).
―― *Drama of Orokolo* (Oxford University Press, 1940).

SOME PERIODICALS

Africa. Journal of the International African Institute (Oxford University Press). Quarterly.
African Studies (formerly *Bantu Studies*). Department of Bantu Studies, Witwatersrand University (Witwatersrand University Press, Johannesburg). Quarterly.
American Anthropologist (American Anthropological Association, Lancaster, Pa.). Quarterly.
American Antiquity. (Society for American Archaeology).
American Journal of Physical Anthropology (Wister Institute of Anatomy and Biology, Philadelphia). Quarterly.
Ancient India. Bulletin of the Archaeological Survey of India (Manager of Publications, Delhi). Biennially.
Antiquity (Gloucester). Quarterly.
Bulletin of the School of Oriental and African Studies. (London University.)
Folk-Lore. Transactions of the Folk-Lore Society (Glaisher, 87 Fetter Lane, London, E.C.4). Quarterly.
Journal de la Société des Africanistes (Musée de l'Homme, Place du Trocadéro, Paris).
Man (Royal Anthropological Institute). Monthly.
Language. Journal of the Linguistic Society of America, Baltimore. Quarterly.
Lingua. International review of General Linguistics (Haarlem, Holland). Quarterly.

Nada (Southern Rhodesian Government). Annual.
Oceania (Australian National Research Council, Sydney). Quarterly.
Polynesian Society *Journal* (The Society, Wellington, New Zealand). Quarterly.
Prehistoric Society *Proceedings* (42 Barton Road, Cambridge). Annually.
Rhodes-Livingstone Papers (Oxford University Press, Cape Town). Occasional.
Rhodes-Livingstone Institute. *Journal: Human Problems in British Central Africa* (Oxford University Press, Capetown). Six-monthly.
Royal African Society *Journal* (Macmillan). Quarterly.
Royal Anthropological Institute *Journal* (The Institute). Annually in two parts.
Royal Asiatic Society *Journal*. Quarterly.
Society of Antiquaries of London *Journal* (Oxford University Press). Quarterly.
Southwestern Journal of Anthropology (University of New Mexico Press, Albuquerque).
Sudan Notes and Records (Editorial Secretary, P.O. Box 282, Khartoum). Biennially.
Transactions of the Philological Society (London). Annually.

INDEX

Abnormalities
  mental, 202
  physical, 109, 202
Abortion, 60, 105
Address, forms of, 99, 192
Adelphic, definition of, 112, 113
Adoption, 52, 152
  ceremonies, 74
  motives for, 73
  of adults, 74
  of children, 73
Adult status, 123
Adultery
  attitude towards, 120
  penalties for, 113, 120, 146
Adzes, 258, 261
Aeolian musical instruments, 326
Affines, 55, 70
  defined, 76
Age
  at marriage, 111
  native criteria of, 59, 67, 199
Age-grades
  definition of, 68
  functions of, 144
Age-set, 59
  definition of, 67
  functions of, 68, 70, 140
  see also Initiation
Agnatic
  descent defined, 71, 75
  see Patrilineal
Agriculture, 160
  land for, 49, 154, 155; *see also* Land tenure
  magic in, 165
  seasons, 49, 248
  terrace-cultivation, 248
Albinism, 202
Alphabet, phonetic, 214

Amulets, 233
Amusements, 334
Ancestor
  cult, 176, 178
  definition of, 71
  totemic, 90, 192
Animals
  as decoys:
    in fishing, 257
    in hunting, 253
  attitude to, 200
  knowledge and legends concerning, 200, 251
  reincarnation:
    in animal form, 179
    in animal form of totem species, 179, 192
  transformation into, 176, 189, 194
  see Domestication of animals
Animism, 38
Anthropometry, 3
  record form, 6, 10
Archæology, 343, 346
  earthworks, 204
  stone monuments, 204
Armfield, Noel, 215
Armour, 270
Arrows, 265
  in fishing, 256
  whistling, 321
Art, 33
  as æsthetic activity, 309
  in relation to society, 309
  objects of, 310
  schools and styles of, 310
  symbolic, 314
  terminology of decorative, 311
Attitude of native to investigator, 29
Avoidance, rules of, 85, 86, 88, 99; *see* Taboo

387

Axes, 258, 261

Baits
  in fishing, 255
  in hunting, 252
Barbs (fishing), 255
Bark-cloth
  bark-canoes, 301
  use of, 285
Barter, 169, 282
Basketry, 254
  coiled basket-work, 273
  plaited basket-work, 272
Beads, 33, 233, 346
Beer, 246
Bellows, 282
Belts, 235
Bend (fishing), 255
Bestiality, attitude towards, 110
Betrothal, 111
Betting, 335
Bibliography, 369
Birth, 105
  rates of, 60, 62
  ritual at, 105, 191
  see also Midwifery
Blake, W. M., 304
Blood
  groups, 8, 16: tests, diagram, 23
  knowledge concerning, 200
  pressure, 3, 7
Blood-brotherhood, 100, 135
Blood-money, 135
  cattle as, 167
  given in lieu of blood-vengeance, 147
  women given in lieu of, 119
Blood-revenge, 74, 141, 143, 147
Blow-tubes, 266
Boats
  carvel-built, 304
  clinker-built, 304
  plank-built, 303
  see also Canoes
Bolas, 251, 268
Bones
  preservation of, 364
  working of, 280, 345
Boomerangs, 264

Boundaries
  disputes, 135, 155
  territorial, 66
Bows, 266
  bowstrings, 267
  musical, 324
Boyd, W. C., 8 (*note* 2), 24
Brass, 283
Breeding, beliefs and customs in regard to, 250
Bride-price, 116
  cattle given as, 150, 167
  see Marriage
Bride-wealth
  responsibility for, 88, 113, 116
  return of, 129
  see also Bride-price, Marriage
Bridges, 298
  maintenance of, 49
Bronze, 283, 346
Bull-roarers, 322
Burial, 126, 127, 178
  places, 129, 347

Cairns, 204
Calendar
  ceremonies of calendrical order, 193
  of seasons, 198
  of work, 162
Calthrops, 269
Camps, arrangements of, 239
Cannibalism, 39, 194, 245
  motives for ritual, 144, 177, 188
Canoes, types of, 301, 302; *see also* Boats
Capacity, measures of, 197
Capital, productive function of, 149, 166
Caravans, 298
Carotenoids, 15
Castes, 61
  definition of, 94
Castration, 228
Catamarans, 301
Cattle as currency, 150, 166; *see also* Livestock
Caves
  as shelters, 237
  in archæology, 348

Celts, 346
Cemeteries, exploration of ancient, 347
Ceremonies
  birth, 51, 105
  concerned with land fertility, 156, 191
  connected with hunting and fishing, 257
  death, 51, 129
  in connection with fire, 69, 183, 191
  in house construction, 238
  initiation, 69, 108
  in pottery, 279
  in warfare, 142
  kingship, 181
  marriage, 51, 120
  procuring of food, 242
  water, 249
Chambers, E. G., 4 (*note 2*), 57
Charms
  defined, 188
  use of, 190
Chastity, attitude towards, 108
Chiefs
  as economic organizers, 165
  authority of, 139, 193
  clan, 90, 138
  female, 137
  investiture of, 193
  meaning of, 139
  paramount, 138
  politico-ritual functions of, 138, 156
  sacred, 181
Chieftainship
  elective, 137
  hereditary, 137
  superimposed, 136
Christianity, influence on social organization, 182
Cicatrization, 230
Cisisbeism, 118
Cinematography, 48, 360
Circumcision, 68, 228; *see* Age-grades, Initiation, Deformations
Cire perdue process in casting, 282
Civil Servant, role of, 30
Clan
  as local unit, 90

clansmen, 78, 90
  definition of, 89
  and totem, 90, 192; *see also* Totemism
  royal, 91
  size of, 58, 90
  solidarity, 90, 136
Clans
  dispersed, 90
  reprisals between, 146
Clanship, 90; *see also* Exogamy, Lineage, War
Clappers, 317
Clarinets, 321
Classificatory terminology, definition of, 77
Class systems, 78, 93
Cleanliness, 99
  personal, 203, 223
Clitorodectomy, 69
Clothing, 99, 234
  use and significance, 236
Club-houses, 65, 75, 100
Clubs (weapons), 260
Coconut vessels, 270
Cognatic
  descent, 70
  defined, 75
Coitus, symbols of, 123
Collection of specimens, 361
Colour, 231, 296, 297
  scales, 5, 15
  see Eye, Hair, Skin
Comb
  in hair, 225
  in weaving, 293
Communication
  means of, 35, 66
  methods of, 195
Communism, primitive, 38
Community, status of members of, 144
Conception, 104
Concubinage, 118
Condiments, 243
Confinement, 105, 202
  practices by husband, 105
Consumption, 160, 165
  analysis of, 171

Continence, ritual, 106, 122
Contraception, 104
Cooking, 242
Copper, 283, 346
Cord, 286
Corpse
  artificial decomposition, 127, 128
  cremation, 126
  disposal in water, 128
  disposal of, 126
  encasing, 128
  exhumation, 128
  exposure, 127
  inhumation or interment, 126
  mummification, 127
  orientation, 127
  preparation of, 126
  preservation, 127
  see also Dead, Mortuary rites
Councils
  of elders, 140
  role of, 139
Cousins
  cross, definition of, 76
  cross-cousin marriage, 54 (diagram), 115
  definition of, 76
  parallel, definition of, 76
Couvade, 105
Crafts (boats), 300
Craftsman, 136, 163
  status of, 222
Cranial deformation, 6, 102, 226
Cremation, 126
Cross-bows, 267
Cross cousin, see Cousins
Cult objects, see under Ritual and Sacred
Cults, 70
  definition of, 180
  individual, 180
  organized, 174
Culture, contact, influence of, on native society, 39, 182
Culture-hero, 90, 179
  historic or mythological, 193
Currency, 170, 197
  cattle as, 150, 166
  metal articles as, 282

Daggers, see Knives
Dams, 254
Dances
  meaning of, 181, 185, 331
  ownership of, 332
  records of, 332
Dancing-ground, 65
Darts, 265
Dead
  abode of the, 179
  fate of the, 179
  for disposal of body see Corpse
  property of, 131
  shrines, relics, memorials, 131
  treatment of
    dying or dead, 125
    enemy dead, 126
  wives to the, 118, 178
Death
  abnormal, 125, 181
  beliefs concerning, 124, 178
  life after, 129, 178
  rates of, 60, 62
  ritual connected with, 191
  see also Mortuary rites
Debts and indebtedness, 167
Decoration
  meaning of personal, 231
  of houses, 310
  of pottery, 277
  of skin, 229
  terminology of decorative art, 311
Decoys, 252
Defloration, 122
Deformations
  attitude to, 202, 229
  dental, 227
  facial, 226
  of genital organs, 228; see also Circumcision
  of head, 6, 102, 226
  of limbs and trunk, 228
  ritual connected with, 233
Demography, 58, 161
  demographic factors, in structure of settlements, 60
Descent
  asymmetrical, 92
  bilateral, 71

Descent—*contd.*
 definition of, 71
 dominant line, 92
 double unilineal, 92
 indirect, defined, 91
 submerged line, 92
 see also Matrilineal, Patrilineal
Descriptive system, 77, 79
Disease
 beliefs concerning, 190
 treatment of, 190, 200
Disparate, definition of, 112
Distaff, 288
Distance, measures of, 196, 297
Distribution, 160
 defined, 168
 principles of, 168
Divination, 189
 methods of, 186
Divorce
 definition of, 121
 rates of, 62
Dobzhansky, T., 8 (*note* 1)
Documentation, types of, 45
Domestication of animals, 200, 250, 300
Dowry, defined, 116
Drainage, 248
Drama, 333
Dreams
 meaning of, 176, 178, 180, 187
 sacred, 181
Drills, 258
Drinks
 manufactured, 100, 246
 natural, 245
 receptacles for, 246, 270
 uses and observances in connection with, 246
Drugs, 33
 knowledge of, 203
Drum
 -language, 210
 types of, 318
Drums, ceremonial, 184, 193
Dual Organization, 79
 beliefs associated with, 193
 described, 92
Dug-out canoes, 302
Dulcimer series, 326

Dwelling, *see* Homestead
Dyeing, 286, 296

Ear
 deformation of, 227
 -training, 213
Earthworks, 204, 270
Eating customs, 183, 243
Economic
 life
  environment, 35, 49
  ritual in, 165, 191
  prerogatives and obligations of chiefs, 130, 165
 system
  as stabilizing factor in social system, 158, 162
  effect of social changes on, 173
Economics, definition of, 150
Education, 101
Elders, council of, 136
Elopement
 marriage by, 119
 sporadic, 120
Embroidery, 286, 294
Endogamy
 definition of, 116
 rules of, 94
Environment
 natural, 35
 reconstruction of ancient, 349, 350
Epidemics, 126, 201
 prevention of, 204
Eriam, 120
Etiquette, 66
 of daily life, 98, 100
 with natives, 34
Eunuchs, 67, 140
Evidence, use of literary, 34
Evil eye, 189, 190
Exchange, 160
 role of, 169
Excision, 229
Excreta, magico-religious attitude towards, 99, 182, 188, 225
Exhumation, 128
Exogamy
 definition of, 115
 in relation to clanship, 89, 90

INDEX

Exogamy—*contd.*
  sanctions of, 146
Expert
  ritual, 180
  social status of, 182
Exposure, 127; *see* Corpse
Eye, colour, 3, 15

Face
  deformation, 226
  height, 12
Family
  compound, 70
  domestic, 71
  elementary, 70
  extended, 72
  family life, 75
  joint, 72
  polyandrous, 70
  polygynous, 70
  status of members of, 72
Fasting, ritual, 181, 185, 232
Father's brother, role of, 85
Fauna and flora, 35, 350
Feasts, 169, 185, 244
Fermentations, 245
Ferries, 298
Fertility, 191
  rate of, 113
Fetishes, defined, 184
Field
  antiquities, 343
  recording, 344
Fire, 98, 281
  ceremonies in connection with, 69, 183, 191
  fire-making, 240
  signal-fires, 209
  traditions and observances, 241
Firearms, 268
Fishing, 35, 160
  methods, 253
  rights, 65, 257
  ritual and social observances connected with, 257
Flail, 261
Flares
  in fishing, 255
  in hunting, 252

Flints, 343
Floats
  for water transport, 300
  in fishing, 255
Flutes, 320
Food, 66, 98, 103
  customs regarding, 203
  exceptional, 244
  gathering, 247
  observances and traditions, 243
  plant cultivation, 248
  preparation of, 203, 241
  prescribed and forbidden, 244
  preservation and storage of, 242
  supply, domesticated animals as, 249
  taboos, 100, 243
Fords, 298
Forehead-bands, 235
Forging, 282
Fostering, 74
  status of foster-parents, 74
Friction instruments, 318
Frigidity, 109
Furniture, 238

Gaffs and other fishing hooks, 256
Gambling, 335
Games, 33
  of chance, 334
  of dexterity, 334
  of movement, 334
  of skill and calculation, 334
  round, 334
Genealogical method
  application of, to analysis of kinship systems, 79
  techniques of, 52
  use of, 42, 50, 61
Genital organs, deformation of, 228
Geography, 48, 199
Geology, effect of, on social organization, 48
Gesture, modes of, 208
Ghost, definition, 178
Glass, 279
Gloves, 235
  in archery, 268
God, 180
Gods, marriage to, 119

INDEX

Gold, 282, 347
Gongs, 317
Gorge, fishing, 255
Goura, 322
Gourds, 270
Grave
   goods, 129, 178, 347
   position of, 64, 127
Grazing rights, 65, 135
Greeting, forms of, 34, 99
Grieve, S. W. and Morant, G. M., 5 (*note* 1)
Growth, 3
Guardian-spirit, 180
Guest-houses, 65, 100
Guest-right, 100

Habitations, 236, 270
   in archæology, 347
   see also Houses
Haddon, A. C., 304, 335 (*note* 1)
Hæmoglobin, 15
Hair
   colour, 3, 15
   magico-religious beliefs concerning, 99, 176, 182, 188
   types of, 5
Hair-dressing, 224
Halting-places for travellers, 298
Hammers, 258
Hapu, 89
Harem, social structure of, 112
Harpoons
   as weapons, 264
   in fishing, 256
Harps, 325
Head
   breadth, 10
   coverings, 235
   deformations, 6, 102, 226
   height, 11
   length, 10
   measurements, 6
Head-hunting
   motives for, 143, 177
   prohibition of, 39
Heddles of looms, 291, 293
Height, sitting, 12; *see also* Stature
Hides, use of, 285

High God, 176, 180
History, reconstruction of, 33, 39, 40, 204
Hoes, 116, 170, 222
Homestead, definition of, 64, 71
Homicide, 146
Homosexuality, 109
Hooks, fishing, 255
Horde, definition of, 65
Hornell, J., 304
Horns, 271
Hospitality, rules of, 99, 100, 139
Hostility, intermittent, between neighbouring tribes, 141, 143
Houses, 221
   appropriation and use of, 239
   ceremonies in construction of, 238
   construction of, 238
   grouping of, 239
   decoration of, 310
   see Homestead
Hrdlička, A., 6 (*note* 1), 11
Human sacrifice, 39, 130, 177
Hunting, 160
   ceremonies connected with, 257
   methods, 252
   rights, 65, 133, 257
   social observances in, 257
Hut, 64, 238
Hygiene, 203

Implements, of stone, 279, 345
Impotence, 109, 188, 202
Incest, 73, 146, 188
   definition of, 113
   incestuous marriages, *see* Marriage
   regulation of, 108, 114, 191
Incision, 69, 228
Indices
   marital, 62
   of mortality, 60
Infanticide, 39, 60, 106
Infibulation, 229
Informants, care in choosing, 31, 37, 43
Inheritance, 75, 79
   rules of, 88, 149, 152, 166
   see also Property
Inhumation, 126

Initiation, 108
  ceremonies in, 69
  fees, 68
  see also Circumcision, Clitorodectomy
Insanity, attitude to, 202
Insect prevention, 203
Insignia
  defined, 233
  importance of, 236
  of chieftainship, 193
Instruments
  anthropometric, 9
  musical, 316
Interment, 126
Interpreter, training of, 42
Investigator
  attitude of native to, 29
  sex of, 30
  status of, 31
Iron, working of, 283, 347
Iron-workers, 44
Irrigation, 248
Ivory, working of, 280

Javelins, 264
Jew-harps, 319
Joking relationship, 86
Judicial
  sanctions, 145
  trials, 147
Justice, 146
  law and, 144

Kakina, 229
Keloids, 230
King
  divine, 138, 181
  killing of, 181
Kinship
  definition of, 75 *see* Matrilineal, Patrilineal, Cognatic
  fictions in social structure, 71, 76
  list of kinship terms, 81
  solidarity of kin groups, 136
  system
    application of genealogical method to, 79
    definition of, 76
  terminology, 76

ties
  as basis for economic relationships, 158, 162
  in relation to structure of settlement, 64
Kites
  in fishing, 256
  in games, 334
Knitting, 287
Knives, 263
Knots, 286
Krogman, W. M., 9 (*note* 1)

Labour
  co-ordination of, 164
  division of, 66, 96, 162, 222
Lace, 287
Lake-dwellings, 237
Land tenure
  changing forms of, 156
  definition of, 154
  security of, 168
Land-marks, 195, 199, 297
Language
  as medium of study, 33, 41, 50
  difficulties with, 42, 211
  grammar, 215
  phonetics, 212
  ritual, 187
  semantics, 217
  sign, 208
  spoken, 210
  use of native terms, 42
Lasso, 251, 268
Law and justice, 144
Laws, knowledge of natural, 35
Leather
  Morocco, 285
  use of, 285
Ledger (fishing), 256
Legends
  concerning death, 125
  counterfeit history, 34, 205
  in connection with clan shrines, 91
  in connection with pottery, 276
Levirate, 153
  defined, 117
Life after death, 178; *see also* Ancestor cult

INDEX

Life-cycle, 59, 104
Lineage
  corporate interests and activities of, 89
  definition of, 88
  group, defined, 89
  matrilineage, defined, 89
  patrilineal or agnatic, defined, 89
  see also Exogamy, Clan
Lines, fishing, 254
Lips, deformation of, 227
Livestock, 116
  as capital, 166; see also Cattle
  species and uses of, 250, 253
Local group, 64
Long-houses, 65
Looms and their parts, 290
Lures
  in fishing, 255
  in hunting, 253
Lute-guitars, 326

Magic
  concerned with economic activities, 165, 191
  in agriculture, 165
  magician, 187
  magico-religious beliefs and practices, 174, 187, 190
  types of, 187, 188
Malinowski, B., 218 (note 1)
Man, beliefs concerning, 176
Mana, 177
Manumission, 97
Maps
  demographic, 47
  economic, 47
Markets, function of, 49, 170
Marks
  owner's, 195, 234, 346
  significance of, 231, 234
  tally, 234
  tribal and personal, 233
Marriage
  adultery, 120
  age at, 111
  by exchange, 117
  ceremonies, 121
  cross-cousin, 54 (diagram), 115

  definition of, 71, 110
  defloration, 122
  divorce, 121
  dowry, 116
  elopement, 119
  endogamy, 116
  exogamy, 115
  extra-marital sexual relations, 120, 123
  group, 120
  incestuous, 68, 108, 114; see also Incest
  marital indices, 62
  matrilocal and patrilocal, 112 (note)
  payment, 116; see Bride-price, Bride-wealth
  regulations, 65, 78, 91, 112, 192
  remarriage of widows, 51, 78, 117
  residential location of couples, 112
  rites connected with, in relation to social structure, 191
  ritual union and ritual continence, 122
  seasons for, 122
  suttee, 124
  to gods, 119
  types of, 53 (diagrams), 112
  see also Monogamy, Polyandry, Polygamy, Polygyny
  unmarried adults, 110
  with dead, 118
Marriages
  preferential, 115
  secondary, 117; see also Levirate, Sororate
Marsh-dwellings, 237
Martin, R., 6 (note 1), 11
Masks, 184
Maternal
  definition, 76
  relatives, 76
Mather, K., 4 (note 2)
Matrilateral, definition, 77
Matrilineal, descent defined, 71, 76
Matrilocal marriage, 112 (note)
Mats, 274
Measures and measurements
  anthropometric, 3, 6, 10
  of capacity, 197

INDEX

Measures and measurements—*contd.*
  of distance, 196
  of surface, 196
  of time, 197
  of value, 197
  of weights, 196
Mechanisms, 258
Medical knowledge, value of, 33, 201
Medicine
  medical treatment, 201
  preventive 190
  ritual in, 189
Medicine-men, 182, 189, 201
Melanin, 8, 15
Melanoid, 15
Membranophones, 318
Menstruation, beliefs concerning, 67, 108, 188, 203, 244
Message sticks and other appliances, 195
Metabolism, basal metabolic rate, 5, 7
Metal
  articles as currency, 282
  working of, 281, 346
Methods
  classical, in physical anthropology, 3
  serological, 3
Metrical measurements, 3
Midwifery, 202; *see also* Birth
Migrations
  rates of, 62
  seasonal, 65
  tracing of, 39
Military organization, 68, 140; *see also* Age-sets, Warfare
Milk, 242
Minerals, presence of, 35, 49
Mining, 283
Miscegenation, 61
Missionary, role of, 30
Moieties, 79, 92, 193; *see* Dual Organization
Moko, 229
Mollison, P. L., Mourant, A. E., and Race, R. R., 24
Money, 170; *see* Currency, Blood-money

Monochords (musical instruments), 324
Monogamy, 112
  definition of, 71
Montagu, M. F. Ashley, 6 (*note* 1), 11
Monuments, historical, 204
Moon, myths concerning, 198
Mortuary rites
  depend on individual's age, rank and sex, 129
  performed before death, 129
  position and orientation of body, 127
  sacrifices at death, 130
  types of burial, 127
  see also Corpse
Mother's brother, role of, 85, 117
Mourning, 129, 130, 178
Mummification, 127
Music, 33
  beliefs and observances connected with, 330
  mechanical records, 331
  musical instruments, 316
  use of, 329
  musical vocabulary, 330
  voices, 330
  written records of, 327
Mutilations, 69, 147–81; *see* Deformations
Myths
  concerning sacred places, 183
  concerning sun and moon, 198
  connected with religious beliefs, 175
  counterfeit history, 34
  functions of, 205
  in connection with incest, 114
  of origin, 90, 138, 200, 205

Nails, human, 99, 176, 182, 188, 225
Nail-violin series (musical instruments), 320
Names
  exchanging, 52
  naming, 107, 176, 179
  ritual connected with naming, 191
  role of, in age-set organization, 69
  taboos on, 51
Narcotics, 246, 335

INDEX

Nasal, height, 11
Nation, definition of, 134
Native
  attitude of, to Europeans, 29
  use of word, 28
Navigation, 300
Needlework, 286
Nets
  in fishing, 253
  in hunting, 252
  netting, 287
Nomadism, 65, 133, 135
Nose, 4
  broadth, 8
  deformation of, 227
Note-taking, 45
Nudity, 99, 234
Nutrition, 3, 7
Nyastaranga, 323

Oars, 305
Oaths, 147
  defined, 186
Oboe, 321
Occupational groups, 65, 196, 163
Odour, personal, 224
Offerings, 185; *see also* Sacrifice
Ogden, C. K., and Richards, I. A., 218 (*note* 1)
Omens, 186
Oracles, 186
Ordeals
  analysis of, 186
  employment of, 147
Ornaments, 246
  as weapons, 260, 269
  attached to canoes, 308
  personal, 223, 233, 283
Ossuaries, 129
Outrigger-boats, 306
Ownership
  definition of, 148, 167
  discrimination between sexes in, 66
  owner's marks, 234, 346
  see also Property

Packing materials, 362
Paddles, 305
Painting
  of objects, 296

of the skin, 229, 297
Pan-pipes, 320
Paper squeezes, 366
Pastoral life, 160, 161, 166, 249
Paternal, definition of, 76
Paternity
  ignorance of, 104
  paternal relatives, 76
  see also Patrilineal, Matrilineal
Paths
  path-finding, 297
  varieties of, 297
Patrilateral, definition of, 77
Patrilineal
  descent defined, 71
  kinship, 75
  lineage, 70, 89
  see also Agnatic
Patrilocal marriage, 112 (*note*)
Payments
  in initiation, 69
  types of, 169
  see Marriage-payment
Peace, truces and treaties of, 142
Pellet-bells, 317
Pellet-bows, 251, 267
Perfumes, 224
Personal
  appearance, 224, 296
  decoration, meaning of, 231
  ornaments, 233
  totemism, 180
Phonetics, 212
Phonograph, 48, 315
Phonology, 211
Photography, 33, 48
  cameras, exposure, development, 353, 355
  storage of negatives, 356
Physical powers, attitude to, 35, 190
Pigmentation, *see* Skin-colour
Pigs as currency, 166
Pile-dwellings, 237
Pile-weaving, 294
Pirrauru, 120
Pitch
  method of ascertaining, 329
  pipes, 328
Plans of villages and houses, 47

Planter
  plant cultivation, 248
  role of, 30
Point (fishing), 255
Poisons, 203
  applied to weapons, 259, 263
  in fishing, 256
  in hunting, 251
Poles (boating), 305
Political organization, 132
Polyandry, 113
  definition of, 70, 71
Polygamy, 112
  definition of, 71
Polygyny, 112
  definition of, 70, 71
Population, estimation of, 59, 60; *see* Demography
Possession by spirits, 181, 189
Potlatch, 172, 194
Potsherds, 343
Potter's wheel, 277, 346
Pottery, 276
  classification of, 346
  decoration of, 277
  firing of, 278
  property-marks on, 346
Prayer, defined, 184
Precedence, rules of, 99
Pregnancy, attitude towards, 104, 244
Priests as informants, 44
Primogeniture, 137, 153
Production
  defined, 160
  ideology of, 163
Prohibition, 73, 113; *see also* Taboos
Property, 72
  destruction of, 172
  types of, 149
Prophets, definition of, 182
Propulsion, method of, 305
Prostitution, status of prostitutes, 110
Proverbs, 163, 206
Psychology, 28
Puberty, 108
  ritual connected with, 191
  see also Initiation
Public life, regulation of, 124, 144

Purification, 233
  after child-birth, 105
  ritual, after contact with corpse, 131

Quarrying, 283

Race, R. R., and Sanger, R., 24
Rafts, 300
Raids, motives for, 141
Rain-makers
  as informants, 44
  as ritual experts, 181
  role of, 199
Rakes, in fishing, 256
Rank, 32, 68, 165
  hereditary, 94
Rattles (musical instruments), 318
Reckoning, modes of, 195, 196
Recognition marks, 233; *see* Marks
Recording, methods of, 195
Reed in weaving, 293
Regalia, 193, 236
Reincarnation, 179
Relatives
  behaviour between, 84, 124
  duties and privileges of, 87, 124
  list of relationship terms, 81
  see also Avoidance, Joking relationship
Religion
  defined, 175
  religious beliefs and practices, 175
Remuneration, 44
Restrictions, *see* Taboos
Rhythm
  in decorative art, 315
  in music, 328
Rites, 175
Ritual
  connected with canoes, 308
  continence, 122
  defined, 175
  forms of, 184
  functions of ritual observances in social structure, 179, 191
  in economic life, 165, 191
  language, 187
  objects, 183
  sanctions, 145

INDEX 399

Ritual—*contd.*
  types of, 175
Rivers, W. H. R., and Haddon, A. C., 335 (*note* 1)
Roads, upkeep of, 49, 66, 297
Rock, 49
  drawings, 348
  paintings, 348
  shelters, 348
Rope
  cord and, 286
  term for indirect descent in Papua, 91
Rudders, 305

Sacred
  chiefs, 181
  dreams, 181
  objects, 183
  places, 183; *see also* Shrines
Sacrifice
  defined, 185
  human, 39, 130, 177
Sacrifices in mortuary rites, 130
Sails, 306
Salt, 33
  as medium of exchange, 284
  use of, 243, 284
Salutations, 99; *see also* Etiquette
Sampling, 56
Sanctions
  judicial, 135, 145
  positive, 145
  ritual, 145, 193
Sanitation, 64, 99, 188, 203, 224
Sansa series (musical instruments), 319
Saws, 258
Seasons
  for marriage, 122
  observations of, 198, 248
Seclusion, 69, 232
Secret societies, 66, 140
  definition of, 68 194
  functions of, 144, 194
Segment, 89, 133
Seniority, 59, 68
  rule of, 70, 81 (*note* 1), 85
Sennet, 286

Settlements
  demographic factors in structure of, 60
  geographical factors in structure of, 48
  in archæology, 347
  kinship ties in relation to structure of, 64
Sewing, 286
Sex
  in dancing, 332
  in games, 335
  puberty and initiation, 108
  sex communism, 120
  sex segregation, 66, 101
  sexual development, 107
  vital statistics of, 60, 61
  see also Chastity, Virginity, **Regulation of Public Life**
Sexes
  division of labour between the, 66, 96, 162, 222
  social distinctions between the, 66
Shadow-plays, 333
Shamanism, 109, 179, 185
  definition, 181
Shank (fishing), 255
Shed
  in weaving, 291
  shed-stick, 291
Shells, 346
  as currency, 166
  working of, 271, 281
Shelters
  natural and portable, 237
  rock, 348
  temporary, 64
Shields, 270
Shoes and sandals, 235
Shrines, 64, 91, 178
  definition of, 130, 183
Shuttles of looms, 290, 293
Siblings
  definition of, 71
  diagram of, 53
  half, defined, 71
Signals, 209
Sign-language, 208
Silver, 283, 347

Simmons, R. T., and Graydon, J. J., 19
Sinnet, 286
Sistrum series (musical instruments), 318
Sitting, height, 12
Skeuomorphs, 346
Skin
  cicatrization, 229
  colour, 3, 7, 15
  decoration, 229
  painting, 229
  staining, 229
  tattoo, 229
Skins
  skin-canoes, 302
  skin-dressing, 285
  skin-vessels, 271
Skulls and bones
  deformations of, 226
  preservation of, 130, 178, 348
  trepanning of skull, 202
Slavery
  definition of slave, 95
  slaves as form of capital, 168
Sledges, 258, 299
Sleep, 98, 103
Slings, 251, 266
Smelting, 282
Snedecor, G. W., 4 (*note 2*)
Social structure
  definition of, 63
  functions of ritual observances in, 179, 191
  kinship fictions in, 71, 76
Sociology, 28
  economic system as stabilizing factor in social systems, 158, 162
  influence of culture contact on native society, 29, 39, 182
  methods, 36
  regulation of public life, 144
  social groups and groupings, 61, 64
  social life of the individual, 98
  terminology, 54
Soils, structure of, 49, 352
Solidarity within kin groups, 136
Songs, 206
  recording of, 331

Sorcery, 175
  definition, 189
Sororate defined, 118
Soul, 107, 179
  definition of, 176
  external soul, 183
  substance, 176
Spears
  as ritual objects, 183, 193
  described, 261
  in fishing, 256
Spear-throwers, 265
Specialists, skilled, 163
Specimens
  collection of, 361
  packing and labelling of, 362
Spindle, 287
Spindle-whorl, 287
Spinning, 287
Spinning-wheels, 288
Spirit
  definition of, 179
  guardian, 180
  helpers, 180
Spirits, alcoholic, 246
Squeezes, 366
Staining
  of objects, 296
  of skin and nails, 229
Stamping tubes (musical instruments), 318
State, definition of, 134
Stature, 7
  measurements, 12
Stebbing, S., 218 (*note 2*)
Sterility, 109, 202
  rates of, 62
Stilettos, 263
Stimulants, 246, 335
Stone
  implements, 279, 343
    collecting of, 345
  vessels, 271
Stories, 206
Stranger, status of, 78, 100, 133
Stringed instruments (musical), 323
String figures
  described, 335
  terminology, 335, 338 (diagram)

Strings
  bow-, 267
  use of, 286
Succession
  definition of, 152
  rules of, 137
Suicide, attitude towards, 125
Sun, myths concerning, 198
Sundials, 197
Surface, measures of, 197
Surgery, 202
Suttee, 124, 130
Swaddling, 102, 226
Swamps, 298
Swords, 263
  in weaving, 293
Symbols
  for males, females and sex unknown, 52
  symbolic art, 314
Sympathetic strings (musical instruments), 327
Syrinx (musical instrument), 320

Taboo
  defined, 185
  on names, 51
Taboos
  for warriors, 143
  imposed on women, 67, 186, 188, 203
  in connection with food, 69, 100, 186, 192, 244
  significance of, in social structure, 66, 100, 186
Tackle, fishing, 254
Talismans, 233
Tanning of skins, 285
Tattoo, 229
Tawing of skins, 285
Technique of investigation, 40
Technology, 33, 49, 160
Teeth, 106
  dental deformations, 69, 227, 228
Tell, in archæology, 348
Tents, 64, 237
Terminology
  of art, 311
  of musical instruments, 317

  of religion, 176
  of social organization, 54
  of weapons, 260
Terrace-cultivation, 248; *see also* Agriculture
Textiles
  collection of, 294
  design, 294
  patterns, 293
Texts, 49
Thread, 286
Throwing-clubs, 264
Throwing-knives, 264
Throwing-spears, 264
Tied cloth, 293
Tildesley, M. L., 5 (*note* 2)
Time, measures of, 197, 199
Titles, use of, 99
Tobacco, 33, 247
Tongue, deformation of, 227
Tools, 166, 257
Topography, 199
Totemism, 38
  ceremonies to increase fertility of totem, 193
  clan marks representing totems, 231
  definition of, 192
  linked, 192
  myths of origin of totem species, 192
  personal, 180
  relation between, and social structure, 90, 192
  totemic animals, 179 192
  behaviour towards, 179, 192
Town, definition of, 64
Towns, arrangement of, 239
Toys, 33, 334
Trade, functions of trader, 30, 170
Trade-marks, 195
Trailers, 299
Training, physical, of children, 101; *see also* Education, Sexual development
Transport
  by land, 299
  by water, 300
Traps, 269
  for fishing, 254
  for hunting, 252

Travel
  by land, 297
  by water, 300
  mode of, 298
Treaties, 135
  peace, 142
Trees
  as dwellings, 237
  exposure of dead on, 127
Tribe, 134
  definition of, 66
Tricks in string-figures, 335
Trident, 256
Trimmers (fishing), 256
Twine, 286
Twined weaving, 293
Twin siblings, diagram of, 53
Twins, beliefs concerning, 106

Ultimogeniture, 137
  defined, 152
Umbilical cord, 105
Uterine
  descent defined, 71
  kinship, 76
  see Matrilineal

Valve instruments (musical), 321
Varna, four, 94
Vegetation, knowledge of, 199
Vehicles, 300
Venereal diseases, 202
Village, definition of, 64
Villages, arrangements of, 239
Virginity, attitude towards, 108, 122

Ward, I. C., 215
Wards, definition, 64
Warfare
  in relation to political organization, 141
  killing of kinsman in, sinful, 141
  ritual concerning, 194
Wars
  accounts of, 142
  causes for, 141, 143
  conduct of, 142
  methods of fighting, 142
  of conquest, 142
  weapons, 143
Warps
  in basket-work, 272
  in weaving, 290, 292
Warriors
  ornaments worn by, 233, 269
  taboos for, 143
Water
  ceremonies connected with, 249
  collection and storage of, 203, 245
  transport by, 300
  upkeep of waterways, 66
  water-clock, 249
  waterworks, 249
Wealth, ownership and consumption of, 172
Weaning, 102
Weapons
  as tools, 258
  history of, 260
  poison applied to, 259, 263
  tally-marks on, 234
  terminology, 260
  used in warfare, 143
Weather, observations of, 198
Weaving, 289 (diagram)
  observances connected with, 295
Wefts
  in basket-work, 271
  in weaving, 290, 293
Weight, 15
  measures of, 197
Weirs, 254
Westermann, D., and Ward, I. C., 215
Wheeled vehicles, construction of, 300
Whistling arrows (musical instruments), 321
Widows
  remarriage of, 51, 78, 117
  status of, 123
Wiener, A. S., 24
Wind instruments (musical), 320
Wine, 246
Witchcraft, 175
  definition of, 189
Women
  as investigators, 30
  captive, 119
  frigidity in, 109